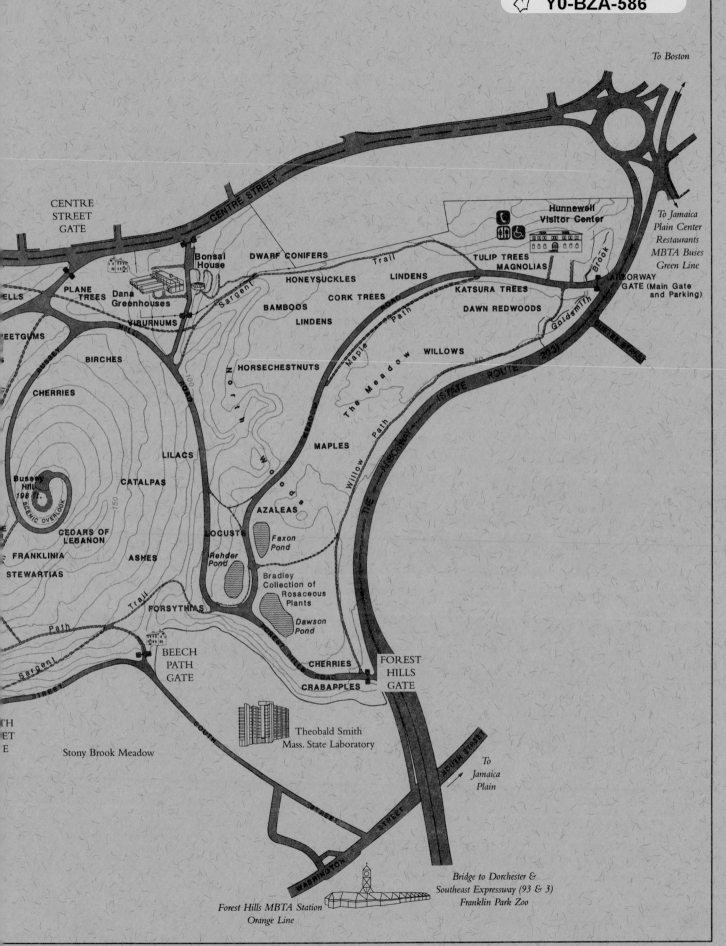

SCIENCE IN THE PLEASURE GROUND

SCIENCE IN THE PLEASURE GROUND

A HISTORY OF THE ARNOLD ARBORETUM

IDA HAY

NORTHEASTERN UNIVERSITY PRESS
Boston, Massachusetts

Library of Congress Cataloging-in-Publication Data

Hay, Ida, 1948–
Science in the pleasure ground : a history of the Arnold
Arboretum / Ida Hay.
        p.   cm.
    Includes bibliographical references (p.   ) and index.
    ISBN 1-55553-201-2
    1. Arnold Arboretum—History. I. Title.
QK480.U52A765   1994
580'.74'474461—dc20   94-29102

Designed by Janice Wheeler

Composed in Bembo by DEKR Corporation, Woburn,
Massachusetts. Printed and bound by Vail-Ballou Press,
Binghamton, New York. The paper is St. Lawrence Matte,
an acid-free sheet.

MANUFACTURED IN THE UNITED STATES OF AMERICA
99 98 97 96 95 94   5 4 3 2 1

*Frontispiece: One of the delights of early spring is the delicate
flowers of pinkshell azalea* (Rhododendron vaseyi), *a native of
North Carolina, here reflected in Faxon Pond along with the
emerging canopy of indigenous oaks in the North Woods behind
them.*

CONTENTS

PREFACE

The Arnold Arboretum is held dear in the hearts and minds of residents of the Boston area as one of the region's most beautiful outdoor spaces. For many it is *the place* to go on the sunniest Sunday in May, when the lilacs bloom and fill the air with fragrance. For others the Arboretum is inviting throughout the year. They come not just to experience the welcome reawakening of spring, but to hear the hum and feel the lushness of summer, to see autumn's glorious color change, and to admire bark and branching patterns and await the swelling of buds in winter. The ranks of Arboretum admirers increase greatly, however, when those who come to make use of its unique exhibits as a *museum* of trees and shrubs are considered. These people come not only from metropolitan Boston but from all of New England, the nation, and the world.

While this wider audience also derives great pleasure from viewing the landscape, its members appreciate as well the ability to observe firsthand so many named and documented trees in one place. They recognize the Arboretum as an institution devoted to scholarship and education as well as aesthetic display. To fulfill its role as collector of hardy woody plants requires a staff with knowledge of many disciplines in plant science. What to grow, what to name it, where to find it, how to keep it healthy, how to display it effectively—these are the sorts of questions arboretum keepers continually seek to answer. To carry out its mission of teaching the knowledge of trees, the Arboretum has arranged its collections according to a classificatory sequence, labeled the plants, produced scholarly and popular publications, and offered instruction to college and university students as well as to the public. Its encyclopedic living collection, and the library and herbarium that have been assembled with it, are a resource for botanists, foresters, horticulturists, professional and amateur growers, landscape designers, and park planners. It is one of the world's centers for learning about trees and shrubs and one of the most extensive collections of hardy plants in the north temperate zone. Yet its landscape is so carefully planned to mimic nature that many who stroll the Arboretum grounds are not aware of its pedagogic role.

At the time of its founding in the late nineteenth century, the Arnold Arboretum was an experiment. No one really knew whether such a collection would indeed add greatly to our understanding of nature. In an era when the vastness of the world's natural resources was still becoming known, and the basis for cataloging plants and animals was shifting from anthropocentric to evolutionary, the idea of bringing all the cold-hardy trees together and growing them in one place seemed promising. This was a time when other collections of natural objects were being gathered into museums. Although trees had been cultivated as individuals for centuries, and although many private collections had been assembled, the Arnold Arboretum founders took the notion further by stipulating an all-inclusive collection and by placing the institution under academic auspices.

While increasing knowledge about the natural world was of great interest to the institution's founders, they equally valued the economic and aesthetic roles of trees in human life. They wished to promote not only conservation but also more knowledgeable use of trees in planted landscapes. Fortunately, just as the new tree garden was getting under way, professionalism in landscape design and urban planning was emerging with the creation of parks in America's cities. Through this movement (namely, with the participation of Frederick Law Olmsted and Charles S. Sargent in laying out the grounds), the tree garden became a model of naturalistic landscape design. Thus, to comprehend the Arboretum landscape, it must be understood as one created by human beings.

Science in the Pleasure Ground has been written to reveal the stories behind the scenery of this intricate garden. A. Lawrence Lowell, president of Harvard, once said of Charles Sargent, the first director, that he built his thoughts into the Arboretum. This applies as well to others who influenced its founding and have developed and maintained it over the years. The story of the Arboretum is made up of the stories of many people. While they had differing personalities and brought a variety of talents and viewpoints to their work, they shared a love of trees and of learning. They all have had the patience, tenacity, and optimism needed to raise trees and complete encyclopedic tasks. The array of motivations and achievements of prior Arboretum keepers provides examples from which to draw future policies and goals.

The creation of this book grew initially from the desire to interpret the landscape and living collections in Jamaica Plain for those who visit the Arboretum. It was stimulated also by staff members' discussions on whether scrutiny of the original design and intent might disclose ways to restore or more effectively manage the collections. During the course of preparation I found that looking at the original plans and early motivations was not enough. Even during Sargent's administration, policies and plantings were changed and added. Full understanding of the place today requires looking at the whole course of its history.

Although my original intent was to tell "only" the story of the grounds in Jamaica Plain, it soon became evident that the development of the research program, the growth of the library and herbarium, and the motives of scientists on the staff have been inextricably bound up with the development of the living collections. In fact, the dynamic tension between science and aesthetics turned out to be one of the most interesting themes of Arboretum history. Nonetheless, my emphasis has been on the living garden. While I have tried to place the Arboretum's work in the context of the history of science, this book could not include the complete story, especially with regard to modern science. The same restriction applies to the Arboretum's role in teaching at Harvard and its relationship to the other fascinating botanical institutions within the university. I could touch only briefly on these subjects here.

The Arboretum was initially arranged to demonstrate plant relationships by a classification scheme while preserving certain existing wooded areas. Although the subsequent changes may seem subtle, the initial arrangement has been altered owing to factors involving growth, space limitations, climate, personal prejudice, and institutional policy. I have traced these changes insofar as they shed light on the collections as they exist today or demonstrate policies and motives differing from those of the present.

Everyone who visits the Arnold (or any other arboretum) has his or her personal interpretation of the place—it is formed by one's intellectual, spiritual, emotional, sensory, or contemplative experiences. This is the best use that can be made of an arboretum: direct contact with trees teaches things that cannot be learned from books. Nonetheless, knowing about the experiences of others who have worked in the tree garden can add dimensions to one's understanding. I hope that by getting to know the people behind the Arboretum readers of this book will gain greater insight into Boston's famous garden of trees, and that they will come to appreciate similar gardens and those who devote their lives to caring for the plants they contain.

ACKNOWLEDGMENTS

The National Endowment for the Humanities supported the research and publication of this and its two companion volumes based on the Arnold Arboretum's living collections. Grateful acknowledgment is made to the endowment for its grants, and to Sheila Connor, who administered them on behalf of the Arboretum. The project also received considerable impetus and encouragement from Peter S. Ashton, director of the Arboretum during the formative stages. Robert E. Cook, present director, has continued to support the publication of this book.

While it is not possible to mention all the Arboretum staff members, past and present, who have given generously of their time and knowledge to the preparation of this book, I extend special thanks to my colleagues there. For guidance in developing the manuscript I particularly relied on Sheila Connor, Stephen A. Spongberg, and Richard E. Weaver. My appreciation for their contributions cannot be adequately expressed. Those who answered my numerous questions and shared reminiscences about the plants and the grounds include John Alexander, Alfred Fordham, Henry Goodell, Gary Koller, Robert Nicholson, Jim Nickerson, Maurice Sheehan, Mark Walkama, R. G. Williams, and Patrick Willoughby. Thanks also go to David Boufford, Michael Canoso, Richard A. Howard, Shiu-ying Hu, Bernice Schubert, and Carroll Wood for their insights on matters relating to botanical research and the herbarium.

I am particularly grateful to those who aided in the preparation of the manuscript. Karen Kane assisted in countless ways, especially by entering the majority of the text into the computer. The type and graphics department of Harvard Student Agencies also assisted with word-processing. Special acknowledgment goes to the late Jennifer Quigley, who cheerfully and generously aided me with the use of computers and the plant records. To those who read and commented on parts of the manuscript I extend my thanks: Cornelia McMurtrie, Stephen Spongberg, and the late Albert Fein. I also appreciate the valuable comments of three anonymous reviewers who read the manuscript.

During the course of my research I benefitted from the assistance of the staff of the botany libraries at Harvard, especially from Judy Warnement and Jean Cargill. For help with the use of the Arnold Arboretum library and archives in Jamaica Plain acknowledgment is due to Sheila Connor and her assistants Carin Dohlman, Karen Kane, and Jennifer Brown. I am also grateful to the Boston Public Library, the Harvard University archives, the Massachusetts Historical Society, the Massachusetts Horticultural Society, the New Bedford Public Library, and the Old Dartmouth Historical Society for allowing the use of their collections.

The majority of images that have been used to illustrate this book come from the extensive photographic archives, slide collection, and library of the Arnold Arboretum. For his excellent reproduction of many of these, I wish to acknowledge the work of Christopher Burnett, photographer to the project. Gustavo Romero also generously aided by photographing some of the Arboretum archival materials, as did the staff of the photographic archives of the Peabody Museum of Harvard University. Many additional people and institutions contributed to the illustrations for this book. For their assistance in locating illustrations and for permission to use material in their care I extend my warmest thanks to the directors and staffs of the archives of the Gray Herbarium; the Harvard University Portrait Collection; the Museum of Comparative Zoology, Harvard University; Mount Auburn Cemetery archives; the National Park Service, Frederick Law Olmsted National Historic Site; the Scott Arboretum of Swarthmore College; and the Society for the Preservation of New England Antiquities. John Alexander, Albert Bussewitz, Bob Howard (Photography Unlimited) Gary Koller, and Stephen Spongberg also generously allowed the publication of their photographs. I am grateful to Carroll E. Wood for permission to reproduce the plate drawn by Dorothy H. Marsh for the Generic Flora of the Southeastern United States project, and to Elizabeth Shaw for her assistance in selecting a herbarium specimen associated with Asa Gray.

To John Weingartner, Senior Editor of Northeastern University Press, I am particularly indebted for valuable suggestions, guidance, and encouragement, especially during the process of revising the manuscript and readying it and the illustrations for publication. I additionally applaud and appreciate the work of Ann Twombly and Diane Levy, who have overseen the production of a beautiful book, as well as Jill Bahcall and Chris Beck, whose efforts to promote it have been inspiring.

A special note of appreciation is extended to Sheila Connor and Stephen Spongberg, my dear friends and co-workers on the sourcebooks project, for their constant help and encouragement. The love and support of my husband, Robert Nicholson, and our son, Charlie, who has never known his mother when she was not working on the book, have been an invaluable resource to me. To them I give my most heartfelt thanks.

SCIENCE IN THE PLEASURE GROUND

1. An inviting park at first glance, seven pillared gateways beckoned visitors to inspect the world of trees once Arboretum construction was complete. The Centre Street gate was one of the first to be finished. This 1917 view shows the black walnut (Juglans nigra) group planted at the junction of Valley and Bussey Hill Roads.

CHAPTER ONE

COLLECTIONS AND LANDSCAPES, THE NINETEENTH-CENTURY TRADITIONS

We want a ground to which people may easily go after their day's work is done, and where they may stroll for an hour, seeing, hearing, and feeling nothing of the bustle and jar of the streets, where they shall, in effect, find the city put far away from them. . . . What we want most is a simple, broad, open space of clean greensward, with sufficient play of surface and a sufficient number of trees about it to supply a variety of light and shade. . . . We want depth of wood enough about it not only for comfort in hot weather, but to completely shut out the city from our landscapes. (Olmsted, 1870, in Sutton, 1971b, p. 80)

The difficulties of making a proper plan for laying out the Arboretum have always appeared very great to me. The site, while offering exceptional beauties, perhaps, for a public park, offers exceptional topographical difficulties for the object to which it is to be devoted; namely, a museum, in which as many living specimens as possible are to find their appropriate positions. In such a museum, every thing should be subservient to the collections, and the ease with which these can be reached and studied. . . . (C. S. Sargent, 1879, p. 100)

In the contrast and the complementarity of these two statements is the essence of the Arnold Arboretum. Its dual nature as a pleasure ground and a scientific collection arose from the interwoven interests of its originators in gardening for enjoyment and in the study of natural history. The Arnold Arboretum was founded during a period when numerous American institutions for education and culture were established. Its initiation was greatly influenced by currents in scientific thought and an ongoing reassessment of humanity's relationship to the natural world. Its vitality among institutions of its kind today is still fed by its nineteenth-century roots.

While the place bears the name of James Arnold and its programs were most thoroughly molded by its first director, Charles Sprague Sargent, its founders included a group of gentlemen who shared interests in natural history and horticulture. In particular, each of them was both student and cultivator of trees. Their desire to increase the knowledge of trees had practical, as well as abstract, motivations. Trees provided the material from which countless essential items were made, from ships to spoons. The dominant aim of life science, as practiced in their time, was identification, the naming and classifying of natural objects. The Arboretum founders also believed that the stately beauty of trees could provide important social benefits. The pleasure derived from viewing a landscape of trees not only was healthful in and of itself but could instill a desire to learn. For the Arnold Arboretum this juxtaposition of scholarly and aesthetic goals has created a valuable tension that has pushed the institution forward many times during its history. The combination has certainly given it the widest possible constituency.

On the Arnold Arboretum's 265 undulating acres thrives a matrix of more than five thousand kinds (species, and their varieties and forms) of accessioned plants. These are interspersed with remnants of ancient woodlands, antedating the Arboretum, that were retained to shelter the planted trees. The mandate of the Arboretum has been to grow and correctly label as comprehensive a collection of plants as possible and to teach the knowledge of trees. Consequently, the plants were set out in a classificatory sequence, by families and genera, in the belief this would be most instructive and practical to manage. Yet, this order does not impose unnaturally on the "genius of the place." The Arboretum designers so skillfully placed the collections and the necessary system for visitor circulation that serenity prevails. Subtlety of design is such that its deliberate plan may go unnoticed.

This tree garden was not created in isolation; from the beginning it bore a relationship to kindred organizations. Locally, it was to complement Harvard University's existing botanic garden, which mainly exhibited herbaceous plants and had scant room for trees. As a living collection, the Arboretum was intended to round out the object lessons available to Boston area citizens at the Boston Society of Natural History Museum and at Harvard's

Museum of Comparative Zoology, where preserved specimens were the norm. Once the Arboretum was under development, its director realized that complete knowledge of the biology and culture of trees would not be obtained until there were gardens for their study throughout the country. Sargent frequently urged a national system of arboreta, encouraged and advised the founding of others, and placed a high priority on the distribution of plant material. The Arboretum has continued to work cooperatively with an international network of gardens and scientific stations.

At its start, however, the Arnold Arboretum was a prototype. While there had been private tree collections and proposals for arboreta under various institutional guises, the Arnold was the first successful American arboretum planned to be open to the public. It was the culmination of notions about the need to know and collect all the kinds of trees, combined with the desire to perfect landscape for human enjoyment. The handful of individuals who produced this subtle demonstration of arboreal wealth were men of science and culture. Benjamin Bussey, on whose land the Arboretum was established, was a successful merchant who, like many of his peers, developed a country estate according to principles known as scientific farming. James Arnold, of New Bedford, Massachusetts, made a fortune in whaling and other enterprises yet pursued his taste for horticulture and natural history with lifelong tenacity. It was George Barrell Emerson who orchestrated the agreement whereby a legacy from Arnold was placed in trust for the creation of an arboretum. Emerson was a pivotal figure in movements to reform science education in the Boston-Cambridge community, as well as a scholar of trees in his own right. Asa Gray, one of the first to profess botany as a field separate from the medical curriculum, promoted scholarship in this subject at every opportunity. He contributed to Darwin's breakthrough theory, authored numerous elementary and advanced textbooks, and greatly influenced the botanical training of Charles Sargent. Sargent, scion of a Boston family of financiers, grew up in an environment of landscaped estates and private plant collections, and he made the Arnold Arboretum his life's work. To aid in its design, he chose Frederick Law Olmsted, foremost American parkmaker and proponent of the profession of landscape architecture. Behind each of these tree advocates were traditions that have shaped the Arnold Arboretum and related institutions.

NATURAL HISTORY AND THE PRIVATE COLLECTOR

During the early nineteenth century, subjects in the life and earth sciences as we know them today were not taught at New England's colleges. Instruction in theology and the classics formed the core of higher education. Biological

2. *Among the many manifestations of the popularity of nature study were ornamental glass cases in which plants and animals were raised indoors. Many a parlor was adorned with a Wardian case, aquarium, or vivarium. The vivarium pictured here was a self-contained environment for observing the life cycles of moths and plants.*

subjects were included in the medical curriculum; botany, for example, was studied to gain practical knowledge of the plants used to treat human ills. But much scientific study was pursued by serious amateurs. Clergymen, lawyers, medical doctors, and holders of political office were the scientists of the time. They collected and studied plants, animals, and minerals, and made meteorological and astronomical observations as well.

As a pastime, the study of nature was thought an especially healthy and morally uplifting activity. It took place out-of-doors and gave people rational amusement and a wholesome occupation. One English author said that for clergymen, especially, an interest in natural history was superior to a taste for gardening because field excursions would give the pastor plenty of chances to visit his parishioners. As part of their religious outlook, many individuals believed that nature provided the best illustration of the design of the Creator. This was natural theology; observation of nature was an exercise thoroughly permeated with the purpose of discovering God's plan. And most biological investigation that was not part of medical study was couched in terms of revealing the Creator's handiwork. Along with the keen interest in observing all of Nature's marvelous "productions" went the human impulse to collect them.

A "cabinet," whether of shells, minerals, fossils, pressed plants, stuffed animals, or an admixture of all of them, was a treasured possession of many New England households from the late eighteenth through nineteenth centuries. It was evidence of a taste for natural objects that was widespread in America as well as in England. Joseph Barrell (1739–1804), successful merchant of Summer Street, Boston, who built his country seat, Pleasant Hill, in Charlestown in 1791, had "an extraordinary cabinet, where birds played in a globe surrounded with a globe of water in which the fish play" (Hammond, 1982, p. 85). The pastor of the First Church in Salem, John Prince, had in his parlor "admirable apparatus and specimens" to teach the "wonders of Astronomy, Optics, Pneumatics, Botany, Mineralology, Chemistry, and Entomology" (Greene, 1984, p. 62). Of a later date is the record of a beautiful collection of shells in mahogany cabinets bequeathed to Harvard College in 1841 by Elizabeth Shaw Craigie

(1772–1841), widow of Andrew Craigie, whose fine Georgian house in Cambridge was later owned by Henry Wadsworth Longfellow.

These collections of butterflies, shells, or rocks, along with seaweed albums, fern cases, birdcages, and aquariums—all inspired by the nature-collecting impulse—were also valued as aesthetic enhancements and status symbols. They adorned parlors and drawing rooms throughout the nineteenth century. Interest in the diversity of nature was furthered by the accessibility of objects from foreign shores provided by New England's involvement in maritime trade. As the nineteenth century progressed, many of these serious amateurs, the collectors of private cabinets, supported the institutionalization of scholarship in natural history within societies and universities.

FOUNDING OF THE HARVARD BOTANIC GARDEN

In the Boston area, many natural-history enthusiasts were graduates of Harvard College. Their wide-ranging interests and desire to strengthen the new Republic drew them together to establish organizations for the communication of knowledge, the acquisition of libraries and scientific apparatus, and the collection of natural-history specimens. At the instigation of John Adams, who knew of Philadelphia's American Philosophical Society (founded in 1743) and had visited scientific institutions in France, the American Academy of Arts and Sciences was founded in Boston in 1780:

And to the honor of our political forefathers, be it spoken, that although the country was engaged in a distressing war, a war most important to the liberties of mankind, that was ever undertaken by any people, . . . they immediately adverted to the usefulness of the design, and incorporated a society . . . by the name of the "American Academy of Arts and Sciences." The purpose of this institution is to promote most branches of knowledge. (Meisel, 1926, pp. 38–39)

Among those who chartered the academy were merchants, professionals, and holders of political office. The study of natural history was lauded as an aid to agriculture,

medicine, and the development of natural resources. Many of the merchants and professionals owned farms on the outskirts of Boston and were very much interested in promoting new knowledge about agricultural practices. The academy soon established a committee on agriculture, which by 1792 was incorporated as the Massachusetts Society for Promoting Agriculture. It sponsored botanical and zoological studies related to agricultural development.

In the Boston area the only lectures in botany were given by Benjamin Waterhouse, professor of medicine at Harvard, who had long hoped to create a botanic garden at the college (he had been teaching there since 1788). After being serialized in the *Monthly Anthology,* his popular lectures were collected in a book entitled *The Botanist* in 1811. It was one of the first botanical textbooks published in America.

The Massachusetts Society for Promoting Agriculture (MSPA) assumed a philanthropic role by giving financial support to institutions that would further the cause of agriculture in its broadest aspects. In 1801, the members voted five hundred dollars for the establishment of the Massachusetts Professorship of Natural History and a botanic garden at Harvard College. As amateur naturalists, many of them recognized that providing instruction in botany could have long-term benefits for the farming community and the general public.

The botanic garden was finally started in 1805, largely through the efforts of a federal district court judge, John Lowell, and his two sons, John, also a judge, and Francis Cabot Lowell, the textile magnate. In addition to the funds they raised, the MSPA convinced the state legislature to grant two townships in Maine (then part of Massachusetts) to the newly formed Board of Visitors of the Massachusetts Professorship of Natural History at Harvard College. For many years, the state provided appropriations for the botanic garden and the professorship. Funds from sale of the land in Maine enabled the board to appoint William Dandridge Peck, who had won prizes from the MSPA for papers on agricultural insect pests, to the professorship. For the first three years of his appointment Peck toured Europe, studying botanic gardens and consulting with naturalists in England and on the Continent before returning to lay out the garden in Cambridge.

3. The Harvard Botanic Garden, founded in 1806, presented its lessons in botany not only to Harvard students but to citizens of Cambridge and greater Boston for nearly 150 years. This view shows the greenhouse and gardens before Charles S. Sargent revised the plantings in the 1870s.

With the support of MSPA members John Thornton Kirkland (president of the university) and John Lowell (member of the corporation), the botanic garden progressed rapidly. Funds were used to purchase land on the Highway to the Great Swamp and Fresh Pond (now Garden Street), and Andrew Craigie donated a parcel from his neighboring estate, bringing the total to more than seven acres. On his return, Peck brought with him a Yorkshireman, William Carter, to be gardener at the new establishment. Carter remained forty years, overseeing the plantings through several changes in leadership.

Peck taught botany to Harvard students until his death in 1822, and the garden became a favorite attraction for Cambridge residents and out-of-town visitors. Peck's catalog of plants cultivated in the garden, published in 1818, listed some 1,040 species. This diversity of plants had been acquired from the private greenhouses of friends of the institution, from "gentlemen who have visited tropical regions in the East and West Indies and in Africa," and from European botanic gardens. Members of the university, original subscribers, the governor and state legislators, "strangers of distinction," and clergymen were admitted free. All others paid admission of twenty-five cents per visit or bought annual tickets. Peck made extensive use of the garden in teaching students and was known as a good field botanist.

After Peck's death, oversight of the Harvard Botanic Garden and instruction in botany at the college were taken over by Thomas Nuttall, an Englishman, who had come to Philadelphia in 1808 to study the natural history of America. By the time he became lecturer at Harvard he had made several excursions in the southeastern and midwestern regions of the continent and had published *Genera of North American Plants with a Catalogue of Species* (1818). Because the appropriations from the state were declining, he was not given a professorship. In 1830 the Massachusetts legislature withdrew its support of the professorship and botanic garden.

After Nuttall resigned from Harvard in 1833 to make an extended journey to the Pacific Northwest, no botanist was left in charge of the botanic garden. Although the Massachusetts professorship and botanic garden funds were not sufficient to cover a salary, prospects improved in 1833 with the bequest of Dr. Joshua Fisher of Beverly, Massachusetts, to support a professor of natural history at Harvard. The college administration wisely decided to let Fisher's bequest accumulate interest for a few years before making an appointment. The botanic garden was left in the hands of William Carter, Peck's original gardener.

THE BOSTON SOCIETY OF NATURAL HISTORY

A common interest in collecting and examining nature's productions also fostered the founding of natural-history societies. The first group in the region devoted solely to the study of natural history was the Linnaean Society of New England, which was founded in 1814, largely through the efforts of Jacob Bigelow, a physician. An 1806 graduate of Harvard, Bigelow spent a few years at the University of Pennsylvania studying medicine and botany with Dr. Benjamin Smith Barton, Philadelphia's most prominent botanist. On his return to Boston, Bigelow set up a medical practice, attended lectures at the botanic garden, and joined the MSPA. The Linnaean Society's main goal was assembling a cabinet of specimens that would spark public interest in the study of natural history. The society printed and distributed directions for the proper preparation of specimens, and zealous collecting ensued. At each meeting members brought specimens and presented papers.

After only about six years of existence, however, in-creasing donations to the collections, located in various rented rooms in the heart of Boston and cared for by volunteers, created a crisis for the society. The members were much too busy with their professions to properly curate the various objects that were rapidly accumulating. Facing an inability to maintain their valuable and diverse collections, they first proposed, but without success, that the Boston Athenaeum acquire them. Finally, in 1823 Harvard College and the board of the Massachusetts Professorship of Natural History were persuaded to accept the cabinet, agreeing to provide a building for it. Unfortunately, there were not sufficient funds to build a museum at Harvard and, despite good intentions, the Linnaean Society's collections languished as the society became defunct, or, more accurately, went into hibernation.

The enthusiasm of some of the original members of the Linnaean Society and a younger generation of Harvard College graduates, many of whom had studied under Professor Peck, could not be quelled, however. In 1830 several of them established a new organization, the Boston Society of Natural History, with much the same purpose as its forerunner. However, the BSNH managed to survive well into the twentieth century. Five of its founders had been active members of the Linnaean Society. One of the newcomers, from the Harvard class of 1817, was George Barrell Emerson, who was to become in later years a moving force in the formation of the Arnold Arboretum.

These men met several times during the next months to draft a constitution and bylaws and to solicit subscribers for a society with the same objectives as its predecessor, namely, to interest the public in the study of natural history. They hoped to fulfill this aim by promoting discourse at meetings, forming collections of natural objects, and offering courses of lectures to the public. One of the original members gave the following description of the state of knowledge of natural history at the time of the society's founding:

At the time of the establishment of the Society there was not, I believe, in New England an institution devoted to the study of natural history. There was not a college in New England, excepting Yale, where philosophical geology of the modern school was taught. There was not a work extant by a New England author which presumed to grasp the geological structure of any portion of our territory of greater extent than a county. There was not in existence a bare catalogue, to say nothing of a general history, of the animals of Massachusetts, of any class. There was not within our borders a single museum of natural history founded according to the requirements and based upon the system of modern science, nor a single journal advocating exclusively its interests.

We were dependent chiefly upon books and authors foreign to New England for our knowledge. . . . The laborer in natural history worked alone, without aid or encouragement from others engaged in the same pursuits, and without the approbation of the public mind, which regarded them as busy triflers. (Bouvé, 1880, pp. 20–21)

With so many of the original members graduates of Harvard, they hoped Thomas Nuttall would be president of the new society. Nuttall, however, declined to accept the office since he would soon embark on plant explorations in the American West. Instead, the members chose Benjamin D. Greene, son of Gardiner Greene, one of Boston's wealthiest citizens, whose fine gardens on Pemberton Hill were well known. The younger Greene graduated from Harvard College in 1812, after which he studied law and medicine (although he practiced neither profession). Having the means, he pursued his real interest, botany, as a capable amateur and liberal supporter of other workers in science.

One of the first undertakings of the BSNH was to sponsor a series of lectures, open to the public for a modest fee; the introductory lecture was given free of charge. The course of sixteen lectures began in October 1830 and continued weekly. Topics included the mineral kingdom and geology, anatomy and physiology of the vegetable kingdom, and anatomy and physiology of the animal kingdom, with separate sessions on mammals, birds, reptiles and fishes, insects, and invertebrate animals. Most of the seven speakers were members of the BSNH, and all were physicians except Thomas Nuttall, who presented four evenings on the vegetable kingdom. The series was held at the hall of the Boston Athenaeum. The *Boston Evening Transcript* reviewed Nuttall's contribution favorably, and reported:

. . . this is a distinctly fashionable assembly of ladies and gentlemen. Surely knowledge hath charms to bring together such

a concourse of intellectual auditors, from all sections of the city on one of the most cheerless November evenings. At each succeeding lecture the hall becomes more closely filled: the necessity for repeating the lectures is very obvious. (Graustein, 1967, p. 271)

Indeed, the lectures became an annual event and provided a small income to the society's funds. Often the introductory lecture was given by a celebrity—the much admired orator Edward Everett opened the series one year, John James Audubon another. In the nineteenth century Americans enjoyed an informative lecture series, much as they avidly watch nature and science programs on television today.

The formation of a cabinet was another project of the society. Although the few remaining specimens of the collection of the late Linnaean Society at Harvard were turned over to the BSNH, the members started collecting geological, zoological, and botanical objects once more. As can be seen in a listing of just a few notable donations to this cabinet in its first ten years, many of the society's enthusiasts had ready access to "natural curiosities" through the masters of sailing vessels:

At one meeting, Park Benjamin presented ninety-two beautifully preserved bird-skins and a box of insects, from Demerara. Joseph Coolidge, forty-five bird-skins, with corals and other objects from Bombay. John James Dixwell, one hundred thirty-three specimens of bird-skins, in perfect order, with many fishes, from the vicinity of Calcutta. J. N. Reynolds, a magnificent collection of between four and five hundred bird-skins; a large collection of botanical specimens; boxes of minerals, organic remains, and of insects; a large and valuable collection of shells; skulls, fruits, and fishes, all from South America, the Islands of Pacific, and the South Shetland Islands. Dr. James Jackson, a valuable Herbarium. . . . (Bouvé, 1880, p. 34)

The availability of exotic items brought home on merchant vessels and from special exploring expeditions certainly abetted the taste for natural history among New England's gentlemen.

At first the society met in a rented room; in 1833 it moved to a larger hall over a new bank on Tremont Street, where members welcomed an important donation to their cabinet, the natural-history collection of the Massachusetts Historical Society. At the first meeting in the new quarters, the Reverend F. W. P. Greenwood of King's Chapel said, "Even the mute representatives from the several kingdoms of Nature, which here in new order surround us, seem to participate in our pleasure, and, rejoicing in their deliverance from the damp and obscure region in which they have been hidden, to bid us welcome to upper air, and the comforts of our present abode" (Bouvé, 1880, p. 22).

Once the collection was in new quarters, those who purchased tickets to the lectures were welcome to visit the society's cabinet one day a week. Despite the public's interest in seeing the collections, maintaining and housing its "mute representatives from the several kingdoms of Nature" was to be a constant financial struggle for the BSNH. But the members' conviction of the value of collected objects in educating the public was unswerving. The society's financial base slowly grew, and in time the collections were opened as a full-fledged museum.

When a geological survey was authorized by the Massachusetts legislature in 1836, the BSNH proposed botanical and zoological surveys for the state as well. Under the auspices of the society, seven naturalists were selected to prepare the reports. Its president, George B. Emerson, was appointed commissioner of the survey, with appropriations for investigation and for publication of the botanical and zoological reports. These came out over the next few years and served as the valuable starting point for many later investigations.

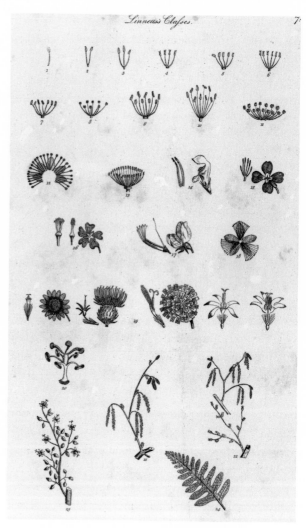

4. To name is to classify. Linnaeus was one of the most successful students of the plant kingdom in devising a simplified scheme to classify plants. In this rendering of his system, the twenty-four classes of plants are represented by single differences in numbers of flower parts.

PLANT CLASSIFICATION, FROM LINNAEUS TO A NATURAL SYSTEM

While botany had yet to receive professional status in the early nineteenth century, the physicians and other thinkers who had been contemplating plants over the centuries had made much progress, especially in the field of plant classification. For medical doctors the ability to identify plants was essential since they supplied many curative preparations, and classification aids identification. It is the long-standing attempt to name plant species and set up some

system of organizing or cataloging them that has resulted in the creation of collections of living plants such as the Arnold Arboretum as well as herbaria containing pressed plant specimens.

By the time the Arboretum founders began their botanical education, the work of the great Swedish naturalist, Carolus Linnaeus (1707–1778), was most widely utilized for identifying and naming plants. In response to the multitude of new kinds of plants and animals coming to the attention of European scholars as a result of world exploration, Linnaeus revamped earlier notions about classification. He created a relatively simple, workable system for flowering plants. To formulate his scheme, Linnaeus counted the numbers of reproductive parts in the flower of each type of plant to place it in a category with others like it; this was the so-called sexual system of classifying plants. Such a scheme would allow the student with a new or unknown plant to count flower parts to find a group (or order) to which it belonged. Even Linnaeus realized this system was artificial and did not always lead to groupings commonly perceived as natural. However, the convenience of the scheme led botanists in Europe and America to adopt it readily.

American books that attempted to name and classify local plants usually were organized on the Linnaean system of classification. One such effort by a New England botanist was Jacob Bigelow's *Florula Bostoniensis,* published in 1814 (the year he founded the Linnaean Society of New England) and dedicated to the trustees of the Massachusetts Society for Promoting Agriculture. As the work was to be a practical reference, it was written in English, although Bigelow was an accomplished Latin scholar. He organized the book on the Linnaean system, with genera and species presented according to the twenty-four categories of the sexual system. Peck's 1818 catalog of plants in the botanic garden at Harvard was also arranged in the Linnaean order.

Long after botanists began to agree that other ways of classification more accurately accounted for the relationships among plants, the Linnaean system was still used in American publications, especially elementary texts, in the belief that the books would gain in popularity over European works because of the simplicity of the Linnaean scheme. More enduring than his classification system was the Swedish naturalist's formulation of a readily referable method for naming plants: binomial nomenclature. This was most thoroughly worked out in *Species Plantarum,* a two-volume catalog of the plant system published in 1753. In this book, long descriptive phrases then used to denominate species were replaced by a two-word combination of genus plus species epithet. For example, to replace the cumbersome phrase-name *Andromeda foliis ovatis ob-* *tusis, corollis corymbosis infundibuliformis, genitalibus declinatus,* Linnaeus coined the binomial *Kalmia latifolia* for the American mountain laurel.

Since even Linnaeus realized that his classification was an artificial one, it is not surprising that students of plants continued to seek a scheme that would better reflect certain natural affinities they perceived. The next phase of taxonomic effort was toward what was termed a "natural system" for plants. This new system was first developed by botanists in France as an alternative to the Linnaean sexual scheme. Soon after Linnaeus published *Species Plantarum,* Bernard de Jussieu (1699–1777) attempted to re-arrange the plants in the Jardin des Plantes, Paris, according to the Linnaean classification. Finding that this did not group the plants that looked alike close together, Jussieu moved the living plants into an arrangement he found more satisfactory. After many years, in 1789 his nephew Antoine Laurent de Jussieu published a book, *Genera Plantarum,* that presented the natural classification scheme as worked out in the royal garden.

These ideas were further developed by the Swiss-French Candolle family, starting with Augustin Pyramus de Candolle's *Théorie élémentaire de la botanique* in 1813. His idea was to look at the whole plant, not just count its reproductive parts, and to look at all the plants that seemed to belong together and see what characters were constant within the group. The ideas in this work were further developed with the aid of his son, Alphonse de Candolle, and published in the seventeen-volume *Prodromus Systematis Naturalis Regni Vegetabilis* from 1824 to 1873. Their approach to the problem of classifying plants was empirical, relying on comparison of many observations, instead of being based on a possibly incorrect assumption about the importance of numbers of flower parts.

Other scientists from abroad whose refinements of the natural system reached American botanists were Stephan Endlicher (1804–1849) and John Lindley (1799–1865). Endlicher was a Viennese botanist who further revised the Candolles' systematic arrangement for his own *Genera Plantarum,* which was published from 1836 to 1840. John Lindley's *Introduction to the Natural System of Botany* (1830) was the first comprehensive explanation of the new ideas on classification to be published in English. Widely used in Great Britain and, eventually, in America, Lindley's work clarified the Candolles' scheme and emphasized the importance of observing the vegetative characteristics of plants as well as their reproductive structures. An officer of the Royal Horticultural Society and professor of botany at University College, London, Lindley published several books that popularized both botany and horticulture, many of which came out in American editions.

The editor of the 1831 American edition of Lindley's introduction to the natural system was John Torrey, a physician and professor of chemistry at the College of Physicians and Surgeons in New York. Torrey was, by this time, the dean of American botany. He had been appointed botanist for the Natural History Survey of New York, and with his protégé, Asa Gray, Torrey had begun compiling a flora of North America, using the natural classification system of his colleagues overseas.

ASA GRAY, PROFESSOR OF BOTANY

Torrey's young assistant was destined to have a great influence on botany as a field of study in America and on the status of the discipline at Harvard. Asa Gray, born in 1810 in central New York, obtained a medical degree in 1831 from the College of Physicians and Surgeons of the Western District, informally known as Fairfield Medical School. As was the custom, Gray's four years at Fairfield were interspersed with spring and summer terms of apprenticeship with a practicing physician. During his rounds in the country, Gray took the opportunity to study and collect plants. This growing passion for botany was furthered by Fairfield's chemistry and materia medica professor, James Hadley, who put young Gray in touch with other botanists in the state, notably John Torrey. Although he received a medical degree in 1831 and with it the capacity to pursue an honorable and financially rewarding career, Gray practiced medicine for only one year. Thereafter, he became a pioneer in establishing the scholarly pursuit of science as an acceptable and remunerative profession in America. For Asa Gray there were ten years of financial uncertainty and false starts before he came to profess botany at Harvard. But these were years of mas-

tering his subject and publishing articles and books that formed the basis of his reputation as America's leading botanist.

John Torrey first employed Gray in the summer of 1833 to collect specimens in New Jersey. In order to publish a flora of the continent, Torrey was accumulating an herbarium to document the plants and a library to aid his researches, all as an aside to a considerable teaching commitment in chemistry for New York's medical school. He welcomed the assistance of bright young Asa Gray to reorganize his herbarium from the Linnaean system to Lindley's natural classification. During the next years Gray stayed at Torrey's home and was paid for his help whenever his mentor's teaching and lecturing could support it. Gray also supplemented his income by sending duplicate sets of his herbarium specimens to botanists who were willing to purchase them. One such patron was Benjamin D. Greene, president of the Boston Society of Natural History.

In 1835 Gray began to write an elementary textbook on botany. In his view the texts in current use—Amos Eaton's *Manual* and Amelia Hart Lincoln Phelps's *Familiar Lectures*—were outdated in their dependence on Linnaean classification and too simplified to truly educate American youth in plant science. Gray did not want to sell his readers' abilities short; he provided them thorough and correct detail in a comprehensible style. Like the members of the Massachusetts Society for Promoting Agriculture, Gray presumed that botany was as useful to future farmers as to future physicians. His *Elements of Botany* was published in the spring of 1836, bringing him favorable reactions from all over the country.

In 1838, Gray accepted an offer from the new state of Michigan to be professor at its embryonic university, still very much in the planning stages. Here was the chance not only to profess botany but to influence the whole educational program of a university, for Gray was called in before a building had even been erected. The proposed curriculum included an emphasis on science, which was unique among American universities at the time. The regents immediately sent Gray to Europe as their agent to obtain books for the new university library.

The trip proved a great opportunity for the American botanist. Gray not only fulfilled his mission for the University of Michigan satisfactorily, but he visited herbaria and botanists, cementing friendships and enlisting cooperation for his and Torrey's North American flora. Foremost among his new botanical acquaintances were the director of the Royal Botanic Gardens at Kew, William Jackson Hooker, and his son Joseph Dalton Hooker. On his return to the United States, however, Gray found that the budding university had suffered a drastic financial setback as a result of the panic of 1837. He had no alternative but to accept Michigan's request that his salary be suspended for the coming year. Gray never went to Michigan. He stayed on with Torrey in New York and worked on the first volume of the flora, which was completed by 1840. Gray also began a series of reviews of botanical works in the *American Journal of Science,* popularly known as Silliman's Journal, after the Yale professor who edited it. In these reviews Gray introduced new European scientific literature to Americans and argued for the adoption of the natural system of plant classification. His reviews of Lindley, Alphonse de Candolle, and Endlicher guided many American botanists.

Hearing of the unfulfilled Fisher bequest for a professorship at Harvard, Gray wrote to Benjamin D. Greene concerning it in 1841. Gray could not have approached a better person, for Greene was the son-in-law of Harvard's president, Josiah Quincy. Greene soon extended Gray an invitation to come to Boston. Although it was his first visit, Asa Gray was already known to the Harvard-Boston scientific community. He had been a corresponding member of the Boston Society of Natural History since 1837, and he and Torrey had been elected to membership in the American Academy of Arts and Sciences in 1840. Gray's numerous publications and textbooks were familiar to the New England botanists. While in Boston, he met many of the area's naturalists and representatives of Harvard.

Soon after Gray's visit, President Quincy began negotiations with him for his appointment to the Fisher professorship. Earlier refusals of the position by two area botanists had been due to their lack of training and interest in the other branches of natural history. In making its

5. *Botanist Asa Gray (1810–1888) joined the Harvard faculty in 1842. He pioneered in the classification and naming of numbers of North American plants, many of which were unknown to European creators of classification schemes until the middle of the nineteenth century.*

offer, the college administration set aside its concern for the whole of natural history and suggested Gray need only teach botany and superintend the botanic garden. Moreover, the presence of the garden justified an emphasis on plant sciences. President Quincy, recognizing the eminence Gray had gained in that single field, wished to bring it to the college. With only botany to teach, Gray would have ample time to continue his research. His appointment coincided with a shift at Harvard toward greater emphasis on research activities.

Asa Gray's appointment was indicative of the beginning of specialization in American science. He had taken the initiative in becoming a specialist and by doing so had surpassed others in his mastery of botany and in his insight into the future direction of the field. At Harvard Gray would have the support to become a leader in North American botany; he would, as well, come to contribute his expertise to the great theory of the nature of life that would be introduced less than two decades later by Charles Darwin.

6. *Asa Gray developed his notion of what constitutes a species by viewing thousands of pressed specimens, such as this* Brickellia baccharidea *from the mountains near El Paso, Texas, collected by Charles Wright in 1849. Dried specimens are compact, portable evidence of the morphology, geographic distribution, and many other characters of plant species. Once mounted, they become enduring documents not only of the plant but of its scrutiny by botanists. When Wright sent the specimen to Gray, he knew only that it was a member of the* Compositae, *a family that was one of Gray's specialties. Once Gray realized that the specimen represented a new species, and after inscribing the name he selected for it, the botanist wrote the abbreviation* n. sp. *(new species) on the specimen label. Thus, the sheet is the type specimen for* Brickellia baccharidea. *Botanists who have subsequently examined the specimen and attached annotations are Benjamin Lincoln Robinson, director of the Gray Herbarium from 1892 to 1934, and David Boufford, who heads the present-day creation of an electronic database of type specimens that includes the use of bar codes.*

Gray was welcomed by the community of naturalists in Boston as he earnestly took up his duties. He participated in the programs of the Boston Society of Natural History, giving the annual address in 1844. But while Asa Gray regularly contributed at the group's meetings and made additions to its herbarium, his duties at Harvard were far too pressing to allow him to become an influence on the society's activities. In Cambridge he instructed students in botany, oversaw the renewal of the botanic garden collections, and attempted to go forward with research on the North American flora.

This last work soon became sidetracked, however, by a constant influx of pressed and dried plants from a vast area, the vegetation of which was unknown and undocumented—the American West. With the collaboration of the physician-botanist George Engelmann of Saint Louis, Torrey and Gray received herbarium specimens from every botanical traveler that the doctor at the "gateway to the West" could enlist. With the promise of potential novelties for the botanic garden, Gray sometimes secured remittance for these collectors from such patrons as John Amory Lowell and Benjamin D. Greene. Then, in the 1850s, the federal government provided for scientists to accompany its surveying expeditions. The Mexican Boundary Survey, from 1848 to 1854, and the great railway surveys of 1853 and 1854 produced a wealth of dried plant materials that made their way to the desk of Asa Gray. The collections contained many new, undescribed species and genera, as well as information that extended the geographic range of known plants—all of it essential for a comprehensive elaboration of the North American flora.

Each collector's booty had to be worked up: the plants identified, compared with specimens of known kinds or with published descriptions, the new kinds carefully described and named. In order that his findings be given priority by other botanists, Gray had to publish accounts of each of the collections as soon as possible. It would not be prudent to wait to include them in the *Flora*, for in the interim someone else might give names to the new plants and negate all Gray's efforts. "Plantae Lindheimerianae" (1846), "Plantae Fendlerianae Novi-Mexicanae" (1848), "Plantae Wrightianae Texano-Mexicanae" (1852–53)—the results of expeditions by Ferdinand Jakob Lindheimer, who botanized for Engelmann and Gray while joining the German Settlements in Texas; by Augustus Fendler, who scaled the southern Rockies in New Mexico; and by Charles Wright, who obtained plants near the Texan boundary of the United States—were among Gray's many papers on the botany of the trans-Mississippi West.

Besides the plants from the North American expeditions, Gray found himself working on those brought home by the United States Exploring Expedition to the South Pacific, the so-called Wilkes Expedition of 1838 to 1842. Essential to Gray's project was a trip to consult European botanists and compare the expedition specimens with those in European herbaria. As the scope of Gray's botanical knowledge took on global dimensions, he received plant collections made on Commodore Perry's mission to open trade with Japan and from the United States Surveying Expedition to the North Pacific Ocean of 1853 to 1856.

THE THEORY OF EVOLUTION

After studying thousands of specimens and determining whether each represented a known species or was new, Gray had more experience with the "species question" than any other American botanist. He realized that species varied much more than was generally supposed. He also observed intriguing patterns in the distribution of species over the world. Why did such a high percentage of eastern North American plants have similar or even identical counterparts in Japan and nowhere in the intervening region? The flora of eastern North America seemed to have greater affinity with that of eastern Asia than with the floras of western America or Europe. What had caused one species to exist only in Japan and in eastern North America, or one genus to have its species distributed only in those two regions?

It was just such questions about geographic distribution that stimulated Charles Darwin's twenty-year probe into the origin of species after the voyage of HMS *Beagle* concluded in 1836. In the Galápagos Islands, Darwin had noticed that many animals were similar but not identical to ones on the mainland and that they varied enough from

island to island to be considered separate species. "It was evident that such facts as these, as well as many others, could only be explained on the supposition that species gradually became modified; and the subject haunted me," Darwin explained in his autobiography (Darwin, 1887, p. 82). In search of more facts on the distribution of species—especially in the plant kingdom, with which he was less familiar—Darwin, at the suggestion of J. D. Hooker, initiated a correspondence with Asa Gray in the mid-1850s. Calling on Gray's knowledge of the floras of several continents, Darwin suggested he compare the North American flora with that of other temperate regions. With Darwin's stimulation, Gray published two benchmark papers on this topic, "Statistics of the Flora of the Northern United States" (1856) and "Diagnostic Characters of New Species of Phaenogamous [i.e., seed-bearing] Plants Collected in Japan by Charles Wright, Botanist of the United States North Pacific Exploring Expedition" (1859). In these papers Gray enumerated the species and genera that occur exclusively in eastern Asia and eastern North America and explained the disjunct distribution as the result of geologic history. He suggested that the flora was formerly a continuous and cohesive one that stretched from North America to Asia across formerly exposed land in the North Pacific, the Bering Strait region. With the advance and retreat of glaciers and the separation of two continents by the sea, the remnants of the former flora that remained in eastern North America and eastern Asia diverged. They retained many similarities, however, since the two regions share similar climates. A quite dissimilar group of plants grew up in the intervening, climatically drier middle- and far-western areas of America.

In the "Diagnostic Characters" paper Gray acknowledged Darwin's theory, which had been presented jointly with A. R. Wallace (who formulated the theory independently) before the Linnaean Society of London in 1858. In fact, it was a sketch of his theory which Darwin had sent to Gray that Darwin used to prove his priority over Wallace. Gray presented his ideas on geographic distribution at meetings of the Cambridge Scientific Club and the American Academy of Arts and Sciences in Boston. Here and in ensuing debates Gray contrasted his and

7. Stewartia pseudocamellia *is an eastern Asian counterpart of the eastern North American mountain* camellia (Stewartia ovata). *The genus* Stewartia *has six species, all of which occur naturally only in these two areas. Asa Gray was one of the first botanists to recognize the relatedness of the floras of these two temperate regions, and his papers on the subject were instrumental in Darwin's development of the theory of evolution.*

Darwin's ideas with idealistic assumptions of immutable species and repeated creations.

When Darwin's seminal *Origin of Species* was published in 1859, Gray introduced it to American scientists with a favorable but critical review in the *American Journal of Science*. With praise for the author's research, Gray argued for a fair hearing of Darwin's well-documented hypothesis. Despite Darwin's lack of explanation for a mechanism to cause variations to be handed down from one generation to the next (the Austrian monk Gregor Mendel's discoveries—the foundation of the science of genetics—were unknown at the time), Gray found that Darwin's theory of derivation by natural selection explained much more about the natural world than prior idealistic constructions. Darwin's theory attributed the geographic distribution and relationship of species to natural phenomena, as he had observed in case after case. To natural theologists, who saw nature study as unveiling a perfect plan, species and their origin were inexplicable. But to empirical scientists like Gray and his colleagues (among them the geologist William Barton Rogers and the zoologist Jeffries Wyman), Darwin's theory at least provided a useful line of research to be pursued. And it certainly lent support to natural systems of classification upon which biologists had long been at work.

Acceptance of this ingenious theory was, nonetheless, difficult for all those natural historians who devoutly depended on nature for their knowledge of the Creator. In America, Louis Agassiz had become their spokesperson. A Swiss comparative anatomist and geologist, Agassiz had made a reputation in Europe with impressive work on fossil fishes and a bold defense of the glacial theory. Agassiz held a professorship at Neuchâtel, but the publication of his five-volume *Recherches sur les poissons fossiles* (1833–43) had left him nearly bankrupt. When John Amory Lowell offered a handsome honorarium to lecture at the Lowell Institute for the 1847 season, he welcomed the opportunity to come to Boston.

Agassiz's infectious enthusiasm for natural history won over all who met him or heard his lectures. After his successful Lowell season, there was such a demand for him on the lecture circuit that Agassiz made the United States his home, eventually taking up citizenship. The Swiss American naturalist became professor in the Lawrence Scientific School, a new department of Harvard backed by the industrialist Abbot Lawrence, and he opened a school of his own for young women.

Agassiz's first lecture before the large Lowell crowd was "The Plan of Creation in the Animal Kingdom." This topic was just what the audience wanted to hear, since their taste for natural history was piqued by piety as much as curiosity. Under the influence of centuries of Christian thinkers, most naturalists believed that the species of animals and plants on earth existed just as the Supreme Being had created them. They had not changed since each was separately created, and they existed in the locations where they had been created. Fossils were explained as previous creations that had become extinct through catastrophic events such as floods or, for Agassiz, ice ages. This way of looking at species—as fixed, distinct, unrelated units—promoted the nineteenth-century collecting impulse. If species were, indeed, immutable and finite, then all one needed was one example of each. Constancy of species was the notion underlying the accumulation of natural-history collections and personal cabinets. The idea influenced the actions of those explorers and scientists who expended more effort on the search for new species than on collecting detailed data about ones already known.

Boston Society of Natural History members eagerly welcomed Agassiz to the community. To them, he was the epitome of a naturalist. In awe of this knowledgeable European who had studied with Cuvier and Humboldt, they had made him an honorary member of their society in 1837. Agassiz, in turn, encouraged their efforts and attended their meetings. He insisted that Americans should have more confidence in their own contributions to science and not always look to Europe for leadership. In truth, promoting the prestige of American science was the greatest contribution of Agassiz's career in the United States.

But when he spoke on the fixity of species, Agassiz met with thorough arguments in support of Darwin from many Boston-Cambridge naturalists. In the Boston Society of Natural History meetings, it was William Barton Rogers who confounded Agassiz with detail after detail of geology that spoke in favor of Darwin. Although the society's president, Jeffries Wyman, took no part in public debates, he shared his vast knowledge of the animal kingdom with Darwin. Again, at the Cambridge Scientific Club and at special meetings of the American Academy of Arts and Sciences, Asa Gray defended the idea of evolution and, most important, maintained that it was not necessarily inconsistent with theistic belief. Gray's published works in support of the derivative hypothesis far outweighed Agassiz's, who seemed never able or willing to commit himself on paper.

Scientific opinion gradually embraced Darwin's revolutionary doctrine, but pre-Darwinian thinking endured among naturalists long after 1859. It required changing one's whole view of nature to see and understand the mechanism of natural selection. For some, sentiment and admiration for Agassiz as champion of God's Plan interfered with acceptance of the derivative hypothesis.

PRIVATE COLLECTIONS BECOME MUSEUMS

The controversy over evolution did nothing to disrupt the museum-building zeal of either the BSNH or Agassiz, however. In fact, this was the beginning of a great period of institutionalization of collections and expansion of science into all levels of education. The Boston Society of Natural History initiated a new building fund in 1854, and by the year of publication of *The Origin of Species* a petition was brought before the legislature requesting for the society a block of the new land being formed in the Back Bay. This proposal was part of a scheme advocated by a group of large-minded citizens to set aside land for institutions of science, industry, and fine art to be housed in one great conservatory. At one point the BSNH considered joining forces with the Massachusetts Horticultural Society to build new quarters. However, the final outcome of negotiations and petitions to the legislature was a grant of one square to be shared by the BSNH and the new Institute of Technology (which would become MIT).

8. The Magazine of Natural History, *said to be the first of its kind, was begun in 1828 by John Claudius Loudon, a Scotsman, who also published works on trees and landscape design. An advocate of educational reform as well, Loudon encouraged the study of natural history by the populace.*

By 1862, the BSNH committee had raised enough money to hire an architect, and construction began; a year later, the three-story brick building was complete. The dedication ceremony in June 1864 included mention of the civil war gripping the nation. Professor Rogers "regarded the interest shown in the Society during these years of war as evidence of the desire of the community for truth." He ended his speech "gratefully referring to those who were struggling through conflict for peace, without which many of the blessings we enjoy would vanish like smoke" (Bouvé, 1880, p. 99). Facing Berkeley Street between Boylston and Newbury, the grand structure would be home to the natural-history collections for the next eighty years. In 1864 the building stood nearly alone on the flat plain of new-made land, but it was soon joined by the building of the Institute of Technology, and later by the Harvard Medical School, the public library, and several churches as the Copley Square area formed a new cultural center for the city. Opening its doors two days a week, the society took on a new role as overseer of a public museum, even as it struggled to raise a permanent endowment to maintain its new building.

Meanwhile, across the river, Agassiz was equally or even more successful in founding a natural-history museum. The Museum of Comparative Zoology, inaugurated in 1860, had as its nucleus his personal collection of zoological specimens, which had been kept in an old boathouse since his arrival at Harvard. Singlehandedly, Agassiz secured an appropriation of $100,000 from the Massachusetts legislature. This grant, together with $71,000 raised by private subscription and a bequest from Bostonian Francis C. Gray, made the new zoological museum one of the best-endowed in America. Under the auspices of Harvard, "Agassiz's Museum" became a center for research and teaching.

9. After maintaining its collections in rented rooms for many years, in 1863 the Boston Society of Natural History garnered the funds to erect this building on Berkeley Street, located on the recently filled area known as the Back Bay. The museum soon drew visitors to view the endless variations on themes that its classified organisms presented. Members of the society also authored surveys of Massachusetts plants and animals.

10. *The cephalopod collection at the Museum of Comparative Zoology, Harvard University. To bring the whole of nature's wonders under one roof and exhibit them in a classificatory order was the goal of such museums as those of the Boston Society of Natural History and the Museum of Comparative Zoology, the latter initiated by Louis Agassiz in 1860.*

Joining the museums movement, Asa Gray took steps to assure the permanent maintenance of his botanical collections. In 1863 he offered his herbarium, comprising some two hundred thousand specimens and a library of more than two thousand works, to Harvard University if a fireproof building could be provided to house them. Funds were raised for a building and for an endowment to maintain and increase the collections. Built up from Gray's own acquisitions and the harvest from his many agents and correspondents across the continent and around the globe, the Gray Herbarium is still one of the nation's most important centers for botanical study. Even in the 1860s it contained a large number of type specimens, increasing its value even further, for "types" document the initial published description of a plant species.

Although the herbarium was by nature not a public museum, the botanic garden was, and Gray maintained it well on limited funds. He made his most valuable contribution to public education, however, by authoring numerous textbooks, including *Elements of Botany; Field, Forest, and Garden Botany; How Plants Grow; How Plants Behave;* and *Manual of Botany of the Northern United States.*

The new museums had a beneficial influence on how natural history was taught. The old method of teaching, by listening to recitation from books, had long been practiced at Harvard, but there was a countermovement, especially among scientists. Jeffries Wyman was celebrated for making his anatomy students find out the nature of a bone by looking at a bone. Agassiz, too, was a master educator; many admired his knack for "teaching by things . . . and not by books, opening the eye to the richness and beauty of nature" (Bouvé, 1880, p. 157). Agassiz would walk into a classroom, place a live insect in each pupil's hand, and allow some moments of observation before beginning to lecture. Opening whole museums of "objects" from which to learn was part of this trend.

The Boston Society of Natural History became a more active educational force soon after opening the Berkeley

Street building. At the instigation of the Reverend Robert C. Waterston, a committee of the society conceived a plan to impart instruction to the teachers of public schools. With a donation for the purpose, in 1865 they began the Teachers School of Science, an illustrated lecture series in natural history expressly for Boston teachers. There was a hiatus after the first, successful year of this program, but by 1871 the Teachers School of Science was in full swing, and the society kept it going for more than ten years with the help of special donations. The public school department took an active interest, introducing natural history into the curriculum. The Institute of Technology, the Museum of Comparative Zoology, the botanic garden, and other institutions cooperated in the project. Each year an average of one hundred teachers attended lectures given by some of the area's noted scientists. They were supplied with specimens, illustrations, and notes to take back to their classrooms. Sometimes there were instructive field trips; always the use of the society's museum collections was encouraged.

Once the BSNH museum was opened to the public in 1864, attendance increased steadily. By the late 1860s, about thirty-five thousand visitors came each year, with the halls open only Wednesdays and Saturdays. Starting in 1869, two city police officers were detailed to duty at the museum on public days. Their presence came to be so necessary to the preservation of order and protection of the specimens that four years later, when the officers were withdrawn, the society shut down the museum. The public outcry was such that the mayor and police chief soon agreed to provide police protection once more. All this outreach caused the members of the society to look again at their collections and decide to make them more comprehensible. The custodian and several members proposed to rearrange the collections according to some easily understood and comprehensive plan that illustrated the general laws of natural science. The collections were labeled uniformly for the visitor's information. In another educational innovation, a special New England section was displayed in each department.

The formation of these sections meant extricating from the vast and heterogeneous collections small groups of objects that would have the most meaning for visitors. At the society's museum, New Englanders could learn larger generalizations about biology from their own flora and fauna. They could study the objects with which they were most familiar or that they were most likely to encounter in the natural world close to home. That the members of the society decided to rearrange the collections to better illustrate the natural sequence and relationship of forms is indicative of acceptance of evolutionary theory as the basis for studying natural phenomena.

SCIENTIFIC FARMING AND FOREST CONSERVATION

While these movements in collecting, cataloging, and teaching about nature's productions were taking place, horticultural developments just as strongly influenced the founding of an arboretum. Early-nineteenth-century landowners began to practice what they termed scientific agriculture, managing their estates as experimental farms. The science of horticulture was widely promoted in the decades before the Arboretum's establishment. And a new concept of landscape design emerged as formal style was replaced by a "naturalistic" one on private estates and in spaces planned for public use.

In addition to advancing the botanic garden, the Massachusetts Society for Promoting Agriculture became an active force for improving farming practices in the region. Few of its founding members were dependent on farming for their livelihood; to the contrary, many were successful merchants. Yet, none of the members was far removed from cultivated ground. Many of them maintained farms in Charlestown, Cambridge, Roxbury, and other surrounding towns. As the society's motto asserted, farming was considered "the source of wealth"—not only of material wealth but of moral and spiritual welfare.

The society awarded prizes for scholarship related to agricultural problems and for successful breeding and introduction of improved livestock and crops. The members published a journal, organized a cattle show for many years, and promoted the founding of local agricultural societies throughout Massachusetts. By facilitating communication and interaction among the state's farmers, the society encouraged excellence in this essential endeavor.

One pursuit advocated by members of the MSPA was raising merino sheep, a Spanish race found to be hardy and to have superior breeding instincts. The introduction of this breed, with its fine, crimped fleece, greatly expanded the woolen-apparel business in the United States, and importing and raising them became a fad among gentlemen farmers. For example, Éleuthère Irénée Du Pont, a French chemist and businessman who emigrated to the Brandywine River valley in Delaware in 1800, also kept merino sheep. Even as his gunpowder mills grew into a great chemical firm, Du Pont maintained gardens and farmlands for the support of his family and his workers. A more famous American farmer, who wrote in 1811 that he was "adding the care of Merino sheep," was Thomas Jefferson. The following year he corresponded with Du Pont about the best method for having cloth manufactured from the fleece of his merinos.

Notable Boston area practitioners of scientific farming included Christopher Gore and Theodore Lyman, both members of the MSPA. After serving in the Revolution and as a diplomat in England for some years, Gore returned in 1804 to create a grand estate and model farm on some three hundred acres in Waltham. Nearby, Lyman, a Boston merchant, also developed his several hundred acres with trees and gardens planted for pleasure, as well as farmland that was screened from the house by an upland clothed with native woods.

This last item is indicative of another practice of these scientific farmers: an early form of woodlot management. The idea of allowing farmland to seed in with trees was, at the time, somewhat radical. Americans had struggled to cut back forests for almost two hundred years. Cleared land, fully used to grow crops and raise animals, with fields defined by walls made from the interfering rocks, was considered a sign of progress. It was during the early nineteenth century that wood shortages were first feared and the woodlot became an accepted part of the well-managed land of an enlightened farmer. Planting trees in shelterbelts or as specimens for their beauty and novelty was also favored by learned agriculturists.

Europeans had been practicing forest management and renewal for decades. In England the warning had been sounded as early as 1664 in a widely read book by John Evelyn, *Sylva, or a Discourse of Forest-Trees and the Propagation of Timber in His Majesties Dominions.* Apparently the country's landowners took Evelyn's criticism of the wastage of forests to heart; by the time of his death, they had planted several million trees throughout the countryside. One of the first to publish a plea for an end to the destruction of American forests was a Frenchman, François André Michaux. In 1819 in his *North American Sylva,* Michaux warned of the need for better management of American forests: "In America . . . neither the federal government nor the several states have reserved forests. An alarming destruction of the trees proper for building has been the consequence" (Michaux, 1819, p. 4).

As the century progressed, forest conservation was taken up in 1864 by a Vermonter, George Perkins Marsh, in his book *Man and Nature, or Physical Geography as Modified by Human Action.* The author was a man of wide interests and experience: lawyer, linguist specializing in Scandinavian languages, designer, Vermont congressman, minister to Turkey and then to Italy. Drawing on sixty-three years of observations of the effects of man's removal of forest in his native Vermont as well as in the Mediterranean region, Marsh put forth the idea that man was a disturbing agent, as important as any geological force in shaping the earth and its climate. He convincingly explained the importance of forest cover to watersheds, and argued that cutting down trees on a large scale had drastically affected the water-retaining capacity of the lands of the Middle East, a process today called desertification. Marsh was the prophet of the modern conservation movement. "With the extirpation of the forest," he remarked, "all is changed. At one season, the earth parts with its warmth by radiation to an open sky—receives at another an immoderate heat from the unobstructed rays of the sun. Hence, the climate becomes excessive, and the soil is alternately parched by the fervors of summer and seared by the rigors of winter. . . . The face of the earth is no longer a sponge, but a dust heap" (Marsh, 1874, p. 301).

In contrast to the eighteenth-century view of man as beneficial agent whose charge was to wrest order out of earth's chaos, Marsh epitomized the nineteenth-century realization that man must learn to harmonize with nature. Man needed to become better informed about the environment and the results his actions could produce. Marsh's volume had a slow reception at first, so much so that he donated its copyright to the United States Sanitary Commission. Friends bought it back for him, however, and in a few years demand for the book was great enough to warrant a revision under the title *The Earth as Modified by Human Action*. This work greatly influenced George B. Emerson and Charles S. Sargent, among others.

HORTICULTURE, A REPUBLICAN FINE ART

Even for New Englanders who did not own farms the desire to cultivate plants was widespread. Whether a cultivator had a modest urban plot, such as the residences on Boston's Summer Street (which were known for their gardens), or many acres in nearby Roxbury, the nurturing of plants for fruit, vegetables, or flowers was an enjoyable avocation for many. Fruit growing was especially popular in nineteenth-century America. Because cider was the most widely consumed beverage, one of the most important productions of New England farms in the early Republic was apples. Orchards were as common on any agricultural tract as plowland or pasture, and even smaller plots were devoted to fruit. Such demand stimulated the search for new varieties. For many growers, the raising of as many varieties as possible became a passion not unlike the impulse for collecting natural-history objects. It became a challenge for these pomologists to grow not only apples but pears, peaches, cherries, grapes, and more. Garden vegetables and ornamental flowers were also widely experimented with. Gentlemen who shared this interest in gardening established the Massachusetts Horticultural Society in 1829. Through weekly shows and competitions during the growing season the society aimed to promote the art of horticulture.

One of the new society's ambitions was formation of an experimental garden, a place to try the seeds of many varieties of vegetables, flowers, and fruits received in ex-

change from colleagues in England and on the Continent and to demonstrate horticultural arts to the public. Another project that interested some MHS members was the creation of Mount Auburn Cemetery in Cambridge, a place intended to combine the sanitary improvement of burial outside the crowded city with the amenity of a landscaped setting. This was instigated by the physician-naturalist Jacob Bigelow. He and the new president of the horticultural society, General Henry A. S. Dearborn, conceived a plan for uniting the experimental garden with the cemetery under the auspices of the MHS.

After the founding of Mount Auburn in 1831, the horticultural society's experimental garden was located in the northeast section, set apart from the cemetery proper by a pond. In 1832 General Dearborn laid it out for the cultivation of fruit and ornamental trees, with beds assigned for vegetable and flower trials. A gardener was hired and provided a cottage on the grounds, as was the custom. For a few years the garden thrived, and its products were proudly displayed at MHS shows and festivals. The connection between the Massachusetts Horticultural Society and the Mount Auburn Cemetery proprietors soon changed, however. The experimental garden did not attract the financial support that the cemetery did, and by amicable agreement the latter was incorporated separately in 1835. The experimental garden was given up and the land turned over for cemetery use. The horticultural society redirected its support of garden arts to promoting exhibitions and competitions in the private sector. Since so many of the society's members were very extensive gardeners in their own right, they probably did not feel the need for the joint experimental garden.

The annual autumn exhibition was the highlight of the horticultural year for these avid gardeners. For example, in 1839 "fruit exhibits were increasing rapidly; indeed, nothing but an examination of the long lists of varieties of apples and pears can give an adequate idea of the labor and time and money devoted at this period to the collection and testing and naming of these fruits. At this exhibition Robert Manning [later secretary of the society, 1876–1902] showed seventy varieties of pears" (Benson, 1929, p. 62).

By the time the Massachusetts Horticultural Society

was twenty years old it had been joined by others in the state. The Worcester County Horticultural Society was founded in 1840. In New Bedford, where many a successful sea captain or whaling merchant gladly spent his leisure time improving his land, puttering in a greenhouse, or coaxing new varieties of fruit to ripen early, a horticultural society was formed in 1847. The New Bedford group met frequently to communicate ideas and promote lively competition in vegetable, flower, and fruit growing.

As a demonstration of the productions of Massachusetts gardens and a gathering of horticulturists from all over the Northeast to mark the success and moral influence of their activities, the twentieth-anniversary celebration of the MHS was unsurpassed. At the show, which was held for several days in early fall, the display of fruit was termed "overwhelming." Robert Manning, for example, exhibited 250 varieties of pears and 80 of apples. For the anniversary dinner, Faneuil Hall was decorated with groves of "large exotics and evergreens" and its columns were wreathed with flowers from floor to ceiling. Tables were laden with flower and fruit arrangements. On decorative banners, the names of time-honored horticulturists, naturalists, and botanists circled the hall, among them Linnaeus, Candolle, Michaux, and Loudon, along with the names of such contemporaries in Boston horticulture as Manning, Lowell, Lyman, and Wilder. Under the full blaze of gaslight, five hundred women and men were seated for dinner, which concluded with ices and, of course, luscious fruits.

Delegates from horticultural societies as far south as Delaware and as far west as Saint Louis, and such political notables as Robert C. Winthrop, Speaker of the House of Representatives, and Josiah Quincy, Jr., mayor of Boston, participated in a long evening of speeches and toasts, interspersed with band music and vocal odes. The president of the society, Marshall P. Wilder, began with an address highlighting the contribution of science to the garden arts and employing tree planting as a metaphor for providing for future generations. The mayor gave a speech, then toasted "the Environs of Boston: the Battle Grounds of Freedom; famous, at present, as the abode of taste and refinement, where as in Eden, woman watches over the flowers, and man finds his most innocent em-

ployments in the culture of the soil" (Anonymous, 1848b, p. 227).

All evening, sentiments were expressed in praise of horticulture as an elevating pursuit. There was plenty of humor and entertainment, yet deep religious and political convictions were revealed in comparisons to the Garden of Eden and in references to horticulture as a pastime available equally to all. At this relatively peaceful and prosperous time, New Englanders could well boast that horticulture was "the Fine Art of common life . . . a Republican Fine Art," as Winthrop expressed it (Anonymous, 1848b, p. 230).

THE ENGLISH STYLE IN LANDSCAPE DESIGN

As important as progress in horticulture was to the interest in forming a tree collection, the corollary evolution of a new style of garden design was probably a greater influence on the creation of the Arnold Arboretum. The new, naturalistic style was developed in England, but Americans eagerly adopted it, especially for large properties and public grounds. This way of composing landscapes used nature as its model. It was especially appropriate for the planting of trees, and in some respects its development was fostered by a concern for reforestation.

It may be hard for citizens of the twentieth century to appreciate that the natural style was a radical departure, since the landscapes of our parks and homes have long been influenced by notions of wilderness and wild scenery that were introduced in the eighteenth century. The late-nineteenth-century American parks movement, of which the Arnold Arboretum is a product, was an important outcome of this new approach to landscape design.

As in many aspects of culture, English ideas influenced Americans in laying out their gardens in the early nineteenth century. And in England a revolution in garden design had begun in the eighteenth century. Until this revolution, the ideal of beauty in gardens had been formal geometrical arrangement. For ease of cultivation and efficient production of food and medicinal herbs, earlier gardeners preferred straight lines and symmetrical designs.

11. *Faneuil Hall was decorated with garlands and elaborate arrangements of plant material for the twentieth-anniversary celebration of the Massachusetts Horticultural Society in September 1848. Banners honored the names of botanists and horticulturists, proclaiming such mottoes as "Your voiceless lips, O flowers, are living preachers."*

12. *Deer park at Inverary Castle, Scotland. The so-called English style of landscape arrangement developed as a revolt against formal, geometric garden designs. With the adoption of a naturalistic style, wooded areas and lawns with widely spaced clusters of large trees became a fashionable part of the manor house grounds. The master of this estate was Archibald Campbell, the third duke of Argyll, who introduced many foreign trees to the grounds at Inverary as well as to the landscape garden he created around his mansion at Whitton, Middlesex, England.*

An orderly, patterned garden was considered beautiful because it represented the ability of humankind to control the supposed chaos of nature. The garden was most often an enclosed area, considered an extension of the house; hence, its design was architectural. This style, now often referred to as formal, was carried over into gardens created for pleasure until the natural style was introduced as an alternative.

From ancient times to the Enlightenment, the parterre, or knot garden, had been the norm. The gardens of Versailles are the grandest example of this type. Envisioned as the setting for the pageantry of the Sun King's court, their rectangular layout with strong central axis flanked by patterned beds proclaimed the royal authority of Louis XIV. Although few English gardens were designed with such self-importance or on such a grand scale, they were usually laid out in geometrical patterns. The great estate owners in England and France often had land beyond their formally planted areas that was forested and cut through with rides for hunting. These deer parks dated back to medieval times.

In eighteenth century England the new, natural style in landscape planning resulted partly from a growing fondness for Italian landscape painting, which showed classical temples amidst rural scenery with groves of trees, lakes and streams, and craggy mountains. Nicolas Poussin and Claude Lorrain, French painters who worked in Italy in the seventeenth century, were the first to concentrate attention on landscapes as subjects rather than as backgrounds. Poussin associated the beauty of nature with moral, religious, or historical themes. His compositions were not exact depictions of a rural or wild place, but rather perfectly proportioned, balanced selections of nature's best, synthesized into an idealized, harmonious setting. Lorrain painted landscapes that showed a greater sensitivity to nature and a more romantic attitude than Poussin's. The landscapes they depicted were idylls of pastoral life, and they became immensely popular.

When discerning English landowners saw these paintings, they were inspired to create landscapes as serene and beautiful as those pictured. The creation of an English landscape garden was not an attempt to replicate wild nature but to create an idealized scene. As the landscape paintings often included poetic or classical references, so did the new garden style. A fresh idea of what was beautiful was communicated through one medium, painting, and then tried in another, garden design.

Writers also expressed new ideas about garden beauty during this period. The satirist Alexander Pope was a passionate gardener who deplored the overly ornate, intricately geometric gardens full of statuary and topiary that were everywhere popular during the reign of William and Mary. In his *Essay on Criticism* Pope exhorted his readers to "first follow Nature, at once the source and end and test of Art." Another advocate of a new landscape style was Joseph Addison, essayist, poet, and statesman. When the vogue was for formal gardens in the French manner, he wrote in an essay: "Our British gardeners . . . instead of humoring nature, love to deviate from it as much as possible. . . . I would rather look upon a tree in all its luxuriancy, than when it is cut and trimmed into a mathematical figure" (Hadfield, Harling, and Highton, 1980, p. 11).

The gradual development of gardens in the natural style in England during the eighteenth century stemmed from the work of a succession of artistically talented men who influenced discerning landowners. One of the first innovations that opened the way for more naturalistic gardens was the elimination of the walls or fences that interrupted scenic views of the landscape. But some kind of barrier was still needed to keep animals, domestic or wild, from straying close to the house and destroying cultivated plants. Early in the eighteenth century a friend of Alexander Pope, Charles Bridgeman, introduced the use of a hidden stockade ditch, or fosse, at the edge of the garden to exclude animals while allowing the garden to appear to extend into outer lands. This new enclosure became known as a "ha-ha wall," because the illusion worked so well that people were alleged to exclaim with surprise at finding an abrupt boundary where none was perceived. The use of a sunken fence meant the landscaped garden was visually fused with more distant pastoral scenery or wooded deer parks. Horace Walpole, who summarized the history of English gardening style in his critical *Essay on Gardening,* published in 1771, said Bridgeman's use of ha-ha walls was "the capital stroke, the leading step to all that has followed."

Another of Alexander Pope's friends, William Kent, an architect and painter, also took up the new profession of landscape design. He had acquired great knowledge of the arts, gardens, and scenery of Italy during nine years spent there studying painting and obtaining objects for patrons. When he began to practice as a landscape designer, Kent's pictorial skills transformed estates into idyllic landscapes said to resemble those in Lorrain's works. He used techniques derived from painting, such as providing an illusion of distance by planting trees with dark-colored, fine-textured foliage and narrowing pathways as they neared the horizon. In Kent's designs the lines of pathways became curved to more closely follow the existing topography of a site, and the shapes of lakes or watercourses were made irregular or "serpentized," for "nature abhors a straight line." The trees placed near the manor house were untrimmed, allowing them to spread branches and show their diverse, natural habit and character.

One of Kent's most important commissions was reworking Stowe House in Buckinghamshire, which Bridgeman had initially laid out for Lord Cobham. Even though it was lauded by contemporaries for its irregular "bewty," serpentized river, and reshaped lakeshore, all bounded by "a ha-ha, which leaves you the sight of a bewtifull woody country" (Clifford, 1966, p. 137), Stowe was abundantly furnished. A temple of Apollo, a triumphal arch, a garden of Venus, and an amphitheater adorned with statues of gods and goddesses, as well as urns, fountains, pavilions, and a tall column that supported a statue of Prince George, were all placed about the grounds. From our twentieth-century perspective these architectural elements in a landscape garden may not seem the least bit naturalistic. It should be remembered, however, that landscape designers were often also architects. Just as landscape paintings included symbolic references to classical heroes and virtues, so did built features in the garden. Such objects as temples and pavilions, grottoes, monumental columns, pyramids, and statues all continued to be included in the great, and even the modest, Englishman's garden, as they had been in the previous formal style. Like the grand landscapes in which they were placed, the monuments and temples were seen to be expressions of the landowner's aspirations and good taste. With its abundance of symbolic architecture, Stowe gained a reputation as the "finest seat in England."

While William Kent was working there, a young man named Lancelot Brown came to Stowe House as head gardener. Brown soon grasped the fundamentals of the new natural layout. After the deaths of his employer, Cobham, and his teacher, Kent, Brown established himself as England's foremost estate designer. Since he often referred to the capabilities of a piece of property to be transformed into a beautiful landscape, he gained the nickname "Capability" Brown. Greatly refining the landscape style, Brown created smooth gradients, gently curving lines, and broad sweeps of short-cut grass, occasionally punctuated with clusters of trees. All traces of geometry vanished, and open, parklike land was brought right up to the house. The simplicity of Brown's transformations gave a majesty to the homes of his aristocratic clients.

Brown's artistic and spiritual heir was Humphrey Repton, who practiced throughout England from the 1780s until his death in 1818. Although he is said to have continued the naturalistic style, Repton was soon modifying the extreme simplicity of Brown's landscape treatment. With the use of more extensive planted areas and a greater variety of shrubs and trees, Repton adjusted to his clients' increasing interest in the exotic plants that were newly arriving from America and the Far East. He created terraced gardens and balustraded walks as a transition zone between house and landscaped park, whereas Capability's pastoral scene swept right up to the manor door.

Although not all of English gardening was patterned by Brown and Repton, by the turn of the nineteenth century their works were well known and widely influential. Their clients were among the most literate and wealthy, and much of the English landscape that was not in farm production had been transformed by their mastery.

AMERICA ADOPTS THE NATURALISTIC STYLE

Accounts of such gardens reached America via firsthand reports and published writings. George Washington and Thomas Jefferson were among the first to introduce the

13. *The Vale, home of Theodore Lyman (1753–1839) in Waltham, Massachusetts. Lyman created a landscape in the "English style," which covered four hundred acres at the time of his death. There were picturesque ponds and meadows gracefully ornamented with planted trees, as well as managed woodlots and a productive farm. Lyman grew grapes and other fruits using many techniques new for his time and raised camellias in a glasshouse he built in 1820. The property is now under the care of the Society for the Preservation of New England Antiquities.*

English style to their estates. At Mount Vernon, Washington's home faced a bowling green bordered by a symmetrical but curved walk planted with trees and shrubs. Ha-ha walls and garden walls separated the landscaped grounds from the other parts of the estate, including a deer park. Jefferson's Monticello also included a lawn on the west front of the house, bounded by a serpentine walk edged with flowers. Thickets of trees formed a transition to the naturally forested surround in which deer were kept. Jefferson's many orchards and fruit and vegetable plots were terraced on a slope that provided a convenient means to separate them from the pleasure grounds. The early influence of the "English style" was apparent in the New England country estates of Theodore Lyman (the Vale), Governor Christopher Gore, and Benjamin Bussey (Woodland Hill).

One of the first books to fully describe the principles of design for "modern" pleasure grounds in America was Bernard M'Mahon's *American Gardener's Calendar,* published in 1806. M'Mahon ran a seed, florist, and nursery business in Philadelphia, as well as a bookstore where he offered a great variety of publications on botanical, agricultural, and horticultural subjects. (Thomas Jefferson was one of his frequent customers and correspondents.) M'Mahon's *Gardener's Calendar* was an encyclopedic work with the chapters arranged by the months of the year. It had a large circulation as a welcome alternative to European gardening books and magazines. Since they were still preoccupied with vegetables and fruits, utility was a stronger motivation than display or ornament for most American gardeners. Yet M'Mahon devoted a section each month to "the district commonly called the pleasure, or Flower-Garden, . . . all ornamental compartments surrounding the mansion; consisting of lawns, plantations of trees and shrubs, flower compartments, walks, pieces of water, etc." He described in detail how to lay out and maintain grounds with all the elements of "modern" design developed by the English landscape gardeners:

In designs for a pleasure-ground, according to modern gardening; consulting rural disposition, in imitation of nature; all too formal works being almost abolished, such as long straight walks, regular intersections, square grass-plats, corresponding parterres, quadrangular and angular spaces, and other uniformities, as in ancient designs; instead of which, are now adopted, rural open spaces of grass-ground, of varied forms and dimensions and winding walks, all bounded with plantations of trees, shrubs, and flowers, in various clumps; other compartments are exhibited in a variety of imitative rural forms; such as curves, projections, openings, and closings, in imitation of a natural assemblage: having all the various plantations and borders open to walks and lawns. (M'Mahon, 1806, p. 55)

With the inclusion of flowerbeds and more varied shrubbery, M'Mahon's recommendations followed Repton's gardenlike schemes.

A Scotsman whose interpretation of the new landscape gardening style reached many Americans was John Claudius Loudon, an energetic and prolific garden journalist who began his publishing career in London in 1803. A reading public greatly in need of guidance on aesthetics and garden design—in the wake of an outburst of publications with an emphasis on plants, such as Curtis's *Botanical Magazine,* that described and listed many plants recently introduced into cultivation—welcomed his all-encompassing *Encyclopaedia of Gardening,* which first appeared in 1822. His practical instructions were especially useful for the middle class, who could not afford to consult a landscape designer.

During his productive career, Loudon published several editions of his *Encyclopaedia of Gardening* and edited the successful *Gardener's Magazine,* as well as journals devoted to natural history and architecture. He also published an edition of Humphrey Repton's *Landscape Gardening* in 1834. Loudon advocated the use of a wide variety of trees, shrubs, and herbs, including the many novel plants that were being introduced into cultivation from foreign lands. Partially as a result of Loudon's activities, there was a surge of botanical interest among gardeners. They desired to grow as many different kinds of plants as they could obtain. This plant-collecting impulse modified the naturalistic style into one termed "gardenesque." Often this meant a return to planting flowerbeds and shrubbery in the "ancient" geometric style close to the house, while the outer property was planted in the parklike manner with lawns and groups of trees or wooded areas.

naturalistic landscape style, referring to works from Kent to Loudon, and concluded by mentioning notable American gardens that exemplified it.

Downing's horticultural and design interests were wide. For a time he and his brother operated a nursery business in their hometown of Newburgh. In *Cottage Residences,* published in 1842, and *The Architecture of Country Houses,* which appeared in 1850, Downing campaigned ardently for the use of Gothic architecture in a simple, tasteful version, which he thought was particularly suited to the naturalistic, unstructured setting. He also authored *Fruits and Fruit Trees of America* (1845), a well-regarded work on that popular subject. He served as a landscape gardener or consultant for many estate owners in his neighborhood of New York and was editor of an Albany-based magazine, the *Horticulturist.* Through editorials and articles in this journal, Downing spoke out for urban parks and initiated the movement that Calvert Vaux (an architect whom he

14. *Andrew Jackson Downing (1815–1852) lived in the scenic Hudson River Highlands of New York. His brief career as nurseryman, landscape designer, and architectural critic greatly influenced each of these movements in America. As an author and as editor of the* Horticulturist *magazine, Downing advocated the adoption of the naturalistic approach to landscape design.*

15. *Pomology was a highly developed and very important aspect of horticulture throughout the nineteenth century. There was great proliferation in numbers of varieties of such domestic fruits as apples, pears, and grapes. A number of cultural methods were tried and improved upon, including raising fruit against protective walls or under glass. Many horticultural periodicals became available to Americans; among them was the* American Journal of Horticulture and Florist's Companion, *published by J. E. Tilton of Boston, in which this picture appeared.*

America's premier advocate of naturalistic landscape gardening was Andrew Jackson Downing. In 1841, he published his first and most influential book, *Treatise on the Theory and Practice of Landscape Gardening.* This work and its subsequent editions were to publicize the ideals of English naturalistic style more widely than ever before. Born in the scenic Hudson River valley, the inspiration for many American landscape painters in the nineteenth century, Downing became the chief proponent and interpreter of Repton's and Loudon's ideas, adapting them to the climate and plants of the New World. The introductory chapters of his treatise sketched the history of the

16. View in Mount Auburn Cemetery, about 1847. It was an innovation when founded in 1830 to create an extraurban burial ground employing natural scenery to soothe the minds of the bereaved. Rural cemeteries became popular places for enjoying nature, and they exemplified the acceptance of naturalistic landscapes for community spaces.

personally recruited in England in 1850) and Frederick Law Olmsted were to carry to fruition.

Andrew Jackson Downing was the first American to call himself a professional landscape gardener. The brilliant Downing met an untimely death at age thirty-seven in a steamboat accident on the Hudson River. Nonetheless, he had done more than any other American of the era to promote horticulture and related arts, especially the practice of landscape design as a profession. At the time of Downing's death, he and Vaux were at work on a commission to design the grounds of the Capitol, the White House, and the Smithsonian Institution in Washington.

FROM RURAL CEMETERIES TO PUBLIC PARKS

The naturalistic style lent itself especially well to a new kind of landscape that was being created: the rural cemetery. The Massachusetts Horticultural Society's support of Mount Auburn Cemetery has been mentioned. This and other rural cemeteries subsequently established in such cities as Worcester (Massachusetts), Philadelphia, and New York became much welcomed places of retreat for an increasingly urbanized public. These naturalistically landscaped cemeteries were the precursors of the nation's city parks.

Consecrated in 1831, Mount Auburn Cemetery was the first of many rural cemeteries created in the United States. In the Boston-Cambridge area the idea started with Jacob Bigelow, whose concern for the health hazard posed by "abuses in the practice of sepulture" in urban burial grounds arose after a yellow fever epidemic in 1822 left his fellow doctors questioning the effect of crowded graveyards on the spread of disease. In 1825, Bigelow began to campaign in earnest for an alternative to urban burial. He called a meeting of twelve gentlemen to consider opening an "extra-urban ornamental cemetery." Bigelow conceived of the proposed cemetery as a new aesthetic space as well as a solution to public health concerns, and the others, many of them prominent Bostonians, concurred. A precedent, known to many Americans who had toured Europe, was Père Lachaise on the outskirts of Paris. This was a garden-cemetery designed according to the theories of English naturalistic style. Unaccustomed as Americans were to the idea of combining garden arts with memorials to the dead, they admired its romantic beauty and expressed surprise that Parisians visited it for Sunday promenades.

A committee was formed to search for an appropriate site, and eventually Bigelow convinced George W. Brimmer, who owned a lovely parcel of land astride the Cambridge-Watertown boundary, of the value of the ornamental-cemetery scheme. Brimmer's property was known to Harvard students as "Sweet Auburn," and they had frequented it for walks at least since the days that Brimmer himself was a student in the class of 1803. It was sometimes used as an informal meeting ground. Brimmer's land was "beautifully undulating in its surface, containing a number of bold eminences, steep acclivities, and deep shadowy valleys" (Bigelow, 1987). There were many large old native trees, as well as ones that Brimmer had planted.

Brimmer decided to offer Sweet Auburn at cost for the cemetery project, and the Massachusetts Horticultural Society agreed to raise the funds. With the aid of a persuasive address by Edward Everett published in the papers, Bostonians were favorably impressed with the idea. Subscribers for the first one hundred lots in the proposed cemetery were recruited, assuring sufficient funds for the project to go forward. Cemetery subscribers who were not already part of the horticultural society became life members.

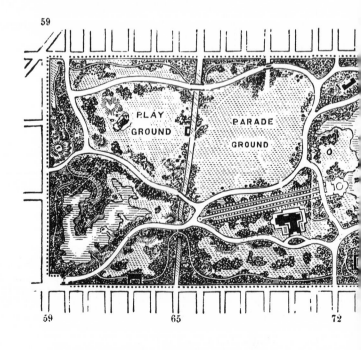

The planning and design for Mount Auburn were carried out by the horticultural society's Garden and Cemetery Committee. Its most active members were Brimmer, Bigelow, and MHS president General Henry A. S. Dearborn, none of them professional architects or designers, for such scarcely existed. A civil engineer was hired to survey the tract and assist in staking out carriageways and footpaths that curved pleasantly with the land's contours. These were named for plants by Bigelow in keeping with the natural theme. General Dearborn was untiring in his labors to supervise the actual work on the grounds in the early years, often taking hoe in hand himself to lead the small crew. He donated numbers of young trees from his own eighty-acre estate in Roxbury to fill out the cemetery plantings. Bigelow designed an Egyptian gateway, a Gothic tower, and a Greek tower.

Almost as soon as the initial construction of cemetery roads and avenues was complete, visitors began to make use of the site, and access had to be regulated to exclude horseback riders and allow only the proprietors of lots to enter in carriages. All others could enter on foot. By 1834 the committee found "the grounds at Mount Auburn were visited by unusual concourses of people on Sundays, and that injuries done to the grounds and shrubbery were far greater on that day than on any other" (Manning, 1880, p. 107). Consequently, it was decided to restrict Sunday access to the proprietors and their guests.

With the success of Mount Auburn, reluctance to break the tradition of inner-city, churchyard interment was overcome by the argument for sanitation, the guaranteed permanence of burial, and the quality of landscape. The cemetery was designed to make the most of the soothing effect that nature would have on the minds of the bereaved. It was a place where natural scenery enhanced contemplation in a manner not possible in the midst of most urban burial grounds. Mount Auburn and the rural cemeteries that were soon started in other cities in America and England exemplified a growing conviction that nature played an important role in one's mental health. And they were examples of the growing use of naturalistically designed landscapes in community spaces.

The two influential landscape writers of the times, Loudon and Downing, expounded favorably on the new rural-cemetery scheme. "Nothing could be easier than to

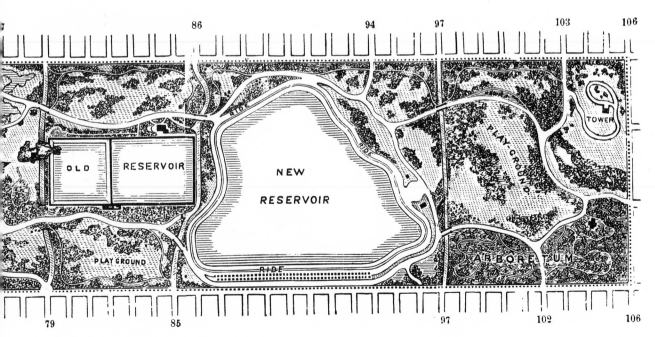

17. *First study of a design for Central Park, 1858, from "Greensward," the description of a plan for the improvement of Central Park by Frederick Law Olmsted and Calvert Vaux. This work was the starting point of their careers as landscape architects and urban planners.*

render every country churchyard in Britain an arboretum and herbaceous ground, with all the trees and plants named, provided the clergyman would give up his right to the grass . . . and the grave digger would be content to acquire a very little knowledge of gardening" (Loudon, 1828, pp. 497–98). As rural cemeteries came into existence in America and in his native Britain, Loudon continued to promote the idea in his magazine.

Urban residents were becoming increasingly eager to enjoy any open space that was available for leisurely strolls and viewing nature. The diversity of trees and shrubs usually included in the cemetery landscapes came to be appreciated by many. Downing published one of his strongest pleas for public parks and gardens in an editorial in the *Horticulturist* in July 1849. Referring to the popularity of rural cemeteries, he asserted that "the true secret of their attraction lies in the natural beauty of the sites, and in the tasteful and harmonious embellishment of these sites by art." In his view, the success of rural cemeteries was proof of the need for "large public gardens." By the 1850s the movement toward public parks, outdoor spaces that were made for and owned by the municipal population rather than private groups, was under way in Boston, New York, and other American cities.

The agitation for public parks first came to fruition in Downing's home state. Five commissioners were appointed by the New York Supreme Court in 1853 to obtain land in the center of Manhattan Island for what would be Central Park. After the acreage was secured, politics and lack of funds allowed only halting development of the site. Advocates of the need for such an amenity kept urging the city to act and the public to support it. In 1856, Robert M. Copeland and Horace W. S. Cleveland, partners in a Boston landscape-gardening firm, published *A Few Words on Central Park,* advocating the development of the New York park in the naturalistic style. Many of their thoughts about parks and landscape design anticipated the ultimate design of Central Park. Citing the need to preserve the inherent features of the land, Copeland and Cleveland stressed the importance of creating a plan that would be usable for centuries into

the future. Since "the love of the beautiful in nature lies at the foundation of all true taste in art," the design should make the most of the "prevailing expression of the natural scene" at the Central Park site. Thereby, the park would have the highest moral influence on the city's people. The design should enhance the "sublime, picturesque and beautiful" in the area's physical features. And they recommended that "opening from different points, a continually varying succession of scenes" be created. They also thought a zoological garden and some provision for the study of botany, arboriculture, horticulture, and floriculture should be included, as well as a popular museum and space for events like military parades and public exhibitions.

In response to such appeals, the commissioners opened a public competition for a plan for the Central Park in 1857. The firm of Copeland and Cleveland entered the competition, but their plan was not successful. Instead, the commissioners chose "Greensward," the design of Calvert Vaux and Frederick Law Olmsted. Vaux, an architect, was determined to carry on the landscape tradition of his deceased partner, Andrew J. Downing. Olmsted, a native of Hartford, Connecticut, had won appointment as superintendent for the new park earlier that year. Previously, he had pursued a varied career as a scientific farmer and as a journalist reporting on social conditions in England and in the Southern slave states. So far, most of Olmsted's work in Central Park had involved surveying and overseeing cleanup. With the adoption of Vaux and Olmsted's plan and the latter's appointment as architect in chief, construction of America's first spacious, naturalistic urban park began. Thus commenced Olmsted's career as landscape architect and city planner, as well as a productive partnership with Vaux.

Central Park's plan included many ideas recently raised about taste in design and the needs of city dwellers that Olmsted was to apply to urban and suburban situations throughout his forty-year practice. In the main, he sought to design parks that would give all classes of people relief from the stresses of urban life by creating scenery that contrasted as much as possible with the city environment.

A spacious, rural park would be a counterpoint to the gridiron of paved and crowded streets that New Yorkers encountered constantly. The sense of freedom and relief that he strove to provide for park users was essential, he believed, to a healthy civilization. It was this effort to reform social conditions, as well as his artistic capability, that set Olmsted apart as champion of the urban parks movement and led to his reputation as the "father of landscape architecture." Behind him was Vaux and the Downing tradition, as well as others who had been refining the naturalistic style and working toward professionalism in landscape architecture. While the Arnold Arboretum would be, in many respects, an exception among Olmsted's numerous works, these traditions in design and social reform would shape its future.

18. *View from Weld Hill (now Bussey Hill) looking south toward the Blue Hills. It was this southwestern plateau and the hill's summit that Benjamin Bussey screened off from his pastures and orchards with a mixed border of lilacs and white pines and other trees.*

CHAPTER TWO

ARNOLD ARBORETUM
FOUNDERS AND FOUNDATIONS

The Arnold Arboretum was officially established in March 1872, when an indenture was signed by which trustees of a bequest of James Arnold agreed to turn the fund over to Harvard College, provided the college would use it to develop an arboretum on land bequeathed earlier by Benjamin Bussey. Mastermind of this scheme was George Barrell Emerson, one of Arnold's trustees. Charles S. Sargent was chosen to manage the new institution and remained in control of it for half a century. The stories of these men—Bussey, Arnold, Emerson, and Sargent—show how each one's personal experience and predilections formed the collective desire to create an institution to provide scientific, educational, and aesthetic benefit for their fellow citizens.

Benjamin Bussey (1757–1842) was a self-made man who acquired an estate on which he practiced scientific farming and scenic improvements during his leisure. An intense regard for the land and for agricultural endeavor led him to leave a large portion of his fortune and all of his property in West Roxbury to Harvard College for the creation of an institution for instruction in farming, horticulture, botany, and related fields. He desired that his beautiful and highly cultivated property be retained perpetually for such purposes. Although the physical evidence of Bussey's tenure remaining in the present-day Arnold Arboretum is subtle, it is hard to imagine a greater legacy for a woody-plant collection in an urban setting.

James Arnold (1781–1868), born to a prosperous Quaker family of Providence, Rhode Island, came of age in time to participate in the success of the whaling enterprise in New England. His religious background predisposed him to beneficence; his personal avocations were landscape gardening, horticulture, and collecting natural-history specimens. Arnold left a large part of his fortune to be used for agricultural or horticultural purposes at the discretion of three trustees. The most influential of the three with regard to the Arboretum was Arnold's brother-in-law, George Barrell Emerson (1797–1881). A schoolmaster and educational reformer, he shared Arnold's pastimes, pursuing an interest in trees to the extent of publishing a scholarly work on them. Emerson was the pivotal figure who made the connection between Arnold's bequest and Harvard College and who interpreted the concept of an arboretum for the founders.

The background that Charles S. Sargent (1841–1927) brought to his appointment was particularly suited for the creation of an arboricultural masterpiece. He was raised in

Brookline, Massachusetts, a neighborhood praised by Downing for its garden landscapes, and he was a frequent visitor at naturalistically styled estates where trees and shrubs were seriously collected. Not the least of these was that of a cousin, an estate in whose design Downing had a hand.

While not all the Arboretum founders knew each other, they shared a love of trees, an interest in plants that ranged from aesthetic to scientific, and a desire to benefit others through education. All were dedicated to the preservation and improvement of natural scenery. During the span of their lives the time became ripe for the founding of an organization to serve these interests; fortunately, their separate actions were unified under the auspices of an institution for higher learning.

19. *Like many of Boston's East India merchants, Benjamin Bussey (1757–1842) sat for the eminent portrait painter Gilbert Stuart. He also shared the Boston merchants' tradition, described by Cleveland Amory in his book* The Proper Bostonians, *of claiming that their great fortunes were started with minuscule amounts of capital. Bussey claimed that the sum of fifty dollars initiated the wealth he ultimately left to Harvard College.*

BENJAMIN BUSSEY, SCIENTIFIC FARMER

Benjamin Bussey was born in 1757 to a family who earned a modest living farming in the colonial Massachusetts town of Stoughton. When he died in 1842, Bussey was one of a handful of wealthy men who resided on country estates on the outskirts of the young nation's prosperous city of Boston. Bussey embraced the changes that took place during his lifetime—from colony to nation, from subsistence farming to "scientific agriculture," from dependence on British manufactured goods to the creation of American industries—and quickly perceived how to turn them to his benefit and the benefit of others.

Benjamin Bussey's interest in agriculture originated from firsthand experience. In an autobiography, written when he was eighty-four years old, Bussey stated that his maternal grandfather, Deacon Joseph Hartwell, was one of the best farmers in the state. Bussey's father, also named Benjamin, was the son of an English farmer. He became a master mariner, spending years away at sea, absent even for his son's birth. Consequently, much of Bussey's early life was spent on the Hartwell farm. Eventually Bussey senior left seafaring, became a country storekeeper, and took up farming. When young Benjamin was not helping in the store or attending school in Stoughton, he worked his father's land.

Bussey's father allowed him to leave the farm at age eighteen. The year was 1775, and the young man joined the Continental army. During three terms of enlistment, Bussey served in Boston and in the northern campaign in New York, attaining the rank of quartermaster. This was a difficult position, as food, arms, and uniforms were seldom readily available to the Continentals. At Saratoga he was stationed with General Benjamin Lincoln's division at Bemis Heights when the British under Burgoyne were forced to surrender in October 1777. This was the first important victory for the Americans, a turning point in the war for freedom and one that Bussey remembered all his life.

After his war experience Bussey "wanted to go into trade," since he had acquired "habits of expense in dress etc. in the army." The duties of quartermaster had given him experience applicable to a business career. With fifty dollars from his grandfather Hartwell, Bussey arranged with a Prussian, a mercenary who had also fought in the war, to learn the goldsmith's business. He mastered this trade in a year, during which time Bussey maintained he worked night and day. At the end of the year, he had earned enough money to buy a house in Dedham. Soon after, in 1780, he married a Dedham woman, Judith Gay. Gradually, Bussey took on other business in addition to the manufacture and trade of gold and silver objects. For one thing, he began to exchange American furs for goods in England.

By 1790 Bussey had garnered enough capital to make the move to the center of New England commerce, buying a home on Eliot Street in Boston. Yet there was still no rest or relaxation—"I worked in Boston as I had in Dedham, day and night" (Bussey, 1899, p. 74). During the next fifteen years he managed to purchase several ships and engaged in many kinds of mercantile ventures. It was a prosperous era for maritime enterprise in New England. From Salem, Boston, and Nantucket vessels sailed to the West Indies, Europe, and China and returned with goods that were eagerly sought in the marketplace.

In addition to these ventures, Bussey shrewdly diversified his investments. He purchased a large part of what became the city of Bangor, Maine (where he had family connections), as well as three thousand acres in the surrounding countryside. Between 1820 and 1860 this settlement on the Penobscot River rapidly became the greatest shipping port for pine lumber in the world. Even in 1832, at the height of Bangor's boom, there was still marketable timber standing within the town limits, perhaps on some of Bussey's lands.

The family's next move was to Boston's fashionable Summer Street in 1798. It was here that he began the pursuit of horticulture in earnest. Many of his equally successful Boston neighbors engaged in this practical and pleasurable activity:

Summer street was for a long time one of the most delightful in the city, and well merited its name from the overhanging branches of ornamental trees and the beauty and fruits of the gardens attached to the mansions of its wealthy occupants.

Here, in the early part of this century, were the residences of Gov. James Sullivan, afterwards of William Gray, Joseph Barrell, Benjamin Bussey, Nathaniel Goddard, Henry Hill, and David Ellis . . . whose gardens were supplied with the fruits and flowers of those days, and where peaches and foreign grapes, and the old pears . . . ripened every year. (Wilder, 1881, p. 9)

Among Bussey's horticulturally inclined neighbors were men who had chartered the American Academy of Arts and Sciences: James Sullivan, James Bowdoin, and John Hancock. Others were founding members of the Massachusetts Society for Promoting Agriculture: Joseph Barrell, Henry Hill, Thomas Russell, and Samuel Salisbury. Bussey joined the agricultural society in 1803 and kept up with its doings and recommendations. Other relatives also joined, and his grandson-in-law, Thomas Motley, became a trustee and was its president from 1870 to 1895. Living in this community of avid horticulturists and amateur scientists fostered Bussey's concern for the application of scientific methods to agriculture. His lifelong interest in farming would influence his future decision on the disposition of his estate.

About the time that the MSPA was establishing Harvard's botanic garden, Benjamin Bussey, then aged forty-nine, was looking forward to a quieter life. He decided to practice farming and live closer to the land. He made the first of several real-estate purchases that would result in an auspicious site for his estate and eventually the Arnold Arboretum.

Bussey bought more than fifty acres from the heirs of Eleazer Weld in 1806. Eleazer had been the seventh generation of Welds to live on the property, part of a several-hundred-acre parcel in "Rocksberry" that had been awarded to Captain Joseph Weld for his personal estate by Governor Winthrop about 1640. Captain Weld had proven valuable to the Massachusetts colony in conflicts with the Indians and presided over a treaty of peace. His land grant lay on both sides of the Highway to Bare Marsh, or Lower Road to Dedham, as South Street in Jamaica Plain was known in colonial times. It included a high hill and both sides of the valley of Stony Brook, a tributary of the Charles River. From then until the time

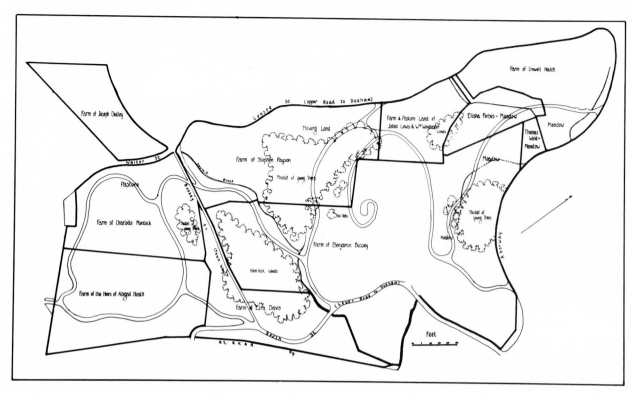

20. *Map of the lands in the Arnold Arboretum (about 1810) showing areas that had begun to seed in with trees when Bussey purchased the Weld estate. As an enthusiast of new ideas in farming, land management, and landscape beauty, he allowed the young trees to grow up and maintained paths through his woods.*

of Bussey's purchase, the Welds added to their holdings. In subsequent acquisitions, Benjamin Bussey obtained more lands from the Welds as well as adjoining tracts. The largest were the farm of Ezra Davis, bought in 1833, and that of Ezra's brother, John, purchased in 1837. Most of the latter property made up the area now known as Peters Hill.

On the south-facing slope of "Weld Hill" (now Bussey Hill), Benjamin Bussey built a large home in the fashionable neoclassic style, always referred to as his "mansion." Located just off the road, it faced the Great Blue Hill, beyond which lay Stoughton, the town of his birth. The road led to Dedham, his wife's family's home and the site of his earliest business ventures.

Although he remained in business in Boston and retained his Summer Street home, after 1815 Bussey lived at his Roxbury estate, which he called Woodland Hill.

He devoted his attention to agricultural pursuits there and to manufacturing at Dedham. One pursuit, advocated by members of the Massachusetts Society for Promoting Agriculture, was raising merino sheep; he was among the first enlightened farmers to import them to Massachusetts.

Extending this interest in sheep, Bussey's business focus turned to woolen mills in Dedham. The War of 1812 and its attendant embargo had forced Americans to start manufacturing goods they previously imported. While woolen manufacturing did not develop as extensively as did that of cotton, textile production was the first major step toward a factory-based economy for the northern states. Small enterprises such as Bussey's were generally welcomed in New England villages in the early nineteenth century as a source of additional earnings for farm families. With his entry into the woolen business Bussey once more seized an opportunity at the right time. As his eulogist,

Thomas Gray, put it: "Mr. B. possessed a quick insight into the best means of accomplishing whatever he purposed; and, after mature deliberation, a persevering and unaltered pursuit of it, by means of which he successfully carried it out. And this union, perhaps, was the key-stone to his fortune" (Gray, 1842, p. 7).

From descriptions of Woodland Hill, we find that Bussey must have ceased working day and night to enjoy country life. He furnished his mansion richly, with mahogany upholstered in red plush, green damask, and French tapestry; Brussels and Venetian carpets covered the floors. He had his family's portraits painted by Gilbert Stuart, the fashionable portraitist best known for his likeness of George Washington. In his hallways and study, busts of the first president and John Adams, a portrait of General Henry Knox, and an engraved Declaration of Independence manifested Bussey's patriotism. He also owned paintings of the masters copied by Rembrandt Peale. Bussey had a coach that greatly impressed his neighbors when he rode to church on Sundays. "On the occasion of President Andrew Jackson's visit to Boston . . . in June, 1833, Mr. Bussey joined the grand procession in his yellow coach, drawn by six horses, richly caparisoned and attended by liveried servants" (Whitcomb, 1897, p. 54).

Bussey greatly loved to read, and he read widely. He owned shares in the Boston library and the athenaeum. Over a lifetime he accumulated a library of approximately eighteen hundred volumes. But it was from improving his land and observing nature that Bussey derived his greatest pleasure. Records indicate that his orchards were highly productive.

Before Bussey's purchases, most of the Roxbury estate was open, cleared for grazing animals or cutting hay, or planted to orchards; there were only small parcels of plowland. An exception was Hemlock Hill. Although its trees had been cut for lumber to supply a sawmill owned by the Gore and then the Morey families in the eighteenth century, the area had never been entirely cleared; its steep, rocky slopes were unsuitable even for pasturage. In addition, two knolls north of Weld Hill, abandoned as orchard and upland meadow sometime between 1770 and 1800, were in young woods or thickets by 1810. Bussey apparently allowed the trees on the knolls and in a few other areas to grow undisturbed, while maintaining mowing land, pasture, and orchard over a large part of his property. He also purchased several outlying parcels in Roxbury for woodlots.

Bussey was well aware of the value of wood for timber and fuel. His Maine timberlands were a significant asset, yet the relatively small parcels on which he allowed forest to regenerate were local examples of what could be done to counter concerns about wood shortages. Bussey also had the luxury and taste to consider his woods "natural attractions" that enhanced the beauty of his property.

In managing his estate, Bussey was undoubtedly influenced by the English landscape gardening style. This was evident in the pathway he created from his mansion to the top of Weld Hill. It was framed by a double row of lilacs interspersed with white pines and other trees. The lilac-and-pine hedge extended in a rectangular pattern around the top of the hill. The northeast and northwest sides of the rectangle had defined the boundary line of the Weld property; subsequently, Bussey obtained the land beyond this boundary. Still, the lilac hedge served to screen the working farm areas from his hilltop observatory, a place for pleasurable activity. Such perennial flowers as lily of the valley, daylily, and periwinkle, which persist in the area today, lined the path up to the observatory. Equipped with two telescopes, couches and chairs, a painted carpet, and paintings adorning the walls, the observatory was a fine place to entertain companions with an interest in natural history. After a pleasant stroll among the lilacs, they might view the surrounding farms and woods, the distant hills, the city, or the celestial bodies in the night sky with the latest instruments from a comfortable enclosure. Owning Weld Hill also had its practical side: an underground reservoir built into the hilltop provided water by means of gravity for the mansion and farm.

BUSSEY'S BENEFICENCE

In 1829, Bussey joined the newly formed Massachusetts Horticultural Society. Although not buried there, he was one of the first proprietors of Mount Auburn Cemetery. It was his wish to support the group and its proposed demonstration garden adjacent to the cemetery, and the improvement of land for community use undoubtedly appealed to him as well. In Bussey's day the need for urban parks was just beginning to be felt. Although cities had not grown large, and the majority of the population was generally too busy for the leisure of country walks, open spaces and wooded areas were visited by Boston's inhabitants. Bussey anticipated the value that natural scenery might have for those who did not own or work on the land. "During Mr. Bussey's life, and for years after, the public enjoyed the freedom of [his] charming grounds. There were lovely wood paths, carefully kept, in all directions. Here was a rustic bridge spanning the jocund brook; there a willow-bordered pond" (Whitcomb, 1897, pp. 52–53). The Jamaica Plain district became accessible to greater numbers of urban residents with the extension of the steam railroad to Forest Hills in 1835 and with the extension of horsecar lines twenty years later. "May's Woods" and Jamaica Pond were two other nearby destinations to which residents of Boston, Dorchester, and Roxbury traveled to enjoy rural beauty in the mid-nineteenth century.

One well-known resident of Jamaica Plain who enjoyed Bussey's grounds was Margaret Fuller (later Margaret Fuller Ossoli), a young woman fired with the theological and intellectual turmoil of the transcendentalists. She was a friend of Ralph Waldo Emerson and editor of the *Dial,* a magazine devoted to the new and freer ideas of their intellectual circle. From 1839 to 1841 Fuller lived in Jamaica Plain, just a short distance from Woodland Hill in what is now the Forest Hills neighborhood. Like her mentor, Emerson, Fuller's reflective temperament was especially receptive to inspiration from nature. In September 1840 she wrote:

I have just returned from a walk this golden autumn morning, with its cloudless sky and champagne air. I found some new wood walks, glades among black pines and hemlocks, openings to the distant hills, graceful in silvery veils. A very peculiar feeling these asters give me, gleaming on every side. They seem my true sisters. . . . For a while I lean on the bosom of nature, and inhale new life with her breath. (Higginson, 1884, p. 99)

This passage could well have been inspired by a walk in "Bussey's Woods," as the estate was then popularly known.

Another active participant in the transcendentalist discussions, William Henry Channing, described a day in the summer of 1840 when he and Margaret became close friends:

It was a radiant and refreshing morning, when I entered the parlor of her pleasant house, standing upon a slope beyond Jamaica Plain to the south. . . .

She proposed a walk in the open air. She led the way to Bussey's wood, her favorite retreat during the past year, where she had thought and read, or talked with intimate friends. We climbed the rocky path, resting a moment or two at every pretty point, till, reaching a moss-cushioned ledge near the summit, she seated herself. For a time she was silent, entranced in delighted communion with the exquisite hue of the sky, seen through interlacing boughs and trembling leaves, and the play of shine and shadow over the wide landscape. (Emerson, Channing, and Clarke, 1852, vol. 2, pp. 31, 34–35)

This description is suggestive of the paths to Hemlock Hill's summit in the present day.

Bussey was eighty-four years old when another of the transcendentalists, George Ripley, began his experimental community at Brook Farm less than two miles west of Woodland Hill. At his mature age and with his lifelong attention to business matters, it is unlikely that Benjamin Bussey would have embraced the new intellectual stirrings. But he was willing to have the "young men . . . with knives in their brains" and shockingly scholarly women partake of his lovely landscape. He was known to be "affable, sociable and kind to everyone."

Hospitality and generous sharing of his assets was long a feature of Benjamin Bussey's character. "No destitute or afflicted applicant ever left his door with an empty hand" (Gray, 1842, p. 8). Bussey also had wide religious tolerance, supporting not only his own church, the Unitarian in Jamaica Plain, but other parishes of other denominations. In Bangor, for example, he donated bells to three different churches. What was important to Benjamin Bussey was that there were churches and that people

21. *"Bussey's Woods" became a popular destination for Bostonians seeking fresh air and natural scenery. Undoubtedly the "glades among black pines and hemlocks" mentioned by the transcendentalist Margaret Fuller were those that clothed Hemlock Hill, at the base of which flowed Sawmill Brook (now Bussey Brook).*

should be allowed freedom of opinion in religious matters. He was aware of the cost of such freedom in terms of war and struggle; during his life he had witnessed changes in religious thought, as Unitarianism evolved from the Congregational faith and was subsequently questioned by the transcendentalists. He realized that only informed people could exercise their right to choose. A self-educated man, he valued formal education. He was aware, as well, of the necessity of an educated population for the effective working of democracy.

In 1835 Bussey wrote his last will and testament. After providing support for his wife, his surviving daughter and her children, and for other relatives, Bussey designated that the remainder of his fortune go to Harvard College, the center of learning in New England. Of his legacy to the schools of law and theology he said:

In a nation whose government is held to be a government of laws, I deem it important to promote that branch of education which lies at the foundation of wise legislation, and

which tends to insure a pure and uniform administration of justice; and I have considered that, in a country whose laws extend equal protection to all religious opinions, that education which tends to disseminate just and rational views on religious subjects is entitled to special patronage and support. (Bussey, 1841, p. 4)

However, the branch of education to which he wished to give even greater support was the knowledge of agriculture and related fields. This, he believed, would "advance the prosperity and happiness of our common country." Bussey died in 1842, but twenty-five years would pass before his trust could be acted on by Harvard. In the interim, there were many changes in the economic and intellectual focus of the country. And other events would influence the ultimate disposition of Bussey's gift. One of these was a bequest to Harvard College from another beneficent man with a special fondness for horticultural and agricultural improvements, James Arnold of New Bedford, Massachusetts.

JAMES ARNOLD,
WHALING MERCHANT

At about the time that Benjamin Bussey began to contemplate retirement to country life, a young Quaker left his native Providence, Rhode Island, to begin a career in New Bedford, Massachusetts. James Arnold entered the whaling offices of William Rotch, Jr., and soon became a partner in this prosperous firm. New Bedford had grown to be New England's whaling capital after 1765, the year the Rotch family brought the industry from Nantucket to the well-protected shore of Buzzards Bay.

Many of the settlers of Dartmouth Township, as the area was known before 1765, were members of the Religious Society of Friends, whose beliefs had been unacceptable to the Puritans and Pilgrims of the Massachusetts Bay and Plymouth colonies to the north. Similarly, many of their denomination had settled in neighboring Rhode Island, where religious freedom was guaranteed. As the Friends of New Bedford and Providence formed a single community, it was natural for James to join the Rotch enterprise and to marry William's eldest daughter, Sarah, which he did in 1807.

Whaling was the economic specialty for nearly all the people of New Bedford. Everyone—the carpenters, caulkers, riggers, and blacksmiths who built the ships; the seamen and captains who pursued whales for months and years at a time; the merchants who backed each voyage— took their pay in shares. The risks were high; the investment needed was great. Chasing whales was a hazardous occupation, to say the least, and the vessels had to be built to endure the ocean's worst. But the potential for financial returns was also great, and for more than half a century New Bedford was one of New England's most prosperous towns. The products of whaling were essential to American society and industry. Whale oil was a primary source of fuel for lighting for America's growing population, and whalebone was used for corset stays, umbrella frames, buggy whips, and countless other items. Few could then envision the devastation that was to come to the populations of the great sea mammals; profit and progress were motivating factors for the industry.

Arnold's sixty-year business career coincided with the steady growth of the New Bedford whaling industry. He also conservatively backed other enterprises from time to time. The hazards of maritime business prompted prudent merchants to organize mutual insurance companies and banks. In 1816 James Arnold became one of the directors of the Bedford Commercial Bank, a responsibility he held until 1854.

As an employer, Arnold had, in his younger years, a reputation for being exacting in his demands. Yet he probably asked no more of others than he expected of himself. "He possessed a strong physical constitution, full to overflowing with vitality, capable of great labor and great endurance. . . . having the large practical understanding, which, combined with an untiring energy and strong will, gave him excellent administrative ability" (Potter, 1868, p. 7).

The amassing of a fortune, which enabled Arnold to generously benefit others, continued until his death at age eighty-seven. During his long involvement with whaling, he owned at least twelve ships and barks. Together with his sister-in-law, Mary Emerson, and three others, Arnold purchased a saltworks that provided steady income for many years. Real estate and railroads, especially railroads in the "Old Northwest," were other reliable financial interests.

Some accounts of the Arboretum's founding have stated that a large part of Arnold's wealth came from investments in Michigan timberland and that this was the basis for his interest in trees. Comparatively speaking, he did not profit greatly from this venture. Arnold did enter into some deals in Michigan with Henry Howland Crapo, and it is true that Crapo eventually turned a profit from the Lapeer pinelands and even went on to become governor of Michigan. However, it was only after Crapo left his home and business in New Bedford and spent more than ten years working incessantly, with ever increasing debt, that he built a financially successful lumber enterprise.

By 1860 it was evident to Arnold that the capital committed to the pinelands was far out of proportion to the rate at which they yielded returns. Apprehensive of further losses, and facing the last illness of his wife, he withdrew from the lumbering partnership and a few years later sold the land to Crapo for less than its original cost. The intricacies of the timber industry were as alien to a longtime Yankee whaling merchant as outfitting a whaling ship would be to a timberman.

In New Bedford, this era was the height of productivity for the whaling industry. The city was famous for its wealth and its stately homes and elegant public buildings. The demand for illuminating and lubricating oil had increased steadily with urban growth and the ever expanding use of machines in factories. It was the trade that New Bedfordites knew better than anyone, and the city boomed. Artificially high prices during the Civil War allowed tremendous profits for voyages that had left New Bedford before the start of the conflict. The industry was, however, doomed. The development of a successful technique for drilling for petroleum in 1859 led to the decline of New Bedford's whaling prosperity. Although the demand for whalebone for corsets kept the industry profitable into the 1890s, New Bedford turned to textile production, eventually becoming second to Lowell, Massachusetts, in this field.

ARNOLD LANDSCAPES HIS GROUNDS

Arnold had interests outside his business. Although he did not receive the training in Latin and Greek that was the core of formal education in the early Republic, he was widely read in the works of English and American writers, and he was a capable speaker who loved a good debate. He was highly respected for this skill by his fellow members of the Dialectic Society, a group of men who gathered for "intellectual culture and social intercourse." The meetings were usually planned debates, but formal lectures by members or outside speakers were also given.

In 1821 James Arnold purchased land from his father-in-law to build a house. It was on County Street at the edge of New Bedford, on farmland that William Rotch, Jr., had obtained from another of the founders of New Bedford's whaling industry. Large elms, lindens, and a landmark oak graced the property. There were woods and some swampy areas in the western part of the eleven acres. County Street followed the crest of a hill overlooking the harbor, where New Bedford's dwellings and workplaces

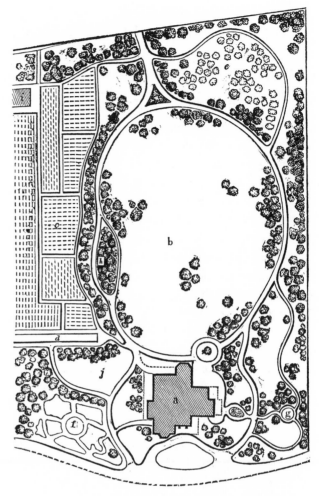

22. This ideal plan of a suburban residence from Andrew Jackson Downing's Treatise on the Theory and Practice of Landscape Gardening *is similar to the layout of James and Sarah Arnold's grounds. Flowerbeds and fruit and vegetable plots were separated by walls and hawthorn hedges from the remainder of the garden, through which tree-shaded walks meandered. Large areas of informally planted trees and shrubs were what the horticulturist Charles M. Hovey thought exemplified the modern, or English, style of gardening at the Arnolds' in 1840.*

were concentrated. Arnold's property sat at the head of Union Street, the town's main thoroughfare. His was the first among many elegant homes and gardens for which the neighborhood became famous. The two-story brick dwelling, built in the Federal style, was symmetrical, with clean lines and simple proportions that reflected Arnold's Quaker heritage. Inside, the house was equipped with conveniences novel for the time: central heating and a bathing room with a water heater.

It was here that the Arnolds would pursue the increasingly popular pastime of creating a garden. For Sarah and James their home grounds became the place for the generous hospitality that they loved to practice. In the design of their gardens and grounds the Arnolds kept up with the latest ideas.

The gardenesque interpretation of the English naturalistic mode was the style in which James Arnold developed his property. Like many of the landed English proprietors after whose estates he modeled his gardens, Arnold opened his grounds to his fellow townspeople. The estate became a place in which citizens of New Bedford took pride. Visiting dignitaries were brought to see the garden, May Day celebrations took place there, and a typical Sunday afternoon for many New Bedfordites included a stroll within its groves.

A detailed account of "the delightful grounds of Mr. Arnold" was given by Charles M. Hovey of Cambridge, nurseryman and editor of the *Magazine of Horticulture:*

Mr. Arnold's grounds are decidedly the most ornamental that we have ever seen, and convey to those who have not a good conception of the modern or English style of gardening, a better idea of what this style consists in than they could learn by reading a hundred descriptions of the same. . . .

The house stands about one hundred and fifty feet from the street; a broad carriage-way enters on one side, and, sweeping by the entrance to the house, in a semicircular form, opens to the street on the opposite side. Between this carriage-way and street there is a fine lawn; this is varied by two or three elegant groups of trees, which break the view of the house from the street, and likewise convey an idea of greater extent. . . . From the approach, on the south side of the house, . . . a walk leads up to the conservatory. . . . In front of the conservatory is a fine flower garden, laid out

with dug beds on turf. This garden is bounded by a wall . . . and by the back of the grapery . . . ; and, to screen the latter building from the eye, a vigorous and luxuriant growth of the woodbine has covered it so completely as to scarcely leave an open space. A rock-work, in a small way, but erected mostly with rare specimens of quartz, &c., and covered with verbenas and petunias, is an interesting feature of this garden.

Passing into a straight walk which leads from the conservatory . . . to the grapery, we entered upon a portion of the pleasure ground: this is laid out with most excellent taste and judgment. . . . Continuing through the winding walks, shady bowers, and umbrageous retreats, through which rustic seats were placed, we arrived at the shell grotto. This is an ingenious piece of work, finely executed under the direction of Mr. Arnold. The roof is supported by columns of rough trunks of trees, the outer part of the roof thatched, and the ceiling elegantly inlaid with shells, quartz, &c. A rustic sofa and table are the only articles in the interior. So secluded is this grotto, that the robin has built its nest and reared its young in some of the niches left for that purpose. From this we pass through other portions of the ground, and enter upon the main walk which leads round the kitchen garden, . . . shut out from view by a hawthorn hedge. . . . We should remark, that between the lawn in front, and the pleasure ground, a belt of trees, composed principally of evergreens running at a right angle with the street, to the grapery, screens the whole from view of any persons entering the house. On the north the grounds are planted mostly with ornamental trees and shrubs. (Hovey, 1840, pp. 363–64)

Since this article depicts a well-established estate, Arnold must have started planting soon after building his house. He included elements from both the formal and naturalistic styles, a hybrid that Loudon would have approved. The careful screening with trees of kitchen gardens from the pleasure ground was something Loudon recommended. The Arnolds traveled in Europe from 1837 to 1839, and gardens seen on this tour inspired their plans for the New Bedford property. Groves of trees, including the indigenous old oak, were an important part of the garden.

The grotto may be most puzzling to gardeners and nature lovers of today. Mentioned in every account of the Arnolds' garden, it is always noted for its surprisingly whimsical quality. The grotto, which recalled classical mythology, was a less important element in American

23. *View of the grounds of James Arnold, New Bedford, as pictured in the second edition of Downing's* Treatise on Landscape Gardening. *The Arnolds' grotto was a secluded, rustic structure ornamented with shells, an emblem of their reliance on the sea for their livelihood.*

estates and gardens than it was in those of England. This feature had many garden precedents, not the least of which was the Grotto of Thetis in André Le Nôtre's original design for Versailles. Befitting the cave of one of the sea nymphs of Greek mythology, Le Nôtre's grotto had a reservoir in the roof that fed dripping and spouting water devices, and its walls were encrusted with shells. In England, Alexander Pope transformed an underground passageway beneath a public road that bisected his property into a grotto by covering the walls with rocks, shells, and chips of glass.

Such a structure, whether an actual cave, a pile of rocks made into a cave, or a wooden structure built to look like a cavern, functioned as a resting place and cool retreat from summer's heat. Grottoes were usually associated with water or ornamented with objects reminiscent of the sea. For James and for Sarah Arnold, who was often named as the one responsible for decorating their grotto ceiling with shells, the sea was certainly significant. Herman Melville, who visited the Arnolds' garden one afternoon, wrote of New Bedford: "All these brave houses and flowery gardens came from the Atlantic, Pacific and Indian Oceans.

One and all they were harpooned and dragged up from the bottom of the sea" (Melville, 1980, p. 50).

Arnold's was not the only fine garden in New Bedford. His father-in-law, William Rotch, Jr., had a beautiful boxwood maze alleged to have been modeled after the one at Hampton Court in England. Many other sea captains and merchants had gardens and were avidly interested in horticulture. When these men decided to organize in 1847, James Arnold joined the founders of the New Bedford Horticultural Society. Grapes and peaches were his forte. He trained peach trees against walls, making use of passive solar heat to protect the flowers from spring frost and to promote the ripening of fruit. There were two graperies, along with the orchard for which the street that now runs between Arnold and Union Streets is named. Arnold's great-niece later recalled a storeroom on the north side of the County Street mansion where a table was heaped with grapes from the hothouses; the scent of fruit always filled that room.

When John Quincy Adams and his son Charles visited New Bedford in September 1835, Arnold's garden was part of the city tour provided by their hosts. Charles wrote in his diary: "We went to Mr. Arnold's, where we stopped. He took us over his garden, which is laid out with much taste. The presence of a female taste is perceptable in it. . . . Mrs. Arnold too is a lady as there are not many." When the Adamses again visited New Bedford in September 1843 they called on Arnold once more. Finding only Mrs. Arnold at home, John Quincy Adams noted that she "received and treated us with a profusion of flowers and of fruits—grapes, pears and peaches" (Old Dartmouth Historical Society, 1919, pp. 19, 21, 22).

24. *Black Hamburg grapes. One of the founders of the New Bedford Horticultural Society, James Arnold enjoyed the challenge of raising the earliest fruits each season. He entered Black Hamburgs, a popular variety, in competition at the Massachusetts Horticultural Society the year he was made an honorary member.*

Arnold began exhibiting at the Massachusetts Horticultural Society shows in Boston soon after his return from Europe. In September 1839 he entered peaches and Black Hamburg grapes in the MHS annual autumn exhibition and later that year was voted an honorary member of the society. At the same show, the brothers Charles and Andrew Jackson Downing exhibited some varieties of pears raised at their nursery in Newburgh, New York. Even though their contributions were small compared to those of some of the regular exhibitors at this increasingly popular fete, Arnold's and the Downings' participation introduced the New Bedford horticulturist to America's premier advocate of naturalistic landscape gardening.

By the time the second edition of his *Treatise on the Theory and Practice of Landscape Gardening* was published, Downing had seen Arnold's garden.

In the environs of New-Bedford are many beautiful residences. Among these, we desire particularly to notice the residence of James Arnold, Esq. There is scarcely a place in New-England, where the pleasure-grounds are so artistically laid out, so full of variety, and in such perfect order and keeping, as at this charming spot; and its winding walks, open bits of lawn, shrubs and plants grouped on turf, shady bowers, and rustic seats, all most agreeably combined, render this a very interesting and instructive suburban seat. (Downing, 1844, pp. 42–43)

In addition to landscape gardening and horticulture, collecting natural-history specimens was a favorite pastime of the Arnolds. In the south wing of their home was a room called "the cabinet," encircled with mahogany cases filled with shells; "a rather dark and awesome room," as their great-niece recollected (Pease, 1924, p. 6). Their rock garden contained rare specimens of quartz and other geologic finds, some from their own excursions. And the ceiling of their garden grotto held another shell collection.

James Arnold's whaling fleet brought back such "treasures" for his cabinet. He was known "for keeping an interest in the progress of science and thought" (Potter, 1868, pp. 7–8). Arnold shared his enthusiasm for natural science with George Barrell Emerson, who became his brother-in-law in 1834 by marrying Sarah's widowed sister, Mary Rotch Fleming.

ARNOLDS AS LEADERS IN NEW BEDFORD COMMUNITY

Although James and Sarah Arnold worshiped with the Unitarians after the Society of Friends divided in the late 1820s over the progressive preaching of Elias Hicks, they maintained their Quaker belief in equality and the dignity

25. While in Rome in 1837, James and Sarah Arnold and their daughter Elizabeth sat for this oil portrait by Robert Scott Lauder (1803–1869), a Scottish painter. (The man in the background has not been identified.) Reputed to be stern in his business dealings and fond of a good debate, James Arnold (1781–1868) mellowed in his later years and took up his wife's program of personal charity after her death. Apparently, he wished to leave his beautifully landscaped grounds to the people of New Bedford for a park, but the city government was not in a position to accept it. Instead, Arnold donated major portions of his whaling fortune to charitable organizations and individuals who were involved in good works.

of human life, which helps to explain their extended philanthropy. Since education has always been an important part of the life of Quaker believers, it is not surprising that James Arnold and his father joined five other gentlemen who put up the funds to found an academy to educate youth in New Bedford. James Arnold served on the board of the Friends Academy from 1812 until his death in 1868.

Typical of their humanitarianism, many New England Friends were leaders in the movement to abolish slavery. In the years preceding the Civil War, Arnold donated funds to several antislavery societies. During the war, his conviction that disputes should be resolved through mediation rather than force prevented him from contributing to the Union army. Instead, he sent ample gifts for medical relief.

Sarah and James Arnold supported many good causes in New Bedford. They were particularly active in the Port Society, which maintained the bethel, or seafarers' church, so vividly described in *Moby Dick*. In 1851, Sarah presented a house she had inherited to the Port Society for use as a mariners' home. This building still stands on New Bedford's Johnny Cake Hill, across from the Whaling Museum. Among the Arnolds' other interests were the Association for Relief of Aged Women, Orphans Home, State Reform School for Girls, and Massachusetts Temperance Union.

Sarah also directly apportioned her Rotch-inherited wealth among New Bedford's "poor and needy" individuals. She had an office at home where she listened to each person's story, then offered sympathy, advice, and gifts to solve immediate problems. Convinced that personal interaction and practical instruction were as valuable to reform as the fortune she shared, Sarah personally investigated the circumstances and kept accounts of each charity case. The Arnold home was known as a refuge for poverty-stricken persons and for fugitive slaves.

James Arnold, who died in 1868, did not live to see New Bedford change from a whaling to a textile-manufacturing center. After the deaths of Sarah and their only child, Elizabeth, in 1860, he withdrew from timber speculation and continued only some of his lifelong business ventures. Much of his last eight years was spent in "taking up . . . the work of home-charity which had dropped from his wife's hands." He carried this work on as a tribute to his wife's memory and continued to support other charitable organizations, such as those that gave support to freed people in the aftermath of the Civil War.

Between 1865 and 1868, Arnold worked on the final revision of his will. With no direct descendants surviving, James divided his large fortune among his and Sarah's relatives and the couple's many friends and colleagues in charitable work. He left his house and the land east of Cottage Street to his nephew, William J. Rotch, who in 1852 became the second mayor after New Bedford's incorporation as a city. Rotch later modernized and enlarged the simple Federal house, adding a mansard roof, which was the style in the 1870s. The garden was maintained, although the acres of meandering woodland paths located west of Cottage Street were sold as part of Arnold's residuary estate. The former Arnold mansion still stands—it is the home of a private club—but nothing remains of Arnold's landscaped grounds.

One of the largest single portions of his residuary estate was entrusted to three of his colleagues for the benefit of the "Poor and Needy in New Bedford who may be deserving." This fund long supported the Union for Good Works, a self-help organization founded in 1870 by the Reverend William J. Potter, Arnold's eulogist and preacher at the Unitarian Church.

A nearly equal portion of Arnold's residuary estate was bequeathed

to George B. Emerson, John James Dixwell and Francis E. Parker Esqrs. of Boston in trust;—to be by them applied for the promotion of Agricultural, or Horticultural improvements, or other Philosophical, or Philanthropic purposes at their discretion, and to provide for the continuance of this Trust hereafter to such persons, and on such conditions as they, or a majority of them, may deem proper, to carry out the intention of the doner [sic], one and one quarter of said 24 parts.

This rather unrestricted item makes no mention of an arboretum, and it was left to the three trustees to interpret the "intention of the doner." One of them, George B. Emerson, was particularly responsible for the creation of the Arnold Arboretum under the auspices of Harvard College. How the idea for the Arboretum developed can be seen by examining Emerson's connections and interests.

GEORGE BARRELL EMERSON, EDUCATOR

At the time James Arnold made out his will, George B. Emerson was retired from thirty-three years of teaching "young ladies" in Boston. This modest and genial man remained active, however, in educational-reform movements as well as in specific projects in natural-history education. He was an advocate of public parks, an enthusiast of the naturalistic landscape style, and had written on the conservation of forest resources. Although he was not a professional scientist, Emerson had written an important work on Massachusetts trees, and his advice had influenced the establishment of botany as a field of scientific study at Harvard College. In addition, he was well acquainted with contemporary developments in botany and the needs of the university in that department. All

26. *George Barrell Emerson (1797–1881), a leader of movements to improve natural-history education at all levels, influenced his brother-in-law James Arnold to leave the bequest that was used to start the Arnold Arboretum.*

these interests were reflected in the advice he gave to Arnold as he planned his bequest.

Emerson, son of the Harvard-educated physician Samuel Emerson, was raised in Wells, Maine (when that state was still part of Massachusetts). He spent much of his boyhood roaming the fields, woods, and seaside and working on the family's farm. Thus, early in life George became familiar with many natural habitats and the plants and animals characteristic of each. After a few years of preparation at Dummer Academy in Byfield, New Hampshire, the young Emerson entered Harvard College in 1813, concentrating in mathematics and Greek. Like many of his fellow students, he spent the winter vacations teaching at country schools.

Besides his enthusiasm for Greek and mathematics, Emerson brought a keen interest in nature to Harvard, where there were good resources for expanding his knowledge. Before entering college he had learned as many of the trees and plants around Wells as he could. His father taught him the Linnaean system of classification. Apparently, the first thing Emerson did after getting settled at college was to visit the botanic garden, hoping to learn from Professor William Peck the names of some plants he had found in Wells that he could not identify. The young student was pleased when Peck recognized them instantly from his descriptions. The summers of farmwork had given George Emerson an appreciation of natural processes as well as the garden arts. All through his adult life he was convinced that active life in the outdoors improved the powers of observation and was beneficial to everyone's learning capacity.

It was an exciting time to attend the college at Cambridge. Among his classmates were young men who would go on to serve as statesmen, presidents of colleges, historians, politicians, and influential clergymen. Although George Emerson graduated in 1817, he kept close ties with his alma mater during this, its "Augustan Age." President Kirkland found him his first job after graduation as master of a private boys' school recently established in Lancaster, Massachusetts, through the generous efforts of a merchant sea captain, Richard Jeffrey Cleveland. Here Emerson was so successful a teacher and leader that the school's enrollment nearly doubled in two years.

In 1820, he took a position as the first headmaster for Boston's new English Classical School, later called English High School. He taught in this position satisfactorily for two years, meanwhile developing many of his own ideas on the best methods of education. At the urging of friends, Emerson left the English Classical School to start his own institution for young women in Boston, which opened in June 1823 with twenty-five pupils. That year he also married Olivia Buckminster, daughter of the clergyman Joseph Buckminster (1751–1812) of Portsmouth, New Hampshire. Olivia frequently helped in the school.

Emerson's belief in the importance of the teaching profession did not remain focused on his own role. In 1830 he joined with other teachers and friends of education to form the American Institute of Instruction. The members' concern for the improvement of schools led them to lobby successfully before the Massachusetts legislature for the creation of a state board of education. The ultimate outcome was the establishment of hundreds of common schools and two normal schools; Emerson served on the boards of the latter in following years.

Emerson also lectured widely and published on such topics as the education of girls and women, moral education, health, home economics, and sanitation. He published texts for schoolroom use and a manual for teachers that was placed in every school by acts of the legislatures of Massachusetts and New York. When the Boston Society of Natural History was founded in 1830, Emerson helped to organize it. He was a very active member, holding several offices, curating one of the collections, and regularly attending meetings.

In the beginning of his second decade as master of his school, George B. Emerson experienced domestic tragedy. Olivia, his wife and assistant in his school, became ill and died in the summer of 1832. George was left with three children, aged seven, five, and three, whose healthy and proper upbringing was a source of concern to him. After two and a half years, in late November 1834, he remarried. Emerson's second wife was a widowed sister of Sarah Arnold, Mary Rotch Fleming. Mary had one daughter by her first marriage, and she and George raised the four children together. Emerson continued to keep school, but moved it to a new location. With assistance from "father Rotch," he purchased a lot and had a house built at Pemberton Square, a new residential development on the site of Pemberton Hill (which had been taken down to fill in a bay on the shores of the Charles River). Home and school were under one roof for the Emersons; there were twin doorways, one of which led upstairs to spacious classrooms located over their living quarters.

With his second marriage, George commenced a close friendship with the Rotch family, including James and Sarah Rotch Arnold. During visits to New Bedford, George and Sarah found they shared an interest in shell collecting, and James led them to neighboring geological sites. Emerson and William Rotch compared their views on religion and reform, discussing such matters as "Dr. Channing's notable book upon slavery."

REPORT ON TREES AND SHRUBS

By 1837, Emerson's home life had settled somewhat and the school was thriving. Its young women benefited from his acquaintances in Boston's intellectual community. For example, he invited the eminent poet Richard Henry Dana to lecture to his students on English literature. Emerson had been elected to membership in the American Academy of Arts and Sciences, and in 1836 he was chosen president of the Boston Society of Natural History.

When the BSNH took on the project of Massachusetts botanical and zoological surveys, Emerson not only acted as commissioner of the surveys but conducted the investigation of trees and shrubs himself. He worked on the project for nine summers, whenever school was not in session. At every chance Emerson traveled throughout the state to look at trees. Relying heavily on his own firsthand observations, he took copious notes while under or near the growing plants. He also consulted the private herbaria of several Massachusetts botanists, while other colleagues sent him specimens for study.

One of the goals of the surveys was to collect information on the economic importance of each subject. To find out more about how Massachusetts trees were used and how forests or woodlots were managed, Emerson sent a circular with twenty questions to some fifty landowners in every part of the state. He obtained much valuable information from their responses. On his own fact-finding excursions, Emerson visited the shipyards in Boston, New Bedford, and other towns, as well as numerous sawmills, machine shops, and manufacturers of furniture, agricultural implements, and other articles using wood. He stopped to make inquiries at farms and woodlots in all parts of the commonwealth.

In addition to direct observation and "networking" with other tree enthusiasts, Emerson's careful scholarship led him to the published works of botanists and agriculturists. After consulting the works of Michaux, Loudon, Endlicher, Lindley, and Torrey, he decided to arrange his report according to the new natural system of classification. By embracing this innovation, George Emerson participated in the advancement of American botany. Moreover, he credited his New England readership with the ability to appreciate and grasp the new system of plant categorization.

Issued in late 1846, Emerson's *Report on the Trees and Shrubs Growing Naturally in the Forests of Massachusetts* turned out to be the most popular of the volumes published in the survey. His ability to present accurate scientific information with lucidity and contagious enthusiasm was universally praised. "It is a work that every intelligent farmer, educated at a New England School, may read and understand fully—and which is at the same time as truly (not pedantically) learned, as if it had been prepared for the Academy of Sciences," reported Andrew Downing's *Horticulturist* (Anonymous, 1847, p. 566).

The main portion of the work consisted of descriptions, which, drawn as they were from firsthand observation, had a freshness and vitality that took the reader out into the woods with the observant schoolmaster. The discussions accompanying the treatment of each species incorporated such facts as the tree's usual habitat, what uses might be made of its wood or bark, its qualities as fuel, how it was best cultivated, what size it usually attained, and where particularly large examples could be seen.

The introduction presented an instructive overview of Massachusetts forests. Emerson summarized the report's chief objective:

27. *George B. Emerson traveled throughout Massachusetts to observe its trees, and he noted particularly large individuals in each species. In Hingham, he admired this old American elm at Rocky Nook. Emerson reported its dimensions as thirteen feet in circumference and sixty or seventy feet in height, with a crown more than ninety feet in breadth. The physician and humorist Oliver Wendell Holmes also liked to measure large trees and sometimes accompanied Emerson in the field.*

A few generations ago, an almost unbroken forest covered the continent. The smoke from the Indian's wigwam rose only at distant intervals; and to one looking from Wachusett or Mount Washington, the small patches laid open for the cultivation of maize, interrupted not perceptibly the dark green of the woods. Now, those old woods are everywhere falling. The axe has made, and is making, wanton and terrible havoc. The cunning foresight of the Yankee seems to desert him when he takes the axe in hand. The new settler clears in a year more acres than he can cultivate in ten, and destroys at a single burning many a winter's fuel, which would better be kept in reserve for his grandchildren. This profuse waste is checked, but it has not entirely ceased. It is, however, giving way to better views. Even since this survey was begun, a wiser economy shows itself. May it be universal. A brief consideration of the general use of forests on a great scale may have a tendency to produce this effect. (Emerson, 1846, p. 2)

What followed was an enumeration of the benefits forests provide for man: improving and holding soil, moderating the climate, providing material for fuel and uncountable necessary objects. Emerson also discussed the nonmaterial, the aesthetic and spiritual, merits of forests and trees. "A single tree by a farmer's house protects it, and gives it a desirable air of seclusion and rest; as if it must be a residence of peace and contentment. . . . While an unprotected, solitary house seems to shiver in the north wind, and we involuntarily wish for the inhabitants a more cheerful home" (Emerson, 1846, p. 9). Massachusetts trees, he argued, could be used not just to supply timber, but, thoughtfully planted, they could beautify many a human environment—dooryards, pastures, roadsides, estates, and public grounds.

In a section entitled "Continuation and Improvement of the Forests," Emerson argued for conservation, management, and restoration of forest resources. Such ideas were just beginning to be discussed in America. There were no governmental authorities to regulate forest use, and conservation organizations did not yet exist. Emerson summarized the experience of many landowners who answered his circular on such topics as how to plant timber trees, when to thin and prune them, how many years each species required to reach suitable size for harvest, and the methods and timing of felling. On these topics, Emerson realized that his report was merely a starting point. Much more scientific study was needed, as well as further development of the fine art of "the best disposition of trees in the landscape."

Emerson was sure that Americans should start to conserve forests and plant trees. Educating them to appreciate trees was a step in the right direction; founding an institution with this role would be another step that Emerson would take.

Several years later, when George Perkins Marsh published *Man and Nature* (revised in 1874 as *The Earth as Modified by Human Action*), George B. Emerson welcomed Marsh's environmental study. Emerson subsequently produced a revised version of his *Trees and Shrubs of Massachusetts*, which appeared in 1874. The most significant addition to the text was mention of *Man and Nature*, "a work richly deserving to be read by every person interested in the . . . preservation, restoration, and care of the

28. Buttonbush (Cephalanthus occidentalis), *which favors sunny locations at the edge of standing water, is one of the commonwealth's many native shrubs inventoried and described by George B. Emerson in his* Trees and Shrubs of Massachusetts. *One of the few woody members of the madder family (Rubiaceae) native to eastern North America, its globose clusters of highly fragrant flowers open in high summer. Emerson argued for the preservation of noteworthy shrubs and trees and for the management of forest resources.*

forests" (G. B. Emerson, 1887, p. xi). In addition, Emerson discussed what European countries were doing to foster conservation and the establishment of forestry schools.

NATURAL HISTORY AND LANDSCAPE GARDENING

Emerson's research into Massachusetts trees widened his contacts and fostered his reputation as a serious scholar. He was offered the Fisher Professorship in Natural History in 1838, but declined to take it. A few years later he supported the appointment of Asa Gray to the post. Soon after Gray's initial visit, President Quincy asked George Emerson's opinion of the young botanist, and he replied in the affirmative. This must not have been a completely easy thing for Emerson to do, as there were others more closely tied to the college and long-standing members of his own Boston Society of Natural History who also might have qualified for the professorship.

The two naturalists commenced a cordial relationship as soon as Gray was established at the botanic garden. Emerson sought the new professor's counsel for his report and found Gray especially helpful when composing the key to identification included in the book. The two men together measured some of the state's noteworthy trees.

When Asa Gray donated his herbarium to Harvard, Emerson was instrumental in raising the fund to endow it. After its transfer to the college, Emerson served on the visiting committee for the herbarium, and Gray turned to him when funds were needed to advance its work. This behind-the-scenes activity is typical of Emerson's ever present support of botanical research and of his interest in education.

His advocacy of Gray's activities notwithstanding, by temperament Emerson was receptive to the interpretation of nature put forward by Agassiz. One of the first to meet the great naturalist, he gave him a tour to see the several species of the tree genus *Carya,* or hickories, then known to occur strictly in North America, that Agassiz knew only from fossils. His mastery as an educator was what appealed most about Agassiz to George B. Emerson. He admired the Swiss professor's knack for teaching by things. Like many naturalists of his generation, Emerson was inclined to agree with Agassiz's idealistic view of the species question, and he tended to ignore the impact of the theory of evolution on that view. As late as 1874, shortly after the death of Agassiz, George Emerson wrote: "During his whole life, while exploring every secret of animal structure, he saw such wonderful consistency in every part, that he never for a moment doubted that all were parts of one vast plan, the work of one infinite, all-comprehending Thinker. He saw no place for accident, none for blind unthinking, brute or vegetable selection" (Bouvé, 1880, p. 157).

Emerson cherished his summers working in the countryside among the trees, and he was impressed by the estates he had seen in the course of his research. Even before the report was complete, he began to search for a rural retreat where his family could gather in summer. Pemberton Square was a fine and fashionable city residence where densely spaced, redbrick houses were shaded by trees in a small central green, not unlike contemporary Louisburg Square. Yet the change, the fresh air, and the stimulation of country living appealed to Emerson, as it increasingly did to many of his fellow Bostonians.

In 1847 he purchased thirty acres of land on the north-eastern side of Chelsea harbor, on a promontory that stretched into Boston Bay. Although the barren site had poor, sandy soil, he was determined to clothe it with trees and anticipated his family's future pleasure in watching them grow. Undoubtedly, Emerson's desire to improve his summer grounds was inspired by the Arnolds' delightful garden and by William Rotch's elegant home grounds. Another garden-estate where the Emerson family were frequent guests was that of Thomas Lee of Brookline, whose wife was a sister of Emerson's first wife. Andrew J. Downing described and praised the Lee place in his *Treatise on Landscape Gardening,* and it subsequently became part of the home grounds of Charles S. Sargent.

The return of his friend and former student Horace William Shaler Cleveland to the Boston area in 1854 also influenced Emerson's management of his property. Cleveland was the son of the founder of the school in Lancaster, Massachusetts, where Emerson held his first teaching post. Having studied civil engineering and spent several years truck farming in New Jersey, Cleveland entered into a landscape-gardening partnership with Robert Morris Copeland of Lexington, Massachusetts. Copeland and Cleveland were very much inspired by Downing's work and publications. They, too, wished to further private and civic improvement through the practice of landscape gardening.

Emerson was one of their first clients. Cleveland credited the *Trees and Shrubs of Massachusetts* with influencing his own endeavors, and the two of them shared an experimental frame of mind with regard to tree planting. On the excessively poor and exposed land they set out many European varieties of oak, beech, birch, linden, maple, elm, ash, mountain ash, and pine to find out whether they were more hardy than the corresponding American trees. Twenty years later, in the second edition of his report, Emerson stated that the European species he planted had performed better than their native American counterparts at his seaside property.

Cleveland conceived of his role as landscape designer with more breadth than it had previously been given. "Landscape Gardening, or more properly Landscape Architecture, is the art of arranging land so as to adapt it most conveniently, economically and gracefully to any of the varied wants of civilization" (Cleveland, 1873, p. 17).

What was important to him was planning to utilize the natural features of land for the best of human occupations, not the mere ornamentation of a piece of property. His was similar to the vision that Frederick Law Olmsted would soon bring to the profession. Not surprisingly, he would eventually collaborate with Olmsted. George Emerson's relationship with Cleveland undoubtedly made the schoolmaster more aware of the goals of the emerging landscape profession.

As a man who always thought in terms of improving the lives of others, Emerson's enthusiasm for trees and for naturalistic landscape gardening could only turn toward public environments. He kept abreast of activities such as

29. *John Claudius Loudon (1783–1843), prodigious horticultural and botanical author, developed the notion that an arboretum should serve both educational and aesthetic purposes. He argued that the plants in such a garden should be laid out in an informal, naturalistic style, but in a sequence that would demonstrate botanical relationship.*

the founding of Mount Auburn Cemetery and became a corresponding member of the Massachusetts Horticultural Society. Developments in Boston and other cities tending toward the establishment of public parks received his support.

ARBORETUM CONCEPT REFINED

Emerson's conception of a public tree collection grew from many sources. His relationship with Asa Gray exposed him to the academic argument for an arboretum. As early as 1844, in an essay on the longevity of trees, Gray condemned the lack of a good living collection of trees and shrubs in America. After a discussion of the contribution of the French botanists André Michaux and his son, François André, he stated:

To these two persons, chiefly, are the French plantations indebted for their surpassingly rich collections of American trees and shrubs; which long since gave rise to the remark, as true at this day as it was twenty years ago, that an American must visit France to see the productions of his native forests. When shall it be said that this statement is no longer true? When shall we be able to point to a complete, or even a respectable, American collection of our indigenous trees and shrubs? (Gray, in C. S. Sargent, 1889, vol. 2, p. 74)

More than once Gray suggested to Harvard's administration that the botanic garden be supplemented by a collection of woody plants.

While the schoolmaster readily agreed with the professor on the need for an arboretum, Emerson's sympathy with the amenity value of trees and landscape widened his vision. From discussions in the horticultural literature and reports of recently established arboreta in England, as well as from unexecuted American proposals, the concept of an arboretum as combining a beautiful space with a scientific function was beginning to emerge.

Just as the naturalistic style of landscape design was introduced from Britain, so too was the formula for an all-inclusive garden of hardy trees and shrubs after which the Arnold Arboretum would be patterned. Most active in this field was John Claudius Loudon, who may have been the first person to use the word "arboretum" in modern times. His dual facility with botany and horticulture allowed him to develop the notion that an arboretum could serve both educational and aesthetic purposes. In turn, Loudon's work was noted by Andrew Jackson Downing, Frederick Law Olmsted, and other American advocates of landscape design.

As early as 1806, in his *Treatise on Country Residences,* Loudon mentioned an arboretum as one of many types of botanic garden that could be formed, noting that such should be arranged either by the sexual system of Linnaeus or by the natural system of Antoine Laurent de Jussieu. After Loudon's reputation had been established with the publication of his *Encyclopaedia of Gardening* in 1822 and the founding of his *Gardener's Magazine* in 1826, he began to report on public gardens on the Continent and to advocate their establishment in the British Isles. He also frequently called for the use of the natural system of plant classification: "If the mode of studying plants in groups, and distributing collections in gardens according to the natural system, were generally adopted, botany would become an easier, more agreeable, and more satisfactory science than it is at present. . . . The Linnean system presents a heap of broken links,—of bricks in a kiln,—or words in a dictionary; the natural system presents a chain, a house, or a discourse" (Loudon, 1827, pp. 300–301).

Loudon's ideas about what an arboretum should be were further developed in articles he wrote criticizing Chiswick, the new garden of the Royal Horticultural Society. He argued that the genera in the arboretum should have been laid out in sequence following the natural system, rather than having been distributed at random. Even an alphabetical arrangement, such as could be seen at the Loddiges' nursery in London, would have been better in allowing students easy access according to Loudon's standards. He also did not think much of Chiswick's landscape composition, finding it neither a good imitation of natural scenery nor a good example of art. From this critique, it is apparent that Loudon expected a pleasingly designed landscape for an arboretum.

Loudon had a chance to put his ideas into practice in 1831, when he was called on to design the Birmingham Botanical Garden in the west Midlands. The supervisory committee desired a plan that would combine a scientific with an ornamental garden, as well as providing growing areas for nursery stock, fruits, and vegetables—a demanding set of requirements for only sixteen acres of ground. For the scientific part of the gardens, Loudon combined the herbaceous planting with the trees and shrubs, these being distributed around the perimeter of the grounds along one side of a main walk. Both the herbs and the woody plants were to be arranged according to Loudon's book *Hortus Britannicus,* which followed the Candolle natural system. In an attempt to make the collections as comprehensive as possible, Loudon recommended no duplicates be planted. The committee, however, disagreed with some of Loudon's recommendations for the rest of the gardens, and he withdrew from the project. Without the unity of composition he had planned, Loudon feared the project would be bungled.

Finally, in 1835, an arboretum was begun for which Loudon had nothing but praise. This was the arboretum at Chatsworth, Derbyshire, the gardens of the duke of Devonshire, designed and executed by his gardener, Joseph Paxton. Although it was a private estate, the sixth duke, William Cavendish, opened it every day in the year and showed it to visitors. With ample acreage, the arboretum was set out on a comprehensive scale. The trees were planted in order according to the Candolle system on either side of a walk that circuited the pleasure ground, curving gently with the land's contours. Some exceptions to the scientific order were made to take advantage of favorable microclimates, as in the placement of the willows around a pond. And the pinetum, which had been started many years earlier, was left in place even though it was out of sequence. This was the first time Loudon noted that the plants were labeled with their scientific names and other useful information.

The year the Chatsworth arboretum opened, Loudon wrote "Remarks on laying out Public Gardens and Promenades" for *Gardener's Magazine,* in which was in-

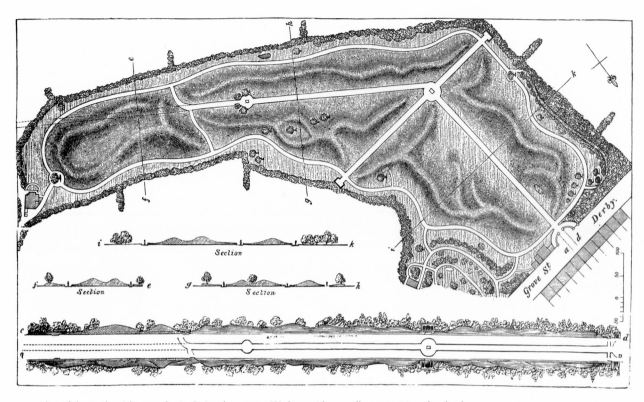

30. Plan of the Derby Arboretum by J. C. Loudon, 1840. Working with a small, unpromising plot, he designed a tree collection and public park for the citizens of Derby at the behest of a wealthy silk manufacturer. The labeled plantings were ordered along a circuitous walk, visually separated from the central promenade by seven-foot-high earthen berms intended to add interest to the level site.

cluded a description of an ideal arboretum. Citing Chatsworth as an example, he stressed that any such garden would need ample space "to allow all the species and varieties of trees and shrubs, foreign and domestic, . . . that will stand the open air in the neighborhood, . . . to attain their full size" (Loudon, 1835, p. 657). He recommended that the plants be arranged according to some classificatory theme, be it geographical, geologic, or botanical. "Whichever system is adopted, one main walk should be conducted through the whole, so as to pass in review every individual tree and shrub, and yet show no plant twice." Loudon reiterated the importance of labeling

the plants with their scientific names and other descriptive information (Loudon, 1835, p. 658).

John Claudius Loudon at last created a full-fledged arboretum in 1840. Not only did this garden, the Derby Arboretum, follow all the principles he had elaborated, but it was expressly intended for the public. It was a gift to Derby from its member of parliament, Joseph Strutt, a wealthy silk manufacturer. Strutt wanted to turn an eleven-acre property into a place of recreation for the inhabitants of the town, and Loudon convinced him that an arboretum would be the most manageable way to fulfill that goal while furthering the interests of education.

The site, a narrow, odd-shaped plot of nearly level ground with no views, was not promising. Strutt had already planted the borders with a thick belt of miscellaneous trees. Loudon retained this planting and used it as a foil for a winding, mile-long walkway, along the interior side of which he arranged the tree and shrub collection in botanical sequence. These were labeled with the same kind of information as the trees at Chatsworth. A straight central walk was intersected by paths connecting to the arboretum walk. Along this central walk were placed the various architectural elements Strutt also required—statuary, urns, and pavilions and gatehouses with visitor amenities. To separate the winding, circumferential arboretum walk from the central, formal avenue and to improve the topography of the site, Loudon raised undulating, gradually sloping ridges of earth from seven to ten feet in height. Strutt provided the land, buildings, and funds to create the arboretum, then turned it over to the Derby town council, specifying that it be maintained with income from a modest admission fee levied five days a week. The arboretum was a success, and it still exists in modified form.

In the creation of the Derby Arboretum, and in all his previous publications mentioning the arboretum idea, Loudon continually emphasized five elements that define this type of garden: it is a tree and shrub collection; it includes only trees and shrubs hardy in the outdoor climate where the garden is located; of these, it is to be all inclusive, with at least "one of every kind" being grown; the plants must be arranged in some rational order, preferably according to the natural system of classification; and the plants must be labeled. He further stressed that the educational tree collection should be accommodated in a pleasing landscape, often suggesting that the best way to achieve this would be to arrange collections along one main path that forms a circuit, so they could be viewed in order by the visitor.

UNEXECUTED AMERICAN ARBORETA

Americans were apprised of the English arboretum activities through reports in the horticultural literature, and the idea was considered and discussed. Before the creation of the Arnold Arboretum there were a few proposals for such gardens in America, but they went unexecuted. These included many of the suggestions put forth by Loudon.

In 1841, the year after the Derby Arboretum opened, the proprietors of the Public Garden adjacent to Boston Common asked Andrew Jackson Downing to develop a plan for its grounds. This twenty-four-acre area had been leased from the city by the private group for use as a botanical garden since 1837.

Downing's plan called for an arboretum scientifically arranged according to the natural system to be sited around the boundary of the property. He prepared lists of the trees to be planted and indicated precisely where they were to be set out. Although this plan was never put into effect, it is noteworthy for introducing the idea of an arboretum to Bostonians at midcentury. Beyond a beautiful planting of trees, his plan included the educational aim of placing them so as to demonstrate current scientific understanding.

Discussion of the arboretum idea and of the need for tree planting and public parks continued in American horticultural journals of the day. In an article entitled "Arboricultural Gossip," John Jay Smith pleaded for increased planting of rare and ornamental trees in addition to fruit trees. Arboreta are "for a country residence, what a museum, or rich collection in natural history or art, is for a town house; a source of interest perpetual and unvarying," he argued. "Many a person . . . to whom trees are only trees—that is, green things in the landscape, in which they perceive little distinction—are immediately struck and interested in a country place by an arboretum" (J. J. Smith, 1847, p. 29). After his trip to England in 1850, Downing published reports on developments like the Chatsworth and Derby arboreta.

The nursery trade was also aware of the potential value of an arboretum or collection of rare trees. Patrick Barry of Ellwanger and Barry Nurseries of Rochester, New York, wrote to the *Horticulturist* proposing an association for collecting rare trees and plants. He was inspired by a visit to England, where he had seen many rare trees native to the American West thriving in its lawns and arboreta. He observed that formerly "our best landscapes have been worked out of the original forest," and, while arboriculture had been mainly restricted to fruits, the growth of cities and towns and the rapid disappearance of natural forests created a new necessity to plant trees ([Barry], 1850, p. 483). With this came the need for organizations that promoted the collection of trees and the dissemination of knowledge about them. Barry's letter was an excellent argument for an arboretum, and helped put the notion abroad in the minds of tree enthusiasts.

Another influential but unexecuted arboretum proposal was included in Vaux and Olmsted's 1858 "Greensward" plan for Central Park. The area available for the arboretum comprised about forty acres. The two designers planned to include only American native trees and shrubs because they thought the space needed for a complete arboretum would be several hundred acres. They also wished to emphasize the great advantage that America possessed in diversity of woody plants. Their "Descriptive Guide to the Arboretum" gave details that harked back to Loudon's many proposals for such gardens:

The principal walk is intended to be so laid out, that while the trees and shrubs bordering it succeed one another in the natural order of families, each will be brought, as far as possible, into a position corresponding to its natural habits, and in which its distinguishing characteristics will be favorably exhibited. At the entrance, . . . we place the Magnoliaceae, associating with them the shrubs belonging to the orders Ranunculaceae, Anonaceae, Berberidaceae, and Cistaceae. The great beauty of these families entitles them, if no other reason prevailed, to a very prominent place on our grounds. In pursuing the path which enters here, we find on our right hand the order Tiliaceae, with the shrubs belonging to the orders Rutaceae, Anacardiaceae, and Rhamnaceae. On each side of the walk groups succeed. . . .

At the next turn of the path, we come upon the Rosaceae. The shrubs of this order being very beautiful, we have placed many of them singley, as well as in thickets between, and over, the large masses of rock, which here occur on both sides. . . .

The oak may be almost called an American tree, as in no other country are the species half so numerous. On this account, as well as for their great beauty, it has been thought proper to give them much open space. A few shrubs of Cupuliferae and Myricaceae form the underwood of the mass which will shut out the view towards Fifth [A]venue, which here passes at an elevated grade.

To these succeed the order Betulaceae, the graceful birches, and Salicaceae, which includes the poplars. Finally, are brought in our various American Coniferae. Only single trees are provided for in this section, as masses of each are elsewhere arranged in the park. (Olmsted and Kimball, 1973, pp. 335–36)

Right down to the order of tree families, this description is prophetic of the Arnold Arboretum, with which Olmsted would be involved nearly twenty-five years later.

In the interim, there was another arboretum proposed for an urban park system by Horace W. S. Cleveland. In 1869 Olmsted engaged him to do some work for Prospect Park in Brooklyn. The following year Cleveland moved to Chicago, where he was placed in charge of South Park and the approach boulevards under development by Olmsted and Vaux. Cleveland projected treatment of a fourteen-mile-long boulevard connecting the city's three parks as an arboretum on a grand scale. He thought that the usual enhancement of natural topography with plantations would not work in Chicago because the area was so flat and featureless. Instead, he suggested, "Let the avenue form in its whole extent, an arboretum, comprising every variety of tree and shrub which will thrive in this climate, each family occupying a distinct section, of greater or lesser extent, according to its importance" (Cleveland, 1869, p. 17). He proposed using masses of each kind of tree in botanical sequence along the boulevard rather than individual specimens, stressing the artistic as well as the educational effect of such an arrangement.

Cleveland felt keenly that the young city would benefit greatly if this large-scale planting were executed. However, Chicago's politics and economics did not allow his vision to be realized. Moreover, the catastrophic fire of 1871 most likely curtailed many of the city's parkmaking efforts.

EMERSON MASTERMINDS THE INDENTURE

In 1855, George Barrell Emerson turned his school over to a nephew and took a journey to Europe with his wife. Although he did not resume his administrative post on his

31. Francis E. Parker (1821–1886), graduate of Harvard College and Harvard Law School, was one of three men directed by James Arnold to administer a bequest to be "applied for the promotion of Agricultural, or Horticultural improvements, or other Philosophical, or Philanthropic purposes at their discretion." Known for his wisdom and fidelity in managing charitable trusts, Parker served for many years on the board of overseers of Harvard.

return, he continued to tutor and counsel former students and stayed active in educational affairs. He also spent more time on philanthropic activity, and during the Civil War he served on a commission responsible for recruiting teachers for special schools for freedmen in the South.

Many affairs—the need for better natural-history education, man's impact on native forests, the importance of trees and naturalistic landscaping in improving public grounds, and the proposals for arboreta—were on Emerson's mind during the 1860s. Happily, his relationship with his alma mater remained strong. In 1859, Harvard awarded him an honorary doctor of laws degree, and in 1862 he was tentatively proposed for the college presidency. James Arnold advised him against this step, convinced that Emerson was more useful as a general philanthropist. Arnold, too, was thinking of philanthropy at this time. He started to revise his will after having lost his wife and only daughter in 1860. He sought Emerson's opinions and the advice of their mutual acquaintance Francis E. Parker, one of Boston's finest trust lawyers.

Parker was a Harvard graduate of the class of 1841. After receiving his law degree from the university in 1845, he developed a remunerative private practice. He was a member of the Boston School Committee and managed some of the city's important charities. Parker also served one term in the state senate and for forty years held the office of public administrator. A lifelong bachelor, Parker traveled to Europe nearly every summer. When home, he was frequently the guest of George and Mary Emerson and of the Arnolds. Once when Emerson gave a dinner party for a group of naturalists including Louis Agassiz, Asa Gray, and a visiting Irish algologist, William H. Harvey, he worried that his young friend, Parker, found the evening dull. But Parker's tastes were cosmopolitan. He knew how to listen to ideas and was skilled in helping others turn their good ideas into permanently funded institutions.

So it was that, although talks with Emerson convinced Arnold that an arboretum was a much needed resource, he left his will sufficiently indefinite to allow his trustees to act. Arnold sought Parker's approval for the final draft of his will and designated him a coexecutor, along with Sarah Arnold's nephew William J. Rotch.

The third trustee named by Arnold to oversee what became the Arboretum bequest was John James Dixwell, a prosperous Boston merchant and president of the Massachusetts Bank. He and Emerson were old friends, long united in their support of the Boston Society of Natural History. Dixwell lived on a ten-acre estate atop a hill that rose between the Bussey farm and Jamaica Pond. There he grew as many kinds of trees as he could obtain. This fondness for trees was shared with the Arnold family, whom he visited often in New Bedford. When James, Sarah, and their daughter were in Boston, Dixwell escorted them to Mount Auburn Cemetery. According to Asa Gray, he also took an interest in the botanic garden.

32. Weeping beech on the estate of John James Dixwell. The third trustee of the Arnold fund, Dixwell (1806–1876) was a friend of both the Emerson and Arnold families. He and F. E. Parker were among the incorporators of the Boston Museum of Fine Arts in 1876. A fancier of trees, Dixwell planted them extensively on his Jamaica Plain estate; located on Moss Hill, it overlooked the property of Benjamin Bussey.

A little more than three years passed from the time Arnold's will was approved and allowed by the court in January 1869 until the trustees, Emerson, Dixwell, and Parker, signed an indenture with Harvard establishing the Arnold Arboretum. With an arboretum in mind, the trustees spent the time weighing how best to carry out their duty. To turn the Arnold fund over to Harvard College, the oldest and most prestigious center of learning in New England, would be a sure way to provide for the continuance of the trust. Both Emerson and Parker were graduates, and all three had close social and professional connections with the college. In 1868 Parker was elected to the board of overseers, a position he held for the next twenty years save one. They all knew of Asa Gray's opinion that a tree collection was needed to complement the herbaceous plantings of the Harvard Botanic Garden.

Over the years Gray had apparently recommended to college authorities that the grounds of the observatory, which were contiguous to the botanic garden, be used for an arboretum. When Gray heard of Arnold's legacy, he reminded the Harvard administration of his idea for the observatory grounds. However, the limited size of this site kept the Arnold trustees from considering it seriously.

On the other hand, the Bussey estate—land already designated for Harvard—must have looked overwhelmingly superior. The Harvard corporation had been considering how best to act on the remainder of Bussey's bequest since 1861. Unlike Arnold, this earlier benefactor gave specific instructions in his will. The college was to retain

Woodland Hill, consisting of over two hundred acres of land, as a place in my judgement well adapted, from the great variety and excellence of its soil, its hills, valleys and water, its great diversity of surface and exposure, and lastly, its high state of cultivation and improvement, for all the objects contemplated. [T]hey will establish there a course of instruction in practical agriculture, in useful and ornamental gardening, in botany, and in such other branches of natural science, as may tend to promote a knowledge of practical agriculture and the various arts subservient thereto. . . . [T]he institution so established shall be called the "Bussey Institution." (Bussey, 1841, pp. 15–16)

Actually, for some years agricultural experimentation had been undertaken at the Bussey farm under the auspices of the Massachusetts Society for Promoting Agriculture. Thomas Motley, whose wife inherited Woodland Hill, participated in the society's program of introducing superior stock of European cattle and workhorses to New England farms. He traveled to Europe several times on the society's behalf to procure animals, often keeping them on his West Roxbury farm until they were evaluated and turned over to Massachusetts breeders.

The Bussey Institution was established soon after Charles W. Eliot began an innovative presidency that would transform Harvard into a modern university. A chemist, he sought to bring the sciences onto equal footing with the classics and humanities and to strengthen the stature of Harvard's professional schools.

In 1870 Mrs. Motley released seven acres from her life tenancy, thus allowing the formal organization of the Bussey Institution. The president and fellows appointed Thomas Motley as instructor in farming and Francis H. Storer as professor of agricultural chemistry. (Storer re-

mained dean of the school for some thirty-five years.) A contract was let for the building about which Benjamin Bussey had carefully expressed his wishes:

I hereby order and direct my Trustees . . . to cause to be erected on the "Plain-field," so called, next easterly of my farm garden, . . . an edifice, with convenient outbuildings, suitable in all respects for said institution: the said edifice to be not less than ten rods from said road; the exterior walls thereof to be built of stone in blocks (not hammered), or to be similar to the front wall of the "Masonic Temple," so called, in said Boston. And I earnestly enjoin it upon my Trustees to have the said edifice constructed and completed, with a proper regard to durability and beauty, and so as best to secure the comfort and convenience of the inmates of said building. (Bussey, 1841, p. 17)

The stone building was completed in 1871 in Gothic Revival style with solid buttresses, heavy lintels, and high slate roofs—substantial and imposing enough to have pleased Bussey, but probably not in the style he envisioned in 1835. With the appointment of more instructors in the spring of 1871, the Bussey Institution opened its doors.

33. The Bussey Institution building was completed in 1871 as directed in the will of Benjamin Bussey. It served as headquarters for an undergraduate school of agriculture until 1907. Subsequently, it housed a graduate school of sciences related to such problems in agriculture as genetics, entomology, plant anatomy, and economic botany. The handsome Gothic Revival structure had to be demolished after a destructive fire in 1971.

At the same time, Emerson was considering options for the Arnold fund. Soon after Arnold's will was probated, Emerson had written college authorities that, although the trust was intended for an arboretum, if most of the money would have to be expended to purchase land other options for the use of the fund would have to be considered. Employing land already in possession of the college would allow all of the Arnold fund to be used to develop the arboretum. The idea seems to have occurred to Emerson and his colleagues early on. Late in 1869, he made recommendations for the education programs of the Bussey Institution to President Eliot, and the two planned to tour the Bussey fields and woods together.

Asa Gray next suggested that the arboretum be located on "Brighton Meadows," a flat parcel along the Charles River on the Boston side that Henry Wadsworth Longfellow was planning to purchase and present to the college. At the poet's urging, Emerson pondered this possibility, but the beauty of Bussey's extensive undulating tract, already partially wooded, must have looked far more promising than the Charles floodplain. Dixwell undoubtedly knew the Bussey parcel well since his own land overlooked it. After careful deliberation, the parties concerned agreed that the West Roxbury acreage was more than the agricultural school could utilize for its purposes and that siting an arboretum on the remainder would be compatible with Bussey's intentions.

Over the next year and a half negotiations between the Arnold trustees and Harvard resulted in a final pact to establish the Arboretum on part of the Bussey property in West Roxbury. In the indenture, signed 29 March 1872, Emerson, Dixwell, and Parker agreed to turn the Arnold fund over to the president and fellows of Harvard College, provided the college allowed some 120 acres of its Bussey estate and the income of the fund to be used for

the establishment and support of an Arboretum, to be called the Arnold Arboretum, which shall contain, as far as is practicable, all the trees, shrubs, and herbaceous plants, either indigenous or exotic, which can be raised in the open air at the said West Roxbury, all which shall be raised or collected as fast as is practicable, and each specimen thereof shall be distinctly labelled, and [for] the support of a professor, to be called the Arnold Professor, who shall have the care and man-

agement of the said Arboretum, subject to the same control by the said President and Fellows to which the professors in the Bussey Institution are now subject, and who shall teach the knowledge of trees in the University which is in the charge of the said President and Fellows, and shall give such other instruction therein as may be naturally, directly, and usefully connected therewith. And as the entire fund, increased by the accumulations above named, under the best management and with the greatest economy, is barely sufficient to accomplish the proposed object, it is expressly provided that it shall not be diminished by supplementing any other object, however meritorious or kindred in its nature.

With the site and an endowment secure, establishment of the Arnold Arboretum achieved many of Emerson's and his colleagues' objectives. Here would be a living collection to augment the "cabinet" of the Boston Society of Natural History. One of every kind of tree and shrub, each labeled and available for study, patterned after Loudon's models—it would be Emerson's report come alive, a living inventory of the region's arboreal resources. Not only would it be a place where those interested in creating parks could determine which trees were suitable for their designs, but it would be a park itself. It would be a place to begin investigation of trees as the components of forests and providers of material wealth. As a station for the study and cultivation of trees hardy in the New England climate, the Arboretum would have a role in the botanical exploration of the world's temperate floras, strengthening the university's stature and resources for botanical research.

Asa Gray was disappointed the Arnold Arboretum was not going to be in Cambridge, an adjunct to the Harvard Botanic Garden. The garden sorely needed additional support, and Gray wished to relinquish the duty of administering it to spend more time on his flora of North America project. As it turned out, he was to benefit from the support generated by the Arboretum's founding. The increased interest of estate horticulturists in botany at Harvard had already manifested itself with a gift from Horatio Hollis Hunnewell to build a laboratory and lecture room adjacent to the herbarium. Additional relief would soon come to Gray with the young man chosen to direct the Arboretum.

CHARLES S. SARGENT, ESTATE HORTICULTURIST

Charles Sprague Sargent was born in Boston in 1841 to Ignatius and Henrietta Gray Sargent. His father, an East India merchant, banker, and successful railroad financier, was some forty years old when Charles was born. Soon after, Ignatius bought land in Brookline for a summer home, the first of several properties near Jamaica Pond that he eventually combined into a family estate. Barely a mile away was the farm of Benjamin Bussey, which had recently been transferred to his heirs. When Charles was six years old the Sargent family moved to Brookline year-round. Ignatius tended to business daily in Boston, but he also began to take an interest in horticulture. In this he was influenced not only by his Brookline and Jamaica Plain neighbors but by two of his kin who were creating grand, landscaped estates, Henry Winthrop Sargent and Horatio Hollis Hunnewell.

Henry Winthrop Sargent (1810–1882), Ignatius's first cousin, was Boston born and graduated from Harvard in 1830. After a ten-year banking career in New York, he moved to a country house on a plateau on the east bank of the Hudson River, nearly opposite the city of Newburgh. Andrew Jackson Downing lived in Newburgh, and the two became friends. Sargent sought the young landscape gardener's advice on the layout of his grounds. "Mr. Downing's leisure moments were never more thoroughly enjoyed than when pacing with [Sargent] around the new grounds, suggesting an effect here, an opening vista through yonder lofty grove, or advising about the hothouses and graperies" ([J. J. Smith], 1856, p. 446).

Through Downing, H. W. Sargent became a devoted student and advocate of the naturalistic landscape style. He named his place Wodenethe, "woody promontory." In fact, it was so thickly wooded that Sargent said his landscape was made mainly by the ax. By judiciously removing some trees, and heavily pruning others to induce thick new growth, he opened views of the Hudson Highlands and Fishkill Mountains framed by lush vegetation.

With his interest in ornamental plants piqued by a

34. *View on the grounds of Wodenethe, home of Henry Winthrop Sargent, at Fishkill Landing, New York. With the guidance of his friend and neighbor Andrew J. Downing, Sargent (1810–1882) created a landscape garden overlooking the Hudson by judiciously thinning and pruning existing woods and planting a great diversity of trees and shrubs. H. W. Sargent became Downing's literary executor and was a great influence on the Arboretum's first director, Charles S. Sargent.*

three-year tour of Europe and the Middle East, Sargent began to plant a great variety of trees and shrubs, many of them exotics and unusual-growth forms, in front of his managed native groves. He placed plants with an artist's eye for color and texture, yet indulged his passion for collecting rarities. Sargent also carefully developed the perimeter:

The boundaries of the place were treated in a similar manner; the original trees reduced to half or two-thirds their height, and, when thick and bushy, faced with ornamental plantations, as an arboretum, with collections of trees in families, and also a portion as a pinetum—each genus being kept by itself—and through which is a walk making a circuit of the place. (H. W. Sargent, 1865, pp. 441–42)

Henry Winthrop Sargent was especially interested in conifers, planting every species and variety he could obtain, experimenting to see how well some of the recently introduced western American, European, and Asiatic plants would withstand his climate. Every few years he wrote up the results of his trials for the *Horticulturist*. Visits from growers and plantsmen were always welcomed, and Wodenethe became known as a fine example of landscape design and an excellent plant collection.

After the tragic death of Downing, Henry W. Sargent became his literary executor, bringing out several editions of the *Treatise on Landscape Gardening,* which he supplemented with notes on new country estates and recently introduced woody plants. He contrasted the creation of Wodenethe with that of the grounds of Horatio Hollis Hunnewell (1810–1902) of Natick, Massachusetts, where a nearly naked plain was transformed into a wooded landscape. Hunnewell and Henry W. Sargent were related by marriage, and the former's horticultural activities were also an inspiration to Ignatius Sargent and his son.

"1843, Sept. Became interested in country life and commenced making improvements" (Hunnewell, 1906, vol. 2, p. 5). So wrote Hunnewell of the start of what was to become one of the finest landscaped plantings in New England. Like Ignatius Sargent, he did not attend college but went right into business, likewise becoming very successful in banking and railroads. Hunnewell made his "improvements" on land that was part of the family estate of his wife, Isabella P. Wells, daughter of one of his business partners. The land overlooked the shore of Lake Waban, in an area that later became the town of Wellesley.

While he made plans for the place, Hunnewell started a much needed program to improve the poor, sandy soil with guano and with peat hauled from an adjoining swamp and began raising trees. By 1847, he figured that he had 1,922 trees that he had planted living on his grounds and another 4,000 seedlings in his nurseries. A few years later he had a grand house built and began creating an Italian garden on a series of terraces in the hillside that dropped from the rear of the house to the lakeshore. Inspired by the topiary at Elvaston Castle in Derbyshire, Hunnewell planted evergreens that he clipped into compact geometrical forms. Since the Irish yew, which was the usual topiary subject in Britain, was not reliably hardy in New England, Hunnewell experimented with shearing other evergreens, notably the native white pine (*Pinus strobus*). The terraces were sodded with fine lawn grasses, and the walks and stairways were lined with balustrades, urns, vases, and fountains. The Italian garden was architecturally and aesthetically linked to the house. Few Americans created extensive gardens such as this in the formal style. While Hunnewell extended and rearranged the area several times, and always maintained it impeccably, other parts of his grounds were more parklike and naturalistic.

Like H. W. Sargent, Hunnewell was a serious collector of conifers, resulting in the establishment of a pinetum in about 1860. He aimed to plant every conifer, native and foreign, that he found to be hardy. Eventually extending over eleven acres, the pinetum was arranged informally and contained more than a hundred different kinds at the height of its development. Another fancy he and Sargent shared was for rhododendrons. These broad-leaved evergreen shrubs were almost unknown in American gardens until both men imported many hybrids from England for trials. The rhododendrons Sargent and Hunnewell found most successful were hybrids involving the southeastern American species *Rhododendron catawbiense,* the first of a group later known as "Iron Clads" for their hardiness. Hunnewell massed his rhododendrons in several places throughout his grounds. He continually tried new varieties, kept records, and reported on their performance in horticultural journals, urging other gardeners to cultivate these elegant shrubs.

H. H. Hunnewell and H. W. Sargent visited each other frequently, and together they toured the estates of fellow horticulturists. Hunnewell was a member of the Massachusetts Horticultural Society, whose Committee on Ornamental Gardening visited the Wellesley grounds periodically and reported on its progress. He also provided support to those engaged in botanical exploration and to Asa Gray at Harvard.

The initial development of the gardens of Hunnewell and H. W. Sargent was nearly simultaneous with the maturing of young Charles Sargent. Visits to Wellesley and Wodenethe, where his relatives were always busy planting, creating scenery, and talking about plants, could not help but impress the young Sargent. Many of the techniques he subsequently used to create the Arnold Arboretum were first observed at these estates.

Several other planters and plant collectors lived in the vicinity of Ignatius Sargent's home. "The whole of this neighborhood of Brookline is a kind of landscape garden. . . . The open gates, with tempting vistas and glimpses under the pendant boughs give it quite an Arcadian air of rural freedom and enjoyment" (Downing, 1844, p. 42). The cottage of Thomas Lee was lauded by Downing for its rhododendrons and mountain laurels. Although it was later purchased by Ignatius Sargent, during Charles's youth Lee was cultivating native and rare shrubs next door.

Another influential neighbor, the historian Francis Parkman, purchased a modest three acres on the shore of Jamaica Pond about the same time the Sargents started living year-round in Brookline. Parkman suffered from several ailments he termed "the enemy," and to have a distraction when pain interfered with his ability to write he took up horticulture. Roses were his first love; he collected and evaluated them as seriously as Hunnewell and Henry Winthrop Sargent did conifers. He published a book on their cultivation, history, and classification. In addition to upwards of a thousand varieties of roses, Parkman planted his grounds with trees and shrubs and many

flowering herbaceous perennials. He became an avid plant breeder, hybridizing lilies, irises, poppies, and other flowers. Parkman joined the Massachusetts Horticultural Society and visited many places as a member of the Committee on Ornamental Gardening. As a man of letters, he strengthened the society's scholarly mission during a ten-year stint as chairman of its library committee.

Near Parkman and Sargent was the estate of the tree fancier John James Dixwell, a distant relation of the Sargents. He and his in-laws, the Bowditch family, owned large tracts on Moss Hill. Some areas of these were wooded, while others were highly cultivated. Beyond the hill, at the Bussey place, Thomas Motley practiced scientific farming and allowed natural woodlands to regenerate; visitors were welcome.

These scenic and cultivated lands shaped Charles S. Sargent's boyhood. His father's grapes waxing almost fabulous in size and weight, the crystal depths of Jamaica Pond, the heights of Moss Hill—wild woodland, pasture, rose garden, grapery, planted grove—all were elements of the boy's landscape. The seriousness with which the adult cultivators pursued their plant collecting and horticultural experimentation deeply influenced Sargent in his future management of the Arboretum. The priority he gave to assembling a library and to publishing was fixed early on by his contact with men of Parkman's inclinations.

Charles Sargent was educated at a private school for boys. His schoolmaster was John James Dixwell's brother Epes Sargent Dixwell, who lived and taught in Cambridge. A brilliant man with interests in music, poetry, and science, as well as in Latin and Greek, Dixwell's curriculum was strictly classical. As a member of the Boston Society of Natural History and a frequent host of the Cambridge Scientific Club, Dixwell would have encouraged any leaning toward natural science in his pupils, however. Along with several of his schoolmates, Sargent entered Harvard College in 1858, graduating after four years. Then, like so many other "sons of Massachusetts," he enlisted in the Union army.

Assigned as aide-de-camp at the headquarters of Nathaniel P. Banks at New Orleans, Sargent saw action in several campaigns in the Gulf states. He was awarded the rank of brevet major for his service in an assault on

35. Charles Sprague Sargent (1841–1927), Paris, 1868. After graduating from Harvard in 1862 and serving the North in the Civil War, he made the grand tour of Britain and Europe, observing gardens and estates at the suggestions of his cousins, H. W. Sargent and H. H. Hunnewell.

Fort Morgan at Mobile, Alabama, which yielded to Northern forces in August 1864. In the summer after the close of the terrible conflict, Sargent was honorably discharged. He rejoined his family briefly, then sailed for a three years' stay in Europe. The trip was a means to recover from war's tragedies; besides, the grand tour was intended to be an important educational experience. On the advice of H. H. Hunnewell and Henry W. Sargent, Charles visited many gardens and country seats.

Having seen the best of British and European gardens, Sargent came home to Brookline in 1868 and took on the management of his father's estate. He was twenty-seven years old and could probably have embarked on a success-

ful business career. Instead, Charles S. Sargent was to be the creative one in his family, the one who accomplished things in horticulture, science, and education. Although he kept a hand in the railroad business, gardens and plant science became his life.

Within a few years of returning from Europe, Sargent was elected a trustee of the Massachusetts Society for Promoting Agriculture. While this group continued its many programs for the improvement of farming, it also supported botanical and horticultural projects. Sargent joined the Massachusetts Horticultural Society and participated on its ornamental gardening and library committees at the same time as Francis Parkman was serving. Undoubtedly, the historian was instrumental in the appointment of Sargent as the one to oversee the Arboretum.

When the Bussey Institution was inaugurated in 1870, President Eliot of Harvard, knowing of Parkman's great experience in the gardening arts, asked him to be professor of horticulture for the new school. Parkman accepted and during his first year oversaw construction of a headhouse, greenhouses, potting sheds, and cold frames. He hired the plant propagator Jackson T. Dawson from Charles Hovey's nursery firm to be the head gardener for the Bussey. Parkman planned to offer courses in propagation and nursery management for the next academic year, but illness, a death in his family, and his historical writing weighed against continuing the professorship; in the spring of 1872, he resigned. Parkman could well have suggested his young neighbor and friend, Charles Sargent, as a replacement. The indenture agreeing to accept the Arnold fund and to establish an arboretum on part of the Bussey farm had just been signed. There would be need in Jamaica Plain not only for a teacher of gardening but for someone to organize and manage the new tree collection. Sargent was appointed professor of horticulture of the Bussey Institution in May 1872 and curator of the Arnold Arboretum in June.

When it came to plant science at Harvard, there were also the concerns of Asa Gray to be considered. The university administration, the Arnold trustees, Parkman, and Hunnewell all knew Gray needed relief from administering the botanic garden and teaching so that he could work on his long-neglected flora of North America. Eliot was unwilling to accept Gray's resignation, however, until someone could be found to take these duties. Sargent, meanwhile, knowing he must become better acquainted with plant classification, spent most of the fall of 1872 in Cambridge studying under Gray. The elder professor at first thought Sargent had "a vast deal to learn," but soon realized his perseverance would stand him well (Sutton, 1970, p. 45).

The Brookline horticulturist caught on to the Cambridge situation quickly and began taking on tasks for the botanic garden, thus alleviating Gray's burdens. Within months further support came through H. H. Hunnewell and Ignatius Sargent. They each agreed to contribute one-half of a yearly stipend for Gray so that he could give up routine duties and work full time on his projected flora. Gray's classes were taken over by George Lincoln Goodale, a graduate of the medical school who had been studying under Gray since 1870. Charles Sargent's administration of the garden became official with his appointments as director of the Harvard Botanic Garden and of the Arnold Arboretum in November 1873. For the next six years he spent as much time renovating the garden as initiating the Arboretum. It would be a period of on-the-job training, coinciding with the time necessary for the Arnold fund to mature.

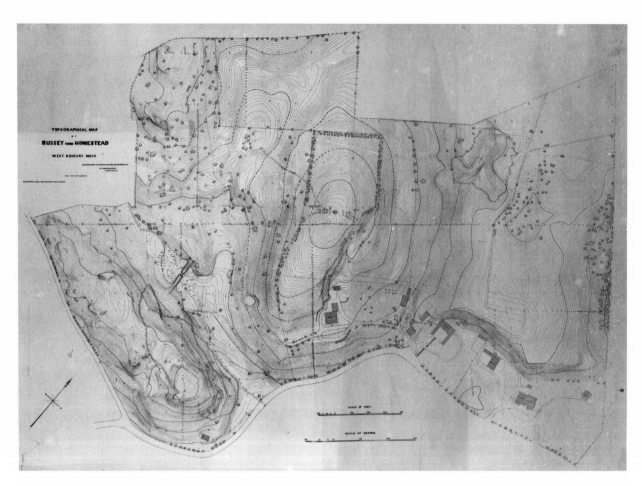

36. *Topographic map of the land on the Bussey homestead to be devoted to the Arnold Arboretum, surveyed in 1879 for Frederick Law Olmsted. Bussey's mansion and the buildings of the Bussey Institution can be seen in the lower right. Also evident are the lilac-bordered walkway to the summit of Bussey Hill, where a hexagon represents the observatory, and the mixed tree planting that separated the hilltop pleasure ground from the working farm, which forms a rectangle on the north and follows the contour on the southwest. Before construction of the Arboretum began, the City of Boston added land to the east, north, and west boundaries of the property pictured here.*

CHAPTER THREE

PLANNING AND CONSTRUCTION,
THE FIRST TWENTY-FIVE YEARS

Although 29 March 1872 marks the founding of the Arnold Arboretum, it took twenty-five years of planning, politicking, collecting, and constructing before the Arboretum was really under way. Nearly half of Charles Sprague Sargent's career as director would elapse before the living collections were in place, roads and paths finished, and a building for administration and research completed. Critical to its successful design was the move to ally the Arboretum with Boston's incipient park system, whereby Frederick Law Olmsted's expertise was gained.

The indenture signed by President Eliot and the Arnold trustees was moderated for an initial term by two conditions: at least two-thirds of the income of the Arnold fund had to be set aside each year until the fund reached $150,000, and the Bussey land had to pass completely into the hands of Harvard. On this account there was a considerable period of planning and policy formulation before construction and planting began. Eliot was pleased with the prospects, however:

An arboretum is intended to educate the public as well as the special students who resort to it. It will, therefore, be laid out as an open park, with suitable walks and roadways, and can hardly fail to become a beautiful, wholesome, and instructive resort, which will be more and more precious as population accumulates about it. From still another point of view, the professorship of arboriculture and the arboretum are substantial additions to the University. The cultivation and preservation of forests will become, in no long time, a matter of national concern. (Eliot, 1872, p. 33)

There never was a groundbreaking ceremony or an official opening day. The Arboretum grew gradually and systematically as a garden should, for its master schemer, Sargent, knew that trees complete their life cycles in increments of fifty or a hundred years. Any time spent on planning would be well worth it in the long run. There were few other arboreta to serve as models for Harvard's Arboretum. Sargent and his collaborators started from scratch. The director took seriously his mandate to gather a comprehensive collection of hardy plants and to set them out in a scientific manner. He also believed it could be accomplished artfully; the Arboretum, if well planned, could be as beautiful as it was educational.

SARGENT'S ON-THE-JOB TRAINING

After his appointments as director for the Harvard Botanic Garden and the Arnold Arboretum, November 1873 marked another important event in the life of Charles S. Sargent. On the twenty-sixth, he married Mary Allen Robeson, daughter of Andrew Robeson, whose family

37. *Mary Allen Robeson (1853–1919) married Charles S. Sargent within days of his change of status from curator of the Arboretum to director of both the Arboretum and the Harvard Botanic Garden in November 1873. At their Brookline home they raised five children and managed beautifully landscaped grounds. During their lifetime together, the elegant hospitality of Holm Lea was extended to visiting botanists and horticulturists, Arboretum supporters, students of landscape design, and many others. Mary Sargent executed a much admired series of watercolors to illustrate the flowers, fruit, and foliage of trees represented in the Jesup Collection of North American Woods at the American Museum of Natural History in New York.*

divided its time between Boston and Tiverton, Rhode Island. A few days after their marriage, the Sargents left for Europe on a combined wedding and business trip. The combination was characteristic of most of their married lives; besides raising five children and managing their home and grounds in Brookline (Holm Lea), the Arboretum was ever their common aim.

On this trip, Sargent observed the European scene with greater purpose. He was welcomed as a garden administrator and an associate of Asa Gray by botanists and horticulturists in Britain and France. Sargent was especially pleased to meet Joseph D. Hooker, who had by this time succeeded his father as director of Kew Gardens. He spent many days with him over the next five months, touring the grounds at Kew and gaining insight into the problems of systematic and evolutionary botany from this knowledgeable botanist. Despite his administrative duties at the royal gardens, Hooker maintained a program of research and publication.

At the time of C. S. Sargent's visit, one of Hooker's projects was a collaboration with George Bentham, a highly esteemed, self-taught botanist who had been working at Kew since 1854. The two were preparing a compendium of the world's genera of seed plants arranged according to a system refined from that of Augustin Pyramus de Candolle. What made their *Genera Plantarum* a landmark in botany was its originality and the quality of its scientific observations. Each genus was precisely described from the two botanists' firsthand observations of living plants in Kew's gardens and of dried specimens in its vast herbarium, freeing their work from errors repeated in earlier literature. Once published, the Bentham and Hooker system of classification was followed by most botanists in Europe and America for the remainder of the nineteenth century.

As to public gardens, Sargent wrote Gray that he admired the Royal Botanic Garden at Edinburgh more than any other, including Kew and the Jardin des Plantes, Paris. The modest extent of well-maintained grounds and arrangement of gardens at Edinburgh was such that Sargent could imagine creating something similar for Cambridge and West Roxbury.

In Geneva, Sargent met Alphonse de Candolle, and in Paris the French botanist Joseph Decaisne. The Sargents toured French nurseries, making contact with potential sources for the Arboretum and its horticulturally inclined supporters. In England once again, Sargent spent the month of March visiting nurseries and private estates, often in the company of English botanists, horticulturists, and authors. He met Harry James Veitch of the family nursery firm with which the Arboretum would later collaborate on plant exploration.

In letters to Horatio Hollis Hunnewell, Sargent described the multitude of plants he saw and contrasted and compared various gardens and nurseries. He raved about a large and thriving example of Japanese cedar (*Cryptomeria japonica*) at one private estate, calling it "a tree to bow down to." In the grounds of Fulham Palace, residence of the bishop of London, Sargent was likewise impressed by the many venerable specimen trees planted more than a century before by Bishop Compton. Indicative of his own preference in landscape treatment was his enthusiasm for another "charming old garden" in Fulham—"I've rarely seen a more delightful old place, simple and dignified and quiet" (AAA, C. S. Sargent correspondence, 16 March 1874). The education gained in discussion with botanists, the inspiration of venerable trees and gardens in many styles, and the contacts made for exchange of plant material were lasting benefits of the journey.

As Sargent began his oversight of the Arboretum he concurrently directed a major renovation of the Harvard Botanic Garden. It was a valuable period of on-the-job training, and the time spent with Gray was to have a great influence. When he started, the botanic garden's seven acres were too heavily shaded by trees and overgrown with masses of shrubbery. The plantings had been erratically added over the years and labels inconsistently applied. Sargent conceived a plan to remodel the taxonomic beds following Bentham and Hooker's scheme. Old and dying trees, and duplicate trees and shrubs, were removed to make room for expanded systematic beds; these were interspersed with grass walks to improve access for students and visitors. Undesirable wet sections were drained, while an artificial bog for moisture-loving plants was im-

proved and replanted. An old rock garden was completely rebuilt and enlarged by three times.

Sargent added a belt of evergreens to screen and shelter the northwest boundary. He made sure that every plant was labeled with scientific and common names and place of origin. The number of species rose from some two thousand to nearly six thousand, and Sargent managed to keep track of all of them during the changes by setting up a new plant record system.

In the course of the rehabilitation, Sargent vastly increased his knowledge of plants. By reworking the botanic garden he became familiar with the characteristics that define plant families and with the current theories of taxonomy. He profited greatly from daily discussions with Asa Gray and learned much from Gray's library and herbarium, which were adjacent to the garden. The professor's understanding of the species question and of evolution, his interest in the relatedness of the floras of eastern North America and eastern Asia, and his esteem for research using herbarium specimens and the botanical literature were all to affect Sargent's policies. In some respects the renovation of the botanic garden was a rehearsal for his more extended work for the Arboretum. Another positive aspect of the time spent on the garden was that it furthered Sargent's reputation among botanical colleagues, the college administration, and interested donors as someone who could get things done.

A turning point in the Arboretum's administration came in 1879. The Arnold fund was complete; it had accumulated to the required amount, so that a larger portion of its income would be available for operating expenses from that time on. Sargent's duties as director of the botanic garden ended, and he was appointed Arnold professor, with his full time to be devoted to the Arboretum. Oversight of the garden was assumed by Professor Goodale, who had relieved Gray from teaching botany in 1874. Although he was reluctant to give up the garden, Sargent told Hooker he was glad "to get really down at the Arboretum, which good or bad must be my life work, and where there will be always more to do than I can ever hope to accomplish" (AAA, C. S. Sargent correspondence, 29 July 1879).

38. *When Charles S. Sargent began his survey of the grounds slated to become the Arnold Arboretum, he found a thicket of American beech (Fagus grandifolia) on the bluff overlooking the then swampy valley of Sawmill Brook. These trees probably grew from suckers at the base of an original parent, a growth habit that distinguishes the American beech from the European (F. sylvatica). After Sargent thinned out extraneous species and pruned the beeches, the stand provided striking contrast to the conifer collection planted on the slopes behind it.*

THE WORN-OUT FARM

When not working in Cambridge, Sargent carefully examined the land meant for the Arboretum. "Undulating and parklike ground, already very finely wooded" was how he described it in a letter to the *Gardener's Chronicle* (Anonymous, 1872, p. 1522). Years later, he revealed his unspoken doubts and dramatized the site's transformation by saying that the Bussey place was "a worn-out farm, partly covered with natural plantations of native trees nearly ruined by excessive pasturage" (C. S. Sargent, 1922, p. 130). This statement was often repeated by Sargent, Arboretum publicists, and the press. The truth was probably somewhere between these two extremes. Neither Benjamin Bussey nor Thomas Motley, who prided themselves on their scientific approach to farming, would have agreed with Sargent's latter statement. Bussey's will stipulated that no living tree be cut or removed and required that the estate be kept in good condition.

Mid-nineteenth-century accounts of the popular use of "Bussey's Woods" mention well-kept pathways, mown meadows, and lush groves. Nevertheless, to the estate horticulturist Sargent it was a farm: its fields were crossed by stone walls and hedgerows that marked old property lines; the moist low areas were inaccessible; abandoned orchards and patches of second- or third-growth timber had not been adequately pruned or thinned.

There was potential, however, in the varied contour and diverse habitats, and charm in the land's rural aspect. At the northern boundary of the property were eskers covered with stands of trees that were about seventy-five years old. Bussey had allowed this area, now known as the North Woods, to seed into trees. Southeast of these ridges was a low, flat area about four acres in extent, now the site of three ponds and the rose-family collection. In the early 1870s there were no ponds; it was open meadow, drier to the east but very boggy where the ponds are today. Immediately southward, the land rose abruptly to a level area, "the plainfield," on which were arrayed the buildings and experimental grounds of the Bussey Institution, fronting on South Street.

At the western end of the institution holdings was the mansion, occupied by Bussey's heirs. Behind this spacious dwelling was the broad mound of Weld Hill or Woodland Hill, now called Bussey Hill, its summit 180 feet above sea level. Bussey's hedgerow of lilacs and white pine and other trees still formed straight lines across the east and north slopes and followed a more natural course surrounding a plateau on the west and south sides of the hill. At the summit the views seaward over Boston and south to the Blue Hills were grand. The hill occupied most of the central portion of the Arboretum property, taking up about one-third of the total area.

On the southwest side of Bussey Hill, culminating in the crest of Hemlock Hill, there were frequent outcrops of puddingstone, also known as Roxbury conglomerate, a bedrock formation peculiar to parts of the Boston basin. This area of rocky terrain contained more trees than the gentler slopes on the north and east. Just below the plateau that extends southeast from Bussey Hill's summit were some ancient white oaks, remnants of the original forest spared by the generations of Welds who first cleared the land. In the valley, a spring issued from below massive puddingstone ledges; there were venerable oaks here as well. Above the spring, extending to the western boundary along Centre Street, was a large upland broken by more exposed rock; it was covered in deciduous woods that dated from the 1780s.

Between these rocky "Central Woods" and the precipitous Hemlock Hill flowed Sawmill Brook (now Bussey Brook), so called for the mill and dam that had already seen many years of service when included in the inventory of Samuel Gore's estate in 1692. In 1872 the brook channel, lost in swamp and alder thicket below the dam site, was joined downstream by the Spring Brook just below a rock promontory. At that point, Sawmill Brook turned and rushed down to cross under South Street, meandered through a pasture used by the Bussey Institution, and joined Stony Brook, a tributary of the Charles River.

The great "hanging wood of Hemlocks" that clothed the southernmost hill was one of the original Arboretum's most celebrated assets. Indeed, it seems mainly to be this

39. An oak on the slope of Bussey Hill, pruned according to the recommendations of Amédée Des Cars, by which old, decrepit trees were rejuvenated by severely cutting back their crowns. After ten or twenty years many venerable Arboretum trees treated in this way were fully recovered; evidence of major cuts was scarcely perceptible.

thickly forested area that midcentury Bostonians thought of as "Bussey's Woods." Often referred to as primeval, its somber, deeply shaded slopes present a primitive, undisturbed appearance even today. This part of the Arboretum probably did come closest to the look of a mature New England woods of cool, northern slopes where hemlocks predominate. A later ecological study showed that hemlock timber had been cut repeatedly before Bussey's occupancy, although the area had probably never been completely cleared. Bussey's trails through the groves and his bridges spanning the brooks remained.

As Sargent took stock of the hills and vales, examining every glade and bole, he counted some 123 species of woody plants growing on that part of the Bussey farm to be devoted to the Arboretum. Getting around on horseback, he scrutinized the soils, topography, and indigenous growth in order to get an idea of what kinds of vegetation could be supported and where to place the future collections. Some 80 of the species already extant were native trees and shrubs, including most of the forest trees that one would expect to find in southeastern Massachusetts. On the uplands and drier sites there were pitch pine, red, white, and black oaks, four species of hickory, and gray birch. In lower areas, where the soils were rich and held more moisture, grew basswood, sugar maple, white ash, butternut, beech, and sweet birch. White pines and American elms towered over the landscape. Smaller understory trees found in the wooded areas included shadbush, flowering dogwood, mountain ash, and hop hornbeam. Hornbeam, red maple, speckled alder, and sycamore edged swamps and streamsides. Besides these natives that had persisted through generations of clearing and farming, or that had sprouted once again when pasture or orchard were given up, there were about forty kinds of planted trees and shrubs. Some of these were the orchard trees that had yielded their fruit in Bussey's day: domestic pears, apples, and mazzard cherries.

The mixed plantation that Bussey placed around the summit of Woodland Hill had many introduced species, such as European larch, horse chestnut, sycamore maple, European ash, wych elm, linden, and lilac. Along the boundaries and around homesites, Sargent found honey locust, tulip tree, catalpa, Scotch pine, Corsican pine, Norway spruce, and several popular garden shrubs.

Although the new curator looked on the spontaneous woodlands as promising features, he immediately perceived that they could be improved. From Henry Winthrop Sargent's experience at Wodenethe, he knew that trees and groves could be greatly improved by the "use of the axe." Since many fine native species were crowded together contending for light and nourishment, Sargent started a program of thinning and pruning. Weak, deformed, and dying trees were cut down; others that interfered with healthy specimens or with rarer kinds were likewise removed. Decrepit trees were given a new lease on life by heavy pruning carried out according to the Des Cars method (after the Frenchman whose book describing the system Sargent subsequently translated).

With limbs cut back, some of the mature trees took on a stag-headed appearance, but to Sargent's eyes this was temporary. He knew the improvement of specimen trees and belts of timber would be a slow process. And, as he predicted, ten or twenty years later new wood had formed on old trees, prolonging their lives. Moreover, selective thinning had allowed the natural copses to retain their diversity and health. To those familiar with today's woodlot-management practices, the pruning and thinning may seem unremarkable. However, these operations were experimental at the time. With the notion of an endless supply of trees still prevalent, Americans had not put much thought into managing what they had. Early on, the Arboretum became a demonstration ground for arboriculture.

Even before going to Europe, Sargent began corresponding with other botanists and institutions to build up the Harvard plant collections. Once back in the United States, he began to accumulate trees and shrubs as quickly as possible. Hooker sent shipments of seeds and plants from Kew's Asiatic and European trees and shrubs. Sargent returned the favor by obtaining some American species of plants for Kew. With the exchange under way, the Jardin des Plantes and the Imperial Botanic Garden at Saint Petersburg became other early international suppliers of seed for the Arboretum. By the time of his first report to the president and fellows of Harvard in December 1874, Sargent had also enlisted some fourteen American botanists, nurserymen, and estate owners who provided seeds of ligneous (woody) plants. Foremost among them was

Dr. George Engelmann of Saint Louis, Asa Gray's man at the gateway to the West. Sargent and Engelmann started a lively correspondence that was to last until the latter's death in 1884. His knowledge of western botany and botanists and of conifers, oaks, and other tree groups, as well as his friendship and guidance, were of great benefit to Sargent and the Arboretum.

Soon, young plants were accumulating from the worldwide exchange. In the first year, the Arboretum raised some 267 kinds of woody plants from seed, grafts, or cuttings received from correspondents—the first accessions for the new garden. At the same time, Sargent was gathering equal numbers of plants for the Harvard Botanic Garden. Realizing these seedlings would soon outgrow their space in Cambridge, he moved propagation and nursery operations for the future Arboretum to the facility planned by Francis Parkman at the Bussey Institution. Jackson Dawson presided there, and Sargent put him in charge of raising plants and handling the exchange of propagating material. Dawson would contribute his propagation genius to the Arboretum for the next forty-odd years.

While rearranging the plantings of the botanic garden, Sargent also contemplated the possible arrangement of the Arboretum. That the trees should be planted in a classificatory sequence was an accepted premise, as indicated in the discussion of the background to the creation of the Arboretum. However, it was one thing to work with the garden at Harvard, where the style was generally formal, the scale small, and the teaching collection composed mainly of herbaceous plants, readily lifted and moved. Displaying trees was another matter: once set in a particular spot they could not easily be moved, and when young enough to be transplanted they would not show their ultimate shape and size. Nor did the undulating Jamaica Plain site lend itself to parallel beds or to row upon row of plants, with each family following its relatives in line. The inherent beauty of hill, wood, and water and the rural character of the place were too precious to ruin with formality.

Another challenge to laying out the Arboretum was that the parcel was nearly inaccessible to a visiting public. Its

40. *Jackson Thornton Dawson (1841–1916), Civil War veteran who allegedly survived one Rebel bullet because it was stopped by the Bible he carried in his pocket, worked for the Hovey Nursery Company in Cambridge before coming to the Bussey Institution as head gardener in 1871. Sargent's need for a propagator soon outweighed the demands of the Bussey school, and Dawson contributed his mastery in starting trees and shrubs for forty-three years at the Arboretum.*

main frontage on South Street was occupied by the school of agriculture, its great Gothic building dominating the view from Forest Hills Station. The portion of Arboretum land that extended to Centre Street, the only other contiguous main street, was rough, rocky upland. Once entered, much of the tract lacked drives or paths. Sargent watched over the place on horseback but did not expect the public would. That the public should be admitted was the assumption that led him and college authorities into a unique partnership with the city.

SARGENT ENLISTS OLMSTED

Fortunately, the creation of a public park system for Boston came just in time to add immeasurably to the Arboretum's design. Since the 1860s there had been much agitation, and halting progress, toward park legislation in the city. Wishing to emulate Central Park, Bostonians successfully petitioned their city council in 1869 to hold hearings on the park issue. At the hearings and in subsequent discussions, proposals were presented by lawyers, engineers, landscape gardeners, and other citizens; many included the notion of a series of parks linked by scenic drives.

In 1870, several interested parties asked Frederick Law Olmsted to comment on proposals or to write letters or articles on the subject of Boston parks. Olmsted declined these requests, however, apparently not wishing to take a stand while planning was in a preliminary state. The one contribution he made to the Boston discussions that year was to give a lecture for the American Social Science Association at the Lowell Institute entitled "Public Parks and the Enlargement of Towns." Although Olmsted made few specific recommendations for Boston, he gave a convincing argument for parks as civilizing components of American cities.

After two failed attempts to implement legislation, park proponents withdrew their initiatives until after Charlestown, Brighton, and West Roxbury voted for annexation by Boston in October 1873. When a new mayor took office in 1874, a special commission was organized to study the parks issue, and once again citizen proposals were invited. At this point it became clear that Sargent and the Arnold trustees were not the only Bostonians who admired the charm of the Bussey farm. Some early planners cast their eyes on the Bussey property, not necessarily as a park, but as scenic open land to be viewed from passing parkways. At the hearings in June, Uriel Crocker, a conveyance lawyer who had long agitated for parks, proposed a series of parks and parkways. One of these routes would cross the Bussey farm to link thickly settled Roxbury with his proposed park system.

Another proposal involving the Bussey land at the 1874

hearings, and one that can aptly be considered the direct antecedent of the Boston park system as designed by Olmsted, was that of Ernest W. Bowditch, a young engineer who had worked with Robert Morris Copeland. The Bowditch plan included a boulevard that extended from Commonwealth Avenue to the Muddy River and along the river to include Ward's Pond and Jamaica Pond. The boulevard would branch at Jamaica Pond, one extension going eastward past the Bussey farm to a large park on the West Roxbury–Dorchester border, the other going west and north to the Chestnut Hill Reservoir. After the hearings the special commission filed a report, and finally, in June 1875, a Boston Park Act establishing a three-man commission was passed by the city council and approved by the voters.

After Boston's park commission had been working for two years, the commissioners were inclined toward including the Arboretum site in their scheme. In their first comprehensive report, issued in April 1876, they described locations for a system of parks connected by parkways that would be accessible to all classes of citizens, as economical to obtain as possible, physically adaptable and beautiful, and designed to use lands and watercourses that would become unhealthy if neglected. One of the proposed connecting links, called the "Bussey Farm Parkway," was to begin at the southwest corner of Jamaica Pond, skirt past the eastern boundary of Bussey farm through a grove of Bussey's trees, and extend to "West Roxbury Park" (now Franklin Park).

The commissioners commented at length on the Arboretum. With praise for the instructive goal of the tree collection, they stated their opinion that an even closer cooperation between the city and the Arboretum than being neighbors might be advantageous to both. They also acknowledged the value of Olmsted's professional advice in choosing the locations presented in the 1876 report, although Olmsted himself felt his participation was handicapped by the preliminary nature of the consultations.

Undoubtedly, Charles S. Sargent's input was behind the park commission's enthusiasm for collaboration with the Arboretum. Just how long he had been following the parks debate is not certain, but by June 1874 he was certainly attentive. The proposals to include the Bussey farm, or at least abut it with parkland, piqued his imagination.

By then, Sargent had probably met Olmsted. His mentor, Asa Gray, had known Olmsted a long time; Gray's wife was a cousin of Olmsted's lifelong friend Charles Loring Brace. Gray, as a Harvard professor with national prestige, was one of several gentlemen whom Olmsted enlisted to write letters on his behalf when he applied for the superintendency of Central Park in 1857. Olmsted was known to the Harvard community as the recipient of an honorary master of arts degree in 1864 for his role on the United States Sanitary Commission. Sargent and Olmsted had an excellent opportunity to meet in June 1873, when the landscape architect visited the great rhododendron show on Boston Common staged by H. H. Hunnewell and Sargent.

While the Boston parks hearings were fresh on people's minds, Sargent wrote to Olmsted on 26 June 1874:

In the general agitation into which the popular mind has now fallen in regard to a public Park or Parks I think I can see some hope for our Arboretum. It has occurred to me that an arrangement could be made by which the ground . . . could be handed over to the City of Boston on the condition that the City should spend a certain sum of money in laying out the grounds and should agree to leave the planting in my hands in order that the scientific objects of the trust could be carried out. (AAA, C. S. Sargent correspondence)

Sargent invited Olmsted to come to Boston and have a look. At the time, Olmsted was still based in New York, at work on Central Park, parks in Brooklyn and Buffalo, the United States Capitol grounds, and many other projects. His reply to Sargent's scheme was ambivalent. He had never seen the ground and was not sure that they would agree on the purposes and policies of an arboretum. He wrote Sargent: "A park and an arboretum seem to be so far unlike in purpose that I do not feel sure that I could combine them satisfactorily. I certainly would not undertake to do so in this case without your cooperation and I think it would be better and more proper that the plan should be made by you with my aid rather than by me

with yours" (AAA, Olmsted file, 8 July 1874). Olmsted quoted his fees for plans, but suggested a trade might be made for preliminary consultation if Sargent visited and gave his opinion of the plan for improving the Capitol grounds.

The Arboretum collaboration was placed in abeyance, however, so long as the question of Boston's parks was still undecided. It is likely that Olmsted did not want to be involved with Boston until there was government backing for its parks. Sargent may have felt that his available moneys, still limited to one-third the income of the Arnold fund, could not be used to pay a landscape architect. The seed of the idea was to germinate in good time, however.

A few years later, in September 1877, Olmsted stopped in Boston, looked over the Arboretum site with Sargent, and offered to work on its plan if the Arboretum paid the office expenses. Perhaps the job now appealed to Olmsted because of its relatively uncomplicated purpose and the promising natural features of the site. It was manageable, in contrast to some of his current projects such as those in New York and Montreal where city politics often hindered progress. With the single-minded, discriminating Sargent behind it, there was a good chance it could be planned and executed quickly and with high standards. It was an opportunity for Olmsted to look the Boston situation over without committing himself there.

Sargent was thrilled, or at least as excited as his Yankee reserve allowed him to be. He wrote Olmsted:

I started yesterday a subscription paper among some of the neighbors to raise the two thousand dollars for the plan of the Arboretum. My appeal has already met with such good success that the whole sum is as good as secured. I place the matter in your hands to make the necessary arrangements for pushing on the park as fast as possible. You did not say that your work in the plan making was thrown in, but I so understand in judging from the figures you named. . . . Certainly nothing could have happened to the Arboretum that could have assisted it so much as the fact that you are to make the plan for laying it out; and now I feel that its future success is assured. (AAA, C. S. Sargent correspondence, 8 October 1877)

In Sargent's appeal for funds "among some of the neighbors," he explained what a coup it would be to have Olmsted design the Arboretum:

Mr. Frederick Law Olmsted, the designer of the Central Park, New York and of the Prospect Park, Brooklyn and the foremost of American Landscape Gardners [sic] has agreed to survey the grounds of the Arnold Arboretum, a tract of some 130 acres forming a part of the Bussey Estate in West Roxbury, and to furnish a plan for laying out and planting it without charge beyond a sum sufficient to pay the surveyors and draughtsmen employed. If the Olmsted offer can be accepted the best possible disposition of the grounds will be secured, and the permanent planting of the immense collection of trees now crowding the nurseries of the Arboretum can be proceeded with. . . . The income of Mr. Arnold's gift for establishing the Arboretum is at the present time inadequate to meet this extraordinary expense, but it is hoped that private generosity will enable me to accept Mr. Olmsted's unhoped for and generous offer. (AAA, C. S. Sargent papers, October 1877)

Twelve gentlemen, including Charles Sargent himself, pledged to make up the fund. Among them were his relatives Ignatius Sargent, Henry Winthrop Sargent, Horatio Hollis Hunnewell, and William R. Robeson. George B. Emerson showed his continuing satisfaction with the Arboretum by a contribution, and the remainder was put up by prominent businessmen from the Boston area. This was a mode of fund-raising that Sargent relied on several more times in his efforts to establish the Arboretum—when money was needed for a worthy project, he made a donation, then convinced others to join him.

Sargent wrote to Hooker of Olmsted's collaboration and the support of the donors: "This I consider the greatest piece of good fortune that could possibly happen to us as there is no one in whom I have so much confidence as in Olmsted in such matters, and the fact that his name is connected with the Arboretum will give it no small éclat with the American public" (AAA, C. S. Sargent correspondence, 12 December 1877).

That winter, as attempts to oust him from oversight of Central Park became apparent, Olmsted's health made a trip to Europe necessary. He returned in April with plenty of notes on parks and gardens seen in England, the Low Countries, Germany, Italy, and France, but not much improved in health or spirits. Prospects for continued relations with the park department in New York looked unpromising. Thinking it best to get away, Olmsted and his family moved to Cambridge for the summer. With Asa Gray nearby for consultation, Sargent and Olmsted began their work on the Arboretum.

CAMPAIGNING FOR THE AGREEMENT

Sargent and Olmsted also had to win the approval of university and city authorities to combine the Arboretum with the park system. Years later Sargent said that "it took five years of exceedingly disagreeable semipolitical work to bring it about." His earliest political maneuver was to credit Olmsted with the notion of joining the Arboretum to the city. Sargent told him in 1877, "I place the matter in your hands to make the necessary arrangements for pushing on the park as fast as possible" (AAA, C. S. Sargent correspondence, 8 October 1877). As an outside expert with considerable experience in urban planning, Olmsted would be a more acceptable proponent of the scheme than his less neutral colleague. In 1878, Olmsted's connection with the park commission was formalized, thus giving him more influence with Boston officials.

Harvard's President Eliot was not particularly enthusiastic at first and expressed reservations to Sargent about partnership with the city in a letter of 1878. He thought that a fence and strategically placed workmen's houses would be sufficient to protect the Arboretum. He assumed the city would provide police when needed, as it had at the Boston Society of Natural History. Nor could he see the need for driveways. Sargent's idea of protecting the grounds was more farsighted:

If the arrangement can be made, the question of taxing the college property in West Roxbury will be forever settled; the danger of the property being entered and crossed by one or more high-ways, and by a rail-road, both of which are probabilities of the near future, will be removed, and the greater

41. Frederick Law Olmsted started his career as a landscape architect by collaborating with Calvert Vaux to design New York's Central Park. His participation in the campaign to include the Arboretum in the Boston park system and his contribution to its design were factors that led him to move his office from New York to Brookline.

evil of the whole Arboretum being taken by the city for a park after the college has done a good deal of work in it will be averted. . . . There are important topographical reasons which make Mr. Olmsted's last plan (the one now before the Corporation) very desirable; and without carriage drives my idea of what the Arboretum ought to be, namely a great museum for public instruction, can never be more than partially realized. (AAA, C. S. Sargent to Eliot, 28 November 1878)

Sargent's view extended to protecting the property from taxation, encroachment, and eminent-domain proceedings. He conceived its mission in the broadest possible sense: as a university museum for research and as an educational resource for the public. Crucial to its role as public educator was the Olmsted plan, realization of which could only be financed if the city was involved. Sargent knew that driveways and pleasing arrangement were needed to draw citizens into the collections to learn:

Left to itself, the Arboretum can never hope to open its collections to the public, except in a limited and unsatisfactory manner. Its income will never be large enough to fully carry out the scientific provisions of Mr. Arnold's bequest; and it will be impossible either to build or maintain carriage-drives for the public convenience. Whatever action is taken by the City of Boston, the public will not be excluded from the Arboretum; but the difference between driving through a broken piece of ground, a hundred acres in extent over a well-graded road, and entering it on foot by the few service paths necessary to the maintenance of the collections, will be so great that it is probable few persons, with the exception of specialists, will ever avail themselves of this privilege; and the usefulness of the Arboretum as a local educator will of necessity be greatly curtailed. (C. S. Sargent, 1880, p. 21)

Apparently his arguments were convincing. The following year Eliot assented to negotiate with the park commissioners. Once they and the university agreed to collaborate, the next step was to obtain permission from the state legislature. This proceeded easily; on 29 March 1880 the Massachusetts General Court, the legislative body, approved an act permitting the parties to enter into an agreement. However, when it came to persuading the city council to appropriate funds, progress slowed considerably.

Olmsted and Sargent started a modest publicity campaign. Several articles appeared in the Boston papers reporting on the Arboretum, detailing its background and progress and extolling the advantages of its inclusion in the park system. In May 1880, Olmsted wrote to his friend Charles Eliot Norton, professor of the history of art at Harvard, outlining the scheme much as he had for the papers:

The college has 118 acres appropriated to the Arboretum. . . . The city is now authorized upon my proposition to condemn 37 acres in parcels one at each end of the college tract. On the 155 acres much the best arboretum in the world can be formed. The scheme is that the city shall lease the condemned land to the college on a nominal rent for a thousand years and that the college shall establish and maintain the arboretum. That the city in good time will lay out a road and walk ($2^{1}/_{4}$ miles) through it and that the public shall be admitted to it under no regulations or restrictions other than such as are usual in well kept public grounds. . . .

This is the whole of the scheme as I would have it. I am sure that it is a capital bargain for both parties. If it fails it will be because of some reserve, caution or lack of push at the college and of indifference and jealousy, growing from ignorance at the city end. (AAA, Olmsted file, 7 May 1880)

Although the park commissioners did their best to urge the city council to authorize the Arboretum agreement, no action was taken in 1880. Meanwhile, Sargent let all parties know the delay was detrimental to the Arboretum's young trees, which were rapidly outgrowing their nursery space. However, the city council was deaf to threats of the loss of rare trees that year.

Finally, on 13 October 1881, the councillors put the Arboretum question to a vote, but less than the necessary two-thirds majority were in favor of the order. As the debate in the common council continued on into November and December the press rallied to the cause with numerous articles and editorials, all insisting it was a first-rate opportunity for the city to gain a park "at one quarter the amount it would otherwise have to pay for it." That the Arboretum would be ready for public enjoyment and instruction sooner than the other parks under consideration and that Harvard College would provide the larger share of its maintenance were further points in favor of the order. Councilmen sympathetic to the idea held out against a few vociferous opponents by voting for establishment of a special committee to further study the Arnold Arboretum.

Without waiting for the special committee to report, Sargent gave matters a push as only he could. On 31 October his supporters circulated a petition in favor of adopting the Arboretum order and collected more than thirteen hundred signatures. Called by the newspapers the most influential petition ever received by city council, its signatories included "nearly all the principal business houses," more than seventy physicians, former mayors, former governors, and several clergymen. Among the most famous signers were Oliver Wendell Holmes and his son, Phillips Brooks, John L. Gardner, James Freeman Clarke, and Edward Everett Hale. The only Arnold fund trustee still alive, Francis E. Parker, signed. (Sadly, George B. Emerson had died in March 1881, too soon to see the

Arboretum secured within the city park system.) In addition, the Massachusetts Horticultural Society, the Boston Society of Natural History, the trustees of the Massachusetts Society for Promoting Agriculture, and the Society of Arts all sent separate resolutions in favor of the Arboretum scheme to the city council during November 1881.

Despite the weighty petition, the sympathetic press coverage, and a favorable report from the special committee, a handful of councilmen attempted to prevent passage of the Arboretum measure by prolonging the debate. Finally, on 22 December, the orders for bringing the Arboretum into the park system were finally passed, along with appropriations for several other parts of the system. The opposition to the Arboretum scheme may have sprung from fear that it would inhibit passage of these other park appropriations. Since Boston's city government still had two houses, the order now needed to be passed by the eight-man board of aldermen. Sargent was only too pleased when they authorized the order with little debate a few days later.

Another year passed before the final agreement was signed by the park commissioners of the City of Boston and the president and fellows of Harvard College, however. During 1882, the city negotiated land acquisitions from adjacent owners. The details of the lease, and a set of rules governing public use, were worked out for the legal document that was to bind the college and the city for the next thousand years. The papers were signed on 30 December 1882. First, the commissioners took some 122 acres from the president and fellows for park purposes and an additional 45 acres from seven abutters. Then the park commissioners and President Eliot executed an indenture to lease most of the land to Harvard for Arboretum purposes for a dollar a year for one thousand years under certain restrictions.

The city retained the land to be made into driveways and the parkway (now the Arborway), as well as two tracts that the parks department was to develop as picnic grounds: one of about five acres at the main entrance, the other of eleven acres on the summit of Bussey Hill. The city was obligated to spend $75,000 to build the driveways as delineated in Olmsted's plan and to keep these in good

42. One of several preliminary studies for an Arboretum drive made by Olmsted in 1878. Like other early attempts to fit the longest possible road into the 120 acres then available, this one called for several hairpin turns to go over the steepest slopes of Bussey Hill. Three circuitous branch roads traversed Bussey Brook valley, Hemlock Hill, and Central Woods before the drive made its return along the north- western boundary. Already apparent, in the upper central portion of the sketch, is the possibility of annexing land from the Adams Nervine Asylum. Also depicted is a possible transverse road to carry traffic from Centre Street to South Street without connecting to the internal Arboretum drive.

condition for the term of the lease. The city promised to deploy police to assure the "preservation of order and good conduct and observance of the rules." Water was always to be available for the Arboretum. The city further guaranteed that no taxes would be assessed against the college for the property and that no street, railroad, or tramway would encroach upon the grounds, nor would any structure be built without the consent of both parties. For its part, Harvard contributed most of the land, promised to fulfill the Bussey and Arnold trusts and to maintain the plant collections and all the grounds not under driveways, and agreed to open the site to the public at all reasonable times.

The "capital bargain for both parties" was struck. The college gained additional land for the Arboretum, the construction of "suitable and dignified approaches and driveways," a guaranteed water supply, policing, and freedom from taxation and future encroachment. The city gained 167 acres of parkland for the cost of 45 plus construction of the main driveway. Both parties had obligations to maintain the property for the thousand-year term of the lease; the lease is renewable on 30 December 2882 if both parties agree.

During the years of negotiation, Sargent annually expressed his regret at the delay in planting the grounds, but he also saw a bright side to the postponement: more thought and experience would go into planning the collections. "Those whose duty it will be to solve the problems of the final arrangement will bring to this work larger experiences and the advantage of wider observation" (C. S. Sargent, 1881, p. 2). In the years that he pushed for the indenture with the city, Sargent was busy with a project that gave him an excellent opportunity to expand his knowledge of trees and to observe the American forests widely: his activity as special agent on forests for the Tenth Census of the United States.

DEVISING THE PLAN

When the city-university agreement was finalized, Olmsted and Sargent had been working on designs for the Arboretum on and off for about four years. Another three years would pass before a configuration for driveways, combined with placement of plant collections, would finally be settled. That plenty of time be devoted to perfecting the plan suited Sargent, since he knew that the final arrangement would be permanent.

Olmsted had worked on plans for an arboretum before, namely, the unexecuted one for Central Park. So he at least had a feel for the problem: to create a circulation system that would cover as much ground as possible, be suitably aligned with the natural topography, and allow room for the plant groups to be distributed along its length. Sargent agreed their goal could be achieved by a one-way drive that would direct visitors by carriage through the collection in the proper sequence, just as John Claudius Loudon had so often recommended. The sequence, decided on with Asa Gray's advice, would be the ordering system of families currently being published by George Bentham and Joseph Hooker in *Genera Plantarum*. The Arboretum drive would have to be conveniently accessible and connect with the rest of the park system. By agreement with the park commission, certain areas were to remain free of scientific collections to accommodate picnicking (during planning, these were termed "city reservations"). A further requirement was that certain large trees and areas already covered with spontaneous woodlands not be disturbed.

One of the first steps was to become familiar with existing conditions. While Sargent knew the place well, the landscape architect needed a graphic depiction with which to work. Under Olmsted's direction the land was surveyed and a topographical map drawn up. Reproduced in quantity, this was the "blank slate" upon which Olmsted sketched several studies for the Arboretum over the summers of 1878 and 1879. In the initial studies the attempt was to make the drive through the collections as long as possible. Each of the early sketches shows an extremely convoluted road that makes a circuit, with few branch roads.

Olmsted also began to work out the relation of the Arboretum's internal road to its surroundings. In one study he tried an entrance/exit point on Centre Street. In most others he linked the Arboretum entry with the proposed parkway between Jamaica Pond and West Roxbury Park (later named the Arborway). Speculating on the possible need to accommodate cross traffic, one sketch included a transverse road isolated by underpasses from the Arboretum visitor carriageway. Olmsted often used some such device to separate disparate transportation modes.

As planning for the Arborway advanced, in the fall of 1878 Olmsted realized the annexation of a few acres adjoining the northeast corner would greatly facilitate connecting the Arboretum with the rest of the park system. Part of this land belonged to the Adams Nervine Asylum, a facility under development that proposed to care for persons with mental illness in a homelike atmosphere. Apparently the administrators of the asylum were willing to sell about twenty acres. An adjoining parcel needed to extend the Arboretum boundary and to locate the Arborway was part of the farm of a Mr. Goldsmith. Most of this land was low and flat, rising only along what became the Adams Nervine border (the area now occupied by the North Meadow and the linden collection). Two branches of a brook entered the area and joined in the low, peaty meadow; both were referred to as Goldsmith Brook. City authorities were sympathetic to adding to the Arboretum acreage, and in subsequent planning Olmsted made use of this space, which allowed a better fit for the return leg of the circuit.

In 1879, he worked on at least seven "distribution

studies," each one a different version of how to arrange the plantings along the circuitous carriage drive. The configuration of the road now began to evolve in response to planning for the living collections.

Placing the plant collections was not a simple matter. While the order of families was more or less fixed, each group had certain traits to be accommodated. Some families had a much larger number of genera than others. Likewise, some genera had a great number of species, others only one. Relying on Sargent and Gray's knowledge, Olmsted had to carefully apportion the area alloted to each group. Habit (i.e., whether the plant was a tree or a shrub) was another variable for which adjustments in the arrangement had to be made. Olmsted also had a choice between distributing each family on both sides of the drive or alternating them from one side to the other; he experimented with both schemes in the studies.

Another character of plants that had to be provided for was each group's preferences in habitat. For example, the best place to site moisture-loving groups like the willows would be in one of the wet meadows or near a water feature. In contrast, most species of oaks thrive on the well-drained soils of hillsides.

In the first sketches of ways to distribute the plant families, Olmsted depicted a road three and a quarter miles in length, of which less than a mile would be level. From the entry point it paralleled the proposed Arborway, then took a broad turn to head up and over Bussey Hill. Traversing the hill from northeast to southwest, the road made five hairpin turns, with a short branch that arced to the very summit. The road then followed the Bussey Brook valley, branching one way to ascend to the long ridge of Hemlock Hill by two hairpin turns, the other way to return to the entry point. The return segment struck up over the rocky Central Woods, where there was a third side branch, then followed the northern boundary along Centre Street and through the Adams Nervine expansion.

The second distribution study showed nearly all of the slopes, as well as the summit of Bussey Hill, unoccupied by the systematic collections. This was an early trial of a city reservation, an important element used to convince the Boston authorities to take up the Arboretum. Later studies allowed for a much less generous city reserve on the hilltop, however. The second distribution study also included the first trial of a pond that was to occupy the low wetland near the main entrance. It, too, was to be part of a city reserve. In these early sketches an exception was made to the Bentham and Hooker arrangement. In following the strict sequence the Cupuliferae, an important tree group including birches, walnuts, hickories, oaks, chestnuts, and beeches, would be followed by the Salicaceae (willows and poplars), and these by another large and important group, the Coniferae. However, in the early studies the Coniferae were placed along the branch road up Hemlock Hill, the Cupuliferae and Salicaceae following them along the return road. This placed the conifers close to and among the native stand of their cohorts, the hemlocks, and the willows at the end of the sequence on the west shore of the pond.

With the fourth attempt to arrange the collections came a simplification of the tortuous carriageway to something like a figure eight with gentler curves; at midpoint, a cross-connection linked the outward and return segments of the circuit. Instead of going over the central mass of Bussey Hill, the main drive skirted the southeast slope, with access to the summit by a branch road that followed the old pathway from Bussey's mansion. Beyond Bussey Hill, the road made a single loop to cover the valley floor before diverging, either to the top of Hemlock Hill in one direction or back toward the entrance along the western boundary, with no further side roads.

As 1879 drew to a close, Olmsted, Sargent, and the park commissioner Charles H. Dalton needed a plan to present to city and university authorities. A presentation-quality copy of the seventh study of distribution was started in mid-December, showing the figure-eight road, subsidiary pathways, and names of major tree groups in block letters. By this time a two-acre plot overlooking the pond near the entrance was assigned as the site for a headquarters, marked "Building Reserve" on the seventh study.

At the same time, Olmsted's office began work on a plan of the Arboretum showing its relation to the parkway and to connecting streets in the neighborhood. This was prepared for printing and showed the area of Jamaica Plain from Jamaica Pond to the site of the future Franklin Park.

Within the Arboretum boundaries, the road was detailed as in the seventh distribution study. A pencil sketch and then an ink version were shown to Dalton and Sargent in November and December. Sometime during the late-year exchange between Olmsted, Sargent, and Dalton, however, an idea occurred to them that thoroughly changed the plans for the driveway.

Dalton apparently thought the drive as then designed would be too expensive to construct because of its length and grades. He considered the plots reserved for park use to be inadequate and not easily approached. Soon after New Year's Day 1880, Olmsted finished a pencil sketch of a completely new road and wrote Dalton:

Upon discussion with Mr. Sargent today it was determined that we should propose to you that the city should buy the body of land adjoining the Bussey Farm on the northwest containing about 14 acres, to be added to the Arboretum.

With this addition we . . . could greatly improve the entire project. It would enable us to meet your wishes as to larger public grounds and additional entrances. . . . We would be able to take better lines and obtain better grades for our roads and build them at less expense. (AAA, Olmsted file, 6 January 1880)

The land in question was a roughly square parcel, contiguous to Arboretum land and bounded by Bussey, Walter, and Centre Streets. It belonged to two families, the Skinners and the Kents. It seemed a good argument to spend city money on additional land rather than on constructing a more costly road in a cramped space. And less of the more expensive Adams Nervine property would have to be purchased if the Skinner and Kent parcels could be utilized.

Olmsted's new road would be a one-way drive, entering at the Arborway and exiting onto Walter Street, in-

43. Detail from a map that showed the proposed Arboretum and its connections to Jamaica Pond and West Roxbury Park (Franklin Park). Although dated 1879, this much simplified version of an Arboretum drive was executed in early January 1880, after Boston's park commissioners agreed to add a parcel of land to the western corner. The addition allowed creation of a one-way road from the parkway (the Arborway) to Walter Street that made far fewer turns and followed a more level course than all previous attempts. This newly configured road would also be much less costly to build.

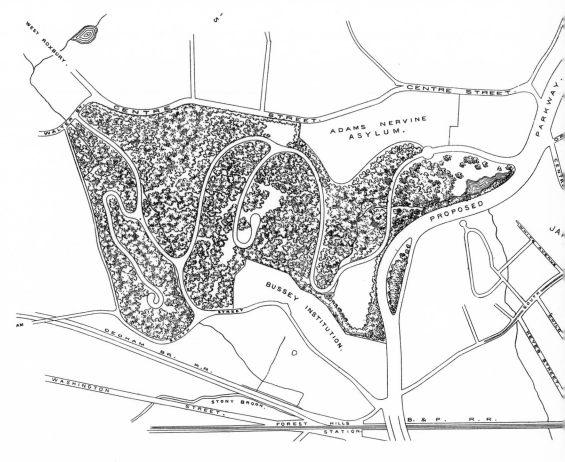

stead of a circuit. An additional entrance from South Street in the valley of Sawmill Brook was provided. The new road followed a more level course, with broad turns around the northern base of Bussey Hill. Access to the summit would be from a branch drive that ascended the more gently inclined northwest slope. The second branch road, up Hemlock Hill, remained as planned previously. The total length was about a mile and a half less than the circuit Olmsted had depicted at the start. This is very nearly the road as it exists today.

An eighth distribution study was begun in early January, enough perhaps to give Olmsted and Sargent a feel for the feasibility of placing the collections compatibly with the new road. Olmsted also completed the Arboretum and neighborhood map and had it lithographed by 31 January 1880. Since this was to accompany the park department's fifth annual report, it was dated 1879, however. Within the Arboretum boundaries it simply indicated the recently revised version of the main carriage drive and symbolic tree covering, giving no further detail. Its purpose was to show college and city authorities the general layout and its relation to nearby parks.

While waiting for some action on the part of the city council, Olmsted and Sargent continued to refine the plan. Four more versions of distribution of plant collections were tried in the spring of 1880; later in the year Olmsted restudied the Arborway entrance, precisely defining the boundaries and figuring the acreage. As a result another map, "Proposition as to a Public Ground to include the Harvard Arboretum," was published in November. The area to be contributed by the college was shaded with light lines, the area to be contributed by the city with dark ones. An important change in the line of what would become Meadow Road is evident in this version of the plan. Previously, this section of the drive had been shown to curve toward the proposed Arborway, skirting east of all the knolls in the North Woods. In the November 1880 proposition, the curve of Meadow Road was made less severe and drawn away from the Arborway boundary by routing it between the oak-covered hummocks. This plan appeared in the newspapers, accompanied by articles favoring the inclusion of the Arboretum in the park system.

In 1881 and 1882, during the intense campaigning and negotiating over this issue, design work seems to have been suspended. After the agreement was signed, however, it did not take long for the city to start construction. In July 1883 work began on what was to be named Valley Road, by then extended to connect two long-established city thoroughfares, Centre and South Streets. Olmsted and Sargent recommended work on the planting plan in 1884, and a study for the roads that shows them very nearly as finally executed was drawn in July. Three more studies for arrangement of plant families were made in October and December of that year. In these, no part of the area east of Meadow Road was designated for collections except a very narrow strip adjacent to the road. Instead, all of the low, flat land was to be occupied by a meandering watercourse. The rest of the land east of Meadow Road was still being used as a nursery. This resulted in all the groups from magnolias to legumes being squeezed along the west side of the drive. The rose-family trees would have been sited on the north slope of Bussey Hill, where forsythia and lilacs are today. Evidently, many details of placement were not yet worked out.

It was not until the following summer that the "final" planting plan was adopted. While Olmsted and his assistants kept trying ways to fit all of the genera of woody plants in sequence, Sargent continually refined his thinking on exactly what to include. The question had to be settled in order to plan the right amount of space for each group. Despite his years of experience, the list of genera and numbers of species to be included seems to have been difficult for Sargent to settle on. For one thing, he had not determined how shrubs were to be accommodated in the sequence of collections.

Once the maple and rose families were moved to sites on the east side of Meadow Road in a sketch made in early August 1885, there was more room for the sequence at this end of the driveway. This change brought the plans much closer to the finally executed version. Details of the changes to the whole sequence were worked out in another distribution study that month, and then a "planting plan," dated 29 August, was drawn up. The fiscal year for the Arboretum ended two days later. In his annual report, Sargent revealed the pressure to finalize the plans:

The progress made by the City of Boston in the construction of its roadways through the Arboretum had reached . . . a point when it was no longer possible to delay the adoption of a definite plan for the permanent arrangement of the collections. The future value of the Arboretum must largely depend upon the manner in which this plan is made; and the development of a scheme of planting scientific in method and practical in scope has long occupied my own attention as well as that of such counsellors as I have been able to call to my assistance. The thanks of the Arboretum are in this connection due in particular to Messrs. F. L. and J. C. Olmsted, who have devoted much time to the study and preparation of the plan which has now been determined on. (C. S. Sargent, 1886, p. 147)

MISSION AND COLLECTION POLICY UNDER DEVELOPMENT

While the plans for the physical layout for the grounds were under development, the Arboretum director was obliged to define policies for the institution's mission and for the content of the living collection. Sargent first publicly outlined his overall policies for the Arboretum in the park commissioners' annual report for 1879. He set out four main functions appropriate to the institution. First, it was to be a museum containing *every* hardy woody plant, along with an herbarium and other collections, to illustrate trees and their products. Its collections, furthermore, should be the subject of research on the classification, cultivation, and economic and ornamental properties of trees and shrubs. Third, it should serve the study of forestry and arboriculture. Finally, the Arboretum should educate the public through "object teaching." To meet these goals, Sargent envisioned that

each species, represented, if possible, by half-a-dozen specimens, will be planted in immediate connection with its varieties, making with its allies, native and foreign, loose generic groups in which each individual will find sufficient space for full development, and through which the visitor can freely pass. Each of these groups will rest on the main avenue so that a visitor driving through the Arboretum will be able to obtain a general idea of the arborescent vegetation of the north temperate zone without even leaving his carriage. It is hoped that such an arrangement . . . will facilitate the com-

prehensive study of the collections, both in their scientific and picturesque aspects. (C. S. Sargent, 1880, p. 21)

Once planning advanced to the stage where a final plant list was needed, it is evident that Sargent was struggling with the idea of fitting all hardy woody plants, especially the shrubs, on the available acreage. In a letter to Olmsted dated 5 February 1885, he asserted that the Arboretum's role as a museum to teach the public would be best fulfilled if it consisted of a selection of well-grown plants illustrating each group or type, rather than being an exhaustive collection. In making the selection, Sargent declared that "we must look ahead not less than a hundred years and try and form some opinion of what the public of those days will want to see." This, he concluded, would be native North American trees. The species would perform most reliably in the Boston climate; since the expansion of urban areas would make natives rare near cities, it would be the Arboretum's increasingly important duty to teach people about their own local resources. He proposed therefore that the dominant feature of the Arboretum's display collection be trees of North America, supplemented by "some of the most important foreign trees."

In addition to this educational, display collection, Sargent outlined two other components. One was a "trial ground," or area for species of doubtful hardiness and garden forms, intended to educate specialists rather than the general public; the other was an herbarium to aid scientific investigation. Sargent did not specifically mention shrubs in this letter, and he promised to give Olmsted a list of the tree species to be included in the main collection.

Sargent expanded on the policy statement outlined to Olmsted in his annual report for fiscal 1884–85. Since the Arboretum was to perform the two concurrent duties of public instruction and facilitation of research, it should have two collections:

1. The permanent collection for display, consisting of a selection of species intended to illustrate . . . the most important types of arborescent vegetation.

2. A collection for investigation which need not necessarily be permanent, and which should be arranged in a manner to permit the admission of . . . new forms and the removal of

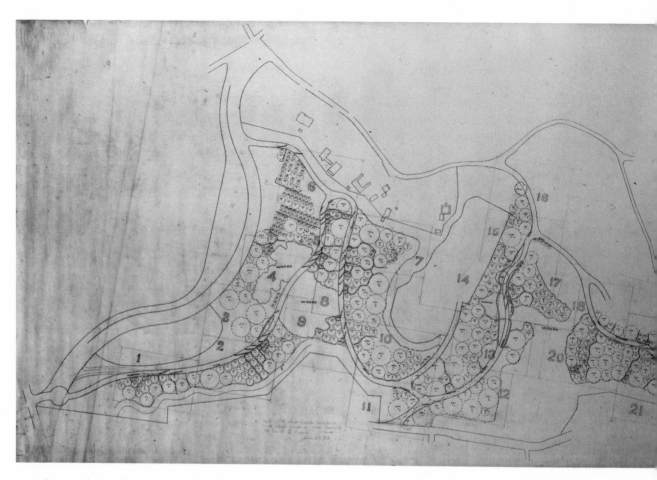

44. *Arboretum planting plan, 29 August 1885. Olmsted and his associates often reversed the orientation of plans ("south" is more or less at the top) when it suited their purpose. In this instance, reversing the compass directions allowed the Bentham and Hooker sequence to be read from left to right. The circles represent expected diameter of crown for each species. Numbers in the circles indicate the name of plant, keyed to a separate list. The lines between groups of circles separate genera or families. That nearly half the acreage was not available for tree collections is apparent from the sizable blank spaces on the plan: the North Meadow and a large part of Bussey Hill were to be "city reservations," and the North, Central, and Hemlock Hill Woods were to remain natural areas. The rectangular grids numbered 1 to 21 may have been added later to define the records maps of larger scale utilized at the Arboretum after planting was complete.*

others which have served their purpose. To this second collection would naturally be joined all minor collections like that of shrubs and other plants of less enduring character than trees. (C. S. Sargent, 1886, pp. 147–48)

On the exhibition collection, Sargent reiterated his argument for emphasis on eastern North American, and especially New England, native trees. For the first time, he detailed the aim of representing each species by an individual planted so that it attains full size as a specimen tree, and by a group of individuals planted closely to show the species' "expression in mass." The mass might include several varieties to show the range of characteristics within the species. These group plantings of selected American trees were to be miniature forestry plantations, intended to show the habit of the trees under nearly natural conditions. Such plantings would also assure the duration of each type in the collection even if something should happen to any one individual. Exotic species, with a few exceptions, would be represented by single specimens. Varieties that are the "most valuable and best fixed" (i.e., known to persist) of some exotics would be displayed in group plantings, as would a few foreign species of exceptional hardiness. In deciding to emphasize native trees, Sargent showed an inclination similar to the Boston Society of Natural History's resolve a decade earlier to display separate New England collections in each of its departments. He was also greatly influenced by his current research on the American forests.

Apparently, Sargent entertained the idea of narrowing the content of the living collections, or modifying the goal to grow *all* the hardy trees and shrubs, in order to create a better, more educative display. He also admitted that a complete collection of trees would require more than a thousand acres, especially considering that the number possible to grow in New England was likely to increase with exploration and cultivar development. Paradoxically, he would later use the argument that the Arboretum *must* be comprehensive to justify its extensive plant-introduction activities. Nonetheless, at the start of planting he preferred to grow a lesser number of plants well than to cram greater numbers into an inadequate space.

An exception was the mass plantings of important American trees intended to exemplify forest conditions.

At this time forestry was scarcely developed as a discipline of study. While it was thought of as one appropriate topic for the Arboretum at the start, this policy would change once government agencies and schools took up forestry at the turn of the century.

Sargent did not go into as much detail on his plans for the second division, the "working or experimental collection," including the shrubs. Neither his annual report nor the plans drawn up that year indicate where this secondary collection was to be located. His inclusion of the shrubs, and statement that the trial ground should allow addition and removal of specimens without disrupting the arrangement, suggests a treatment similar to the interim shrub beds recently created on the flat land below the Bussey Institution next to the nursery.

All the shrubs had been moved into an adjoining area and arranged in systematic order in parallel rows as a measure to relieve overcrowding in the nursery. This, Sargent declared, would be temporary, "until the completion of the roads by the city on the east side of the grounds make it possible to permanently group the different shrubs upon a more picturesque and natural plan than can now be adopted" (C. S. Sargent, 1885, p. 155). With the ground drained and prepared in late 1884, eleven hundred different kinds of shrubs were set out in the 1885 season.

According to recent drawings, however, this area was the ultimate site for the rose family. In 1885, the rows of shrubs were still thought of as a temporary solution. As the years passed the satisfactory disposition of shrubs and other special research collections was never entirely resolved. The shrub collection became a feature of the Arboretum for more than a hundred years, despite its expedient origin. The notion of research plantings took on a variety of forms and locations over the years.

PLANTING AND CONSTRUCTION; THE ARBORETUM TAKES FORM

Although some planting was done before the overall design was finished, the installation of the majority of the collections took place in stages between 1886 and 1895, once the plan was accepted. Planting usually followed completion of each road by the city, although there were

45. *Rock elms* (Ulmus thomasi) *on the northeast slope of Bussey Hill about twenty years after planting. Still made up of youthful trees in this photograph, the elm collection was one of the first eleven genera installed in the Bentham and Hooker sequence. For the important American forest trees, Sargent planted an individual, providing ample space to allow its fullest growth as a specimen. Near it he placed a closely spaced group of the same species to demonstrate its habit when grown in forestry plantations. The elms were planted in 1886, before Bussey Hill Road was constructed.*

times when it was necessary to establish some tree groups before the construction of the drive nearest them began. Because construction of the roads started in the center of the Arboretum property, the installation began in the middle of the Bentham and Hooker sequence. The realities of construction and planting also occasionally called for revision of the original plans.

The first planting on the grounds in Jamaica Plain was made in 1876, when 3,181 young trees were planted out from the overflowing nurseries on various portions of the Bussey estate. Since there was no plan for the collections then, these initial plantings were made along the boundaries, to serve as visual screens and shelterbelts; they consisted largely of native New England species. Remnants of these original border plantings can be seen today in several large specimens of native oak that line the bridle path and the fence to the southwest of the Dana Greenhouse complex—the old boundary line between the Arboretum and the Adams Nervine Asylum. The grove of red pines and white pines that overtop shrub masses on the southern slope of Bussey Hill overlooking the South Street gate was another of these original boundary plantations.

Construction of the Arboretum roads by the city engineer's force began in early July 1883, commencing with the "drive between Centre and South Streets," the section known today as Valley Road. It traversed land entirely within the original Bussey tract and would allow early access to the new parkland from two city streets; all the other proposed entries would be through newly acquired land or connected to the as-yet-unbuilt Arborway. As the line of Valley Road was greatly limited by the channels of Spring Brook and Bussey Brook, it had long been settled on by Sargent and Olmsted. Its connection to an entrance on Centre Street was sketched in on a plan made in May 1880, at a time when further needs of the city were being considered. The Centre Street entrance would be convenient to the eleven-acre city reservation on the summit of Bussey Hill. Sargent revealed his enthusiasm for the start of construction in a letter to Joseph Hooker: "The Arboretum is getting on at last . . . roads, belts, grade, etc. are making grand progress and I really begin to see daylight ahead" (AAA, C. S. Sargent correspondence, 3 October 1883).

Planting of the main collection of trees was begun at last in the spring of 1886. Sargent and Jackson Dawson oversaw the placement of seventy thousand trees and shrubs that year. Some were additions to the boundary plantations begun years before, but most were in several important groups for the "type collections": pines, firs, spruces, larches, ashes, elms, catalpas, birches, hickories, hop hornbeams, and beeches. The ground to be occupied by the oaks, walnuts, and chestnuts was also prepared for spring 1887 planting. Trees of these fourteen genera, Sargent felt, would require the longest time to reach maturity. When these plantings were made the only road completed was Valley Road, along which the walnuts, hickories, oaks, chestnuts, hop hornbeams, and beeches were sited. The other groups had to be positioned to accommodate future road construction.

By this time, Sargent's confidence in the plan was resolute, and he prepared the ground for collections that were to be maintained for a thousand years. He desired to give the trees the best possible growing conditions and feared the Arboretum soil was thin and exhausted. For the individuals that were to grow singly and develop as specimen trees, pits twenty-five feet square were dug; for the others, pits ten feet square. Rock, gravel, and sandy soil were removed from these pits to a depth of three feet and replaced with loam and peat. Ample space was allowed between planting holes for fully mature trees. At the end of the season Sargent remarked, "Trees have never been planted with better promise of undisturbed old age" (C. S. Sargent, 1887, p. 123). The newly planted specimens were going to "appear needlessly remote from each other" for some time, but Sargent felt this a much wiser strategy than planting densely to achieve instantaneous effect.

The director had settled on one notion about the role of shrubs in the Arboretum landscape. For many years he had sent Jackson Dawson on excursions into the eastern Massachusetts countryside to collect native shrubs. Wild roses, sweet fern, bayberry, mountain laurel, nannyberry, sweet pepperbush, and more were brought back in a hired horse and wagon. As each succeeding section of Arbore-

46. Construction of Hemlock Hill Road, near the Walter Street entrance. Originally planned to cross Bussey Brook twice in order to accommodate the junction of a branch road ascending the hill, the course of this section of the Arboretum drive was modified as the city workforce began construction in 1889. The idea of building a branch to the summit was abandoned and the main road was moved away from the brook valley to skirt the base of Hemlock Hill, leaving the stately evergreen grove undisturbed.

tum driveway was constructed, masses of native shrubs were set out to line the edges, to fill the strip between drives and walks, and to form occasional undergrowth for the tree collections. Often referred to as roadside plantations, nearly every year thousands of plants were installed to subtly ornament the drives and provide a foil for the generic tree groups. The roadside shrubberies rapidly imparted lushness to the landscape, as the naturally slower growing trees took years to develop. Since their ultimate role was to be a beautiful but unobtrusive foreground to the all-important tree collections, just as the existing natural woods were to serve as background, Sargent considered that wild forms and not garden varieties were most appropriate for this planting.

Sargent also intended the roadside shrubberies and the natural woods to serve an equally important ecological function in his garden of trees. Since much of the land would eventually be laid down to grass and be subject to soil compaction by visitor use and turf maintenance, the thickly vegetated woods and shrub masses would be areas where soil would remain open and receptive to rains and where leaves would collect and form natural mulch. Moreover, the woods and shrubberies would modify the Arboretum microclimate by checking winds, raising the relative humidity, and lowering summer temperatures through transpiration.

Although Sargent was eager to continue planting, he was at the mercy of the parks department when it came to road construction. After the 1884 mayoral election, the three-man park commission was reappointed, and the resultant upheaval stopped Arboretum construction altogether for the 1886 season. Despite this, Sargent carried out the plans to install the chestnuts, walnuts, and oaks in the spring of 1887. The parks department finished construction of the section of road to the top of Bussey Hill that year, but then did no more roadwork until fall 1889.

Despite the halting road construction, the director carried on. For his 1887 annual report he had his nephew, Henry Sargent Codman, who had become an apprentice to Olmsted, draw up a plan of the Arboretum showing the progress of the work. In addition, Codman drew up a series of detailed maps (on a scale of twenty feet to one inch) on which he plotted the exact location of each plant, thus creating the first system for mapping collections. Each plant was designated by an individual number, which was then used on the metal plant labels and on the records of origin of the plant that were bound in manuscript form with each map. In this way the mapping and numbering system served as a cross-reference and as a backup record to the card file that had been used up to that time.

47. Thousands of native shrubs were planted along the edges of Arboretum drives and walkways as soon as construction was complete. These roadside shrubberies provided an immediate lushness while young trees slowly matured. Pictured here, wild roses softened the verge of Forest Hills Road in 1900. The building at the top of the slope in the background served as a dormitory for Bussey Institution students.

If the 1886 halt in road making discouraged Sargent, by the early 1890s he had learned to live with the uncertainties of partnership with a parks department that could only rely on yearly appropriations from the city fathers. For two years "almost nothing" was constructed in the Arboretum. Exasperated once more by delays beyond his control, Sargent again threatened the loss of many hard-won young trees if they could not be planted out from the nurseries in the proper time. He had already done as much planting in advance of the road making as he could.

Relief came in the autumn of 1889, when the city workforce returned to start grading for the driveway leading from Valley Road to the Walter Street entrance, eventually named Hemlock Hill Road. Sargent's spirits revived with the prospect of opening up "the most picturesque and attractive part of the Arboretum." It was at this point, however, that the proposed branch road to the summit of Hemlock Hill was abandoned. Some sketches by the Olmsted firm made in October 1887 still included plans for a driveway to ascend through the hemlock woods. In order to create a graceful junction for the summit branch, the main road would have to swing wide and cross Bussey Brook twice. As construction commenced, Olmsted and company drew up revised plans, changing the line of the main road and omitting the branch to the hilltop. While the decision was more than likely driven by the city's desire to curtail expenditures, it had the happy consequence of conserving the natural state of Hemlock Hill and the beautiful view of the brook valley.

For the next four years construction of the Arboretum proceeded at an even pace. In 1890 a contract was let for the remainder of Bussey Hill Road and all of Meadow Road. As the work of subgrading this section began in 1891, the contract was amended to include grading for the final segment. Referred to then as a road "leading to the Parkway near South Street," it is known today as Forest Hills Road. By the spring of 1891, construction had progressed sufficiently to allow modest planting by the Arboretum. Along the nearly finished Hemlock Hill Road, Sargent placed the alders and six more genera that would round out the conifer collection: *Taxodium, Taxus, Tsuga, Pseudotsuga, Sciadopitys,* and *Pseudolarix.* At the op-

posite end of the property, the maples were planted on the east side of the unfinished Meadow Road. By 1892, Hemlock Hill Road was finished and surfaced like the others with layers of hard-rolled, crushed stone. The work on the junction of Meadow Road and Bussey Hill Road had progressed enough so that Sargent could plant trees of the legume family on the slope within the great curve of the drive.

Across Bussey Hill Road from the legumes, he planted "a few genera of the Ash family" (C. S. Sargent, 1893, p. 169). Since the ash trees had been among the first plantings of 1886, this somewhat cryptic remark undoubtedly refers to their relatives, forsythia and lilacs, neither of which have species native to North America. Even though this spot on the slope of Bussey Hill was where the ash family belonged in the botanical sequence, Sargent had serious misgivings about the placement of these two shrubby genera, which he voiced in a note to Olmsted:

Isn't it a mistake to plant forsythia, syringa and other showy flowered garden shrubs on the Arboretum Hill?

I should be afraid that they would not harmonize with the general scheme of planting we have adopted to the Arboretum and that they would invite unlawful picking of flowers.

The slope of the hill is as you know a conspicuous object from the road in the valley. How will a mass of bright colored garden flowers look rising above the softer first tints seen everywhere else in the Arboretum? (AAA, C. S. Sargent correspondence, 22 April 1888)

Ironically, it would be one group of these "showy flowered garden shrubs," the lilacs, that would become the Arboretum feature most loved by citizens of Boston. Lilac Sunday, that day in mid-May when thousands throng to see and inhale the fragrance of lilacs, has become as much a Boston tradition as baked beans or cod. But in the early 1890s Sargent was still making up his mind on the best deployment for shrubs, and he favored the completely naturalistic effect of native shrubs along the Arboretum drives. Subsequently, his conservative attitude toward exotic shrubs changed.

In addition to the parks department's steady progress on the roads, several improvements that were outside the city's commitment were accomplished under Sargent's perfectionist direction in 1892. Severe drainage problems were corrected in several spots, rendering more land suitable for tree planting. The channel of Bussey Brook was deepened and widened; drains were set and grades improved in about fifteen acres of the north side of Bussey Hill where some of the earliest plantings had been made; wide channels were cut through the twenty-five-acre North Meadow in an attempt to drain it into an improved outlet of Goldsmith Brook that had been constructed by the city. In all, some 4,667 feet of stone drains and 2,689 feet of ditches were installed that year. The construction

of Meadow Road itself by the city required considerable grade change to bring it above the high-water level. Culverts were installed over the two branches of Goldsmith Brook. The one closest to the main gate was surrounded by a puddingstone arch and rockwork similar to Ellicott Arch and other structures designed by the Olmsted firm in Franklin Park.

With the ground thus prepared, planting of the final groups to complete the Bentham and Hooker sequence took place in the 1894 season. Ironically, these last "type

48. View of the hill during construction of Bussey Hill Road, from near Forest Hills gate (looking west), 1890. The shrub-collection rows, then considered a temporary measure, can be seen in the foreground. The large pile in the middle of the photograph is gravel, crushed on the site from rocks thrown up during grading, to be used to finish the road surfaces. This is located roughly where the lilac collection was placed. Old field boundaries and Bussey's lilac and white pine hedgerow are still evident on the hillside.

collections" to be planted were those planned to be seen first by visitors entering the main gate. They included the magnolia, rue (cork trees), linden, sumac, horse chestnut, and rose families.

The parks department's construction of the Arboretum drives and the posts for the Walter Street gate was completed in the 1893 season. There remained only the Forest Hills and the Arborway gateways, finished in 1894. Construction of the Arborway started in 1893 and had progressed sufficiently to allow final boundary plantings along that perimeter in 1895.

ADDING THE PETERS HILL TRACT

Even before construction and planting were complete, the Arboretum administrator began to feel the pinch for land. Despite the planning, Sargent did not find enough space in the original allotment to place all the genera he deemed important. In groups already sited, such as the conifers, recent discoveries yielded more potential taxa than were known when planning started. The shrub collection was so successful that more land rightly could be devoted to it. It is the nature of gardens that plans are a starting point. Adjustment and revision are inevitable once the attempt is made to work with the land. In a garden with a mandate to be comprehensive, room for expansion became a must.

In 1891, Sargent received permission to plant the poplar collection on some of the area west of Bussey Street that Harvard retained for the use of the Bussey Institution. The three acres granted were in the corner of a roughly seventy-acre upland parcel used as hayfield and pasture by the Bussey Institution horse- and cattle-boarding farm (which brought a small income to the school).

The following year, Sargent proposed that the Harvard corporation add to the Arboretum a considerable part of the hill that came to be called Peters Hill. Although the Bussey Institution had been operating a farm utilizing some of this land for ten years, it rarely did better than break even. Two factors of change since the days when Benjamin Bussey dreamed of a school of agriculture in the 1830s were causing the institution to have trouble holding its own. One was that the country's expansion westward

and improved transportation had made farming much less important in the economy of Massachusetts. Second, between the time Bussey wrote his will in 1835 and the institution opened its doors in 1871, the Morrill Act, passed by Congress in 1862, initiated the nation's system of land-grant colleges. Granted federal lands in each state, these colleges were established primarily to educate citizens in agricultural and mechanical arts. They more than adequately fulfilled the need Bussey had seen. By the 1890s, the Jamaica Plain institute felt competition for students from the Massachusetts Argricultural College at Amherst, founded in 1862. The suggestion was made that sale of some land might correct the finances of Harvard's agricultural school.

After two years, the Harvard administration judged that Benjamin Bussey's desire for the college to "forever retain" his West Roxbury lands precluded their sale even if it were to benefit the agricultural school. The corporation decided that the Arboretum's occupancy of the tract would be compatible with Bussey's wishes, just as the original 120 acres designated for tree collections were.

The college assented to Arboretum use of the hill west of Bussey Street on the condition that the same agreement with the city as held for the original Arboretum be applied to the new territory. A further year and a half passed before the second contract with the city was made. The park commission readily agreed to join with the university to develop an extension of the Arboretum, which would tie in well with the commission's plans to connect with their proposed West Roxbury Parkway. As evidenced by the concurrent development of the greater metropolitan park and parkway system, both the city and the state were in favor of keeping as much land open within the urban area as possible.

As a further incentive, Sargent offered the city use of a puddingstone quarry on the tract if the city would enter into an agreement similar to that of 1881. Stone from the quarry could be used for future construction of park roads and walls. Sargent also requested that the city relinquish control over three areas within the original Arboretum that were not included in the first lease. The first was a four-and-a-half-acre parcel, the remainder of the Skinner and Kent properties, which had been recently purchased by the park commission and already informally turned over to the purposes of the Arboretum. Located at the junction of Centre and Walter Streets, the area was contiguous with the conifer collection. It would round out the Arboretum boundary to adjacent public streets. A century later the upper conifer area is still informally referred to as "Kent field."

Sargent had strong opinions on the second area, the eleven-acre "city reservation" on the top of Bussey Hill. He felt that the city was not maintaining it to the high standard he held for the rest of the grounds, and he had recently agreed to replant it at city expense in an attempt to improve its appearance. "I believe this piece of ground was reserved by the city when the [first] lease was made to insure the view at the top of the hill being kept open, but, as we are as much interested in the view as the City of Boston can be, it would probably be as safe in our hands as in yours" (AAA, C. S. Sargent to Charles F. Sprague, 23 November 1894).

The third area he wished added to the lease was a five-acre portion of the North Meadow, kept by the city because a lake was to be placed there. At the time, it seemed desirable that the construction and care of such a body of water should be carried out by the park commissioners. The lake was never constructed, however. Sargent was so determined to use as much of the North Meadow as possible for tree planting that he had spent considerable sums to trench and install drains in the area. Instead, ponds were created near where the junction of Meadow and Bussey Hill Roads was to be placed. During the construction of the summit section of the Bussey Hill Road in 1887, excess substrate not needed on the hill was deposited across that boggy area. Between 1887 and 1891, two ponds were created as peat was dug from either side of this fill and used for soil enrichment during planting. These are now known as Rehder and Dawson Ponds. The idea of a lake for the North Meadow must have been abandoned by then. In 1894, Sargent argued that since the piece of ground had been turned over to the Arboretum already, its tenure should be assured by a lease.

The second indenture between the city and Harvard's president and fellows was signed 22 April 1895. It extended the agreement of 1882 to the Peters Hill territory

and the other three parcels. The city pledged to construct suitable driveways and, as part of its duty to protect the grounds, to maintain boundary fences or walls around the whole of the Arboretum. The 67.6-acre parcel had much the same boundaries as it does today, except Bussey Street was then an unimproved lane.

The broad mound was largely treeless except for the puddingstone ledge along Bussey Street and a spring-fed depression on the southeastern slope. In 1895 a few ancient white oaks dotted the southeastern slope. These, like the ones on the western side of Bussey Hill, may have been remnants of the original forest left when the land was first cleared for pasture. Walnuts, elms, maples, and oaks lined the Mendum and Walter Street sides, suggesting old plantations from Bussey's or earlier farmers' times. The southwest corner bordered on an ancient burying ground dating back to about 1712, the time of the first religious services at the nearby Walter Street Meeting House. Reaching 237 feet above sea level, Peters Hill was the second highest point in Boston: "From the summit . . . fine views are obtained; the wooded portions of the Arboretum lie at its base; to the south all the Blue Hill range is in view; the waters of Massachusetts Bay are seen to the south-east, and to the north and west a broken well-wooded country" (C. S. Sargent, 1895b, p. 292). It was a fine spot from which to view the entire Arboretum and remains so to this day.

Even before the second indenture was signed, Olmsted and his assistants began work on circulation plans for the new area, and Sargent deliberated about what plant groups could be placed there. As in planning for the original part of the Arboretum, several possibilities were sketched before an acceptable configuration for the carriage drive was worked out. Most utilized a circuit of the hill with two or three approaches and a branch drive to the summit.

A version dated 10 December 1894 delineated a "classic Olmstedian" separation of ways of transportation. This plan was published with the twentieth annual report of the park commissioners. For Arboretum users intent on continued scrutiny of the tree collections by carriage, a drive left Hemlock Hill Road and crossed Bussey Street east of the Walter Street gate. Once in the new tract, the approach drive turned to make a great circuit of the base of the hill at an elevation of 110 feet. A short, straight branch to the summit departed the circuit on the west, opposite the cemetery. There was an approach to the circuit from South Street at the junction of South and Bussey Streets, as well as an entrance from Mendum Street located near the ancient cemetery. In order that Arboretum viewers would not have to hazard cross traffic on Bussey Street, and to improve Bussey Street as an efficient route for those not interested in tree viewing, the landscape architects proposed a "transverse road" to the west

of Bussey Street that would be depressed under the Peters Hill approach roads and hidden by the quarry ledges. Had it been executed, the transverse road would have replaced Bussey Street and would have agreeably separated distinct transportation functions.

The plan that accompanied the 1895 agreement was not much changed from the version just described, except that the Mendum Street entrance was omitted, most likely because of the cost of purchasing the required house lots. As preparation of construction specifications got under

49. *This plan of roads for the Peters Hill extension of the Arboretum by Olmsted, Olmsted & Eliot for Boston's parks department shows the proposal to create a transverse road below the level of the connecting links to the existing Arboretum drive and South Street. The transverse road, screened by puddingstone outcrops and the old quarry, would have replaced Bussey Street, but it was never built. When the interior circuit road was built, the Mendum Street entrance was located closer to Fairview Street than on this plan, and the summit branch was not included.*

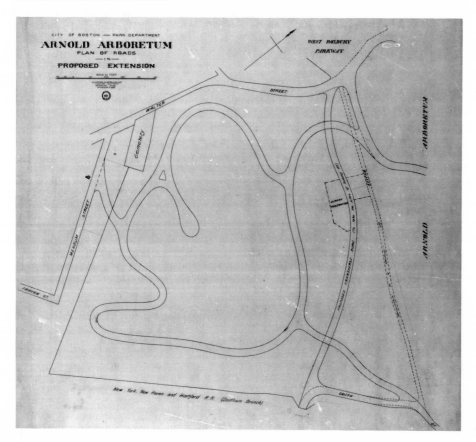

way in 1896, Sargent and the Olmsted partners found they disagreed on the grading:

The roads for the steep and difficult Peters Hill were first designed to have a maximum grade of six per cent; but the Director of the Arboretum objected so seriously to the extensive grading of side slopes which was involved, that we were instructed to re-draw the plans in such a way as to reduce the side slopes to the narrowest possible limit. This was done; but the resulting roads will necessarily have maximum grades of eight per cent; in other words, they will be steeper than any other roads of the Boston parks, and in our opinion steeper than any much-frequented pleasure driveways ought to be. (Eliot, 1902, pp. 705–6)

Apparently Sargent was more intent on leaving the terrain suitable for planting than on providing easy driving conditions for visitors. On the revised plan, grading for an extensive system of paths, separate but roughly parallel to the drive, was added. In recent years innovations in the main Arboretum had also been made with pedestrian, rather than carriage, access in view. Namely, a sketch for additional walks was drawn up by Olmsted, Olmsted & Eliot, and in 1896 Sargent reported that several mown-grass paths had been cut through the woods and collections to facilitate their use and enjoyment.

The parks department finished construction of the Peters Hill circuit drive and three entrances in the last three years of the century. The spring on the southeast slope necessitated installation of extensive tile drains. Access from Roslindale was finally placed where Fairview Street met Mendum, requiring a more modest land acquisition than the earlier proposal. Whether from lack of funds or from Sargent's desire to limit carriage traffic, the branch drive to the summit was not built. (The present paved summit drive was constructed in 1963.) As will be seen, Sargent's eventual use of the Peters Hill tract suggests that he came to think of it as more of an "experimental ground" than the remainder of the Arboretum.

During the five years of planning and construction for Peters Hill, the parks department was active in the original Arboretum as well. The city workforce cleared and graded the Skinner and Kent parcels, which had included homes and outbuildings, and built a boundary wall along the Walter and Centre Street margin. Once Arborway construction was near enough to completion, in 1896, a contract was made for a "substantial but inexpensive" stone wall to enclose the Arboretum on its eastern boundary. Although massive granite gateposts framed by short lengths of wall had been built as successive drives were completed, it was not until 1899 that Sargent reported that the city had installed "handsome and substantial iron gates" at five entrances. Two more entrances, those on the Peters Hill extension, were provided with iron gates the following year.

While the city finished construction of roads, walls, and gates in the last half-decade of the nineteenth century Sargent perfected the plantings of the Arboretum. In 1894 the ground around the recently erected museum was graded and converted into a shrubbery. Broad groups of barberries, viburnums, and other shrubs were thickly planted along walks and drives in the vicinity. Twenty-five thousand plants were used that year to surround the museum and to extend the fruticose profusion along the drives elsewhere in the Arboretum. Trees of the rose family—cherries, crabapples, mountain ash, pears, and hawthorns—were installed around the area of the shrub collection. The border belt of trees defining the Arborway side was extended with the installation of willows on the eastern edge of the North Meadow.

In 1896, with the placement of another twenty-five hundred feet of roadside plantations, Sargent commented that such undergrowth had now been established at the sides of all the finished roads. During the subsequent few planting seasons, Arboretum work consisted of filling in gaps in the shrub collection and in the generic tree groups. It was not until 1899 that any planting on Peters Hill commenced. Between then and 1901, wide belts of trees were established to surround and protect the area. To reflect the coniferous plantings and Hemlock Hill, a swath of white pines and related cone bearers was arrayed along Walter Street. In harmony with the existing woods, Sargent placed mixed plantations consisting primarily of native oaks and mixed native shrubs in the vicinity of the South Street entrance and along the boundary from the southeast corner to the old burying ground. Genus collections of spruce, fir, willow, poplar, and crabapple were initiated on the annexed tract in 1901.

50. View of Peters Hill from above the conifer collection, winter 1904 (looking south). A few ancient white oaks were the only trees on the north slope when the Arboretum took over the area. A supplementary conifer collection and the hawthorn plantation were soon added to these. The weeping tree in the foreground was probably a remnant of ornamental plantings on the remainder of the Skinner or Kent homesteads, lands added to the Arboretum by the city at the time the Peters Hill parcel was annexed.

EDUCATION IN THE NEW TREE GARDEN

By the turn of the century the Arboretum's framework and planting were complete, although the landscape and collections would be constantly improved and modified. Programs to extend the institution's commitment to education were started before it entered the twentieth century, however. Through publications, lectures, and classes a start was made toward teaching the public about many aspects of trees and shrubs. As visitors became more frequent, the Arboretum administration was heartened to see they sought both knowledge and pleasure.

The first educational effort of the new institution was communication on the importance of tree planting. Sargent's concern about the destruction of forests and its effect on soils, climate, and the economy was fired by George Perkins Marsh's *Man and Nature* and by his correspondence with the environmentalist. George Barrell Emerson also imparted his observations and conclusions on forest practices to Charles Sargent and urged the Arboretum to take up this subject. In 1875, Sargent gave an address that was subsequently issued as a pamphlet, *A Few Suggestions on Tree Planting;* its distribution was supported by the Massachusetts Society for Promoting Agriculture.

The pamphlet gave persuasive arguments for halting the destruction of forests and for planting trees. Convinced that thousands of acres of nonarable land could support the growth of trees if the right species were used, Sargent also thought plantations would yield a profit when harvested. Knowing the thrifty bent of the Yankee mind, he brought out the profit motive for those who would not be convinced by the abstract benefits of protecting the microclimate, soils, and water resources. Sargent stressed that planting the best kind of tree for each situation would maximize the advantages, concluding with a species-by-species discussion of the requirements and attributes of the most important trees. He also explained the need for much more information and experimentation in the field of tree culture, in which he felt the Arboretum could play a leading role. George B. Emerson was so pleased with the pamphlet that he wrote Sargent, "If the Arboretum had never produced or would never produce anything else, I should be richly paid for all I have done for it" (AAA, G. B. Emerson correspondence, 9 March 1876).

The tree-planting recommendations were revised and reissued a few years later. Once Sargent started investigating American forests as an agent for the tenth census in

51. *The museum building was designed by the architect Alexander Wadsworth Longfellow and constructed in 1892 to serve as administrative headquarters and to house library, herbarium, and educational exhibits. In 1894 the ground around the building was graded and planted heavily with viburnums, barberries, witch hazels, and other native shrubs.*

the 1880s, he continued to urge landowners to plant trees. The writings of the Arboretum's first staff member, John Robinson, a botanist from Salem, Massachusetts, were other early attempts to capture popular interest in trees and shrubs. During the few years that he took charge of the herbarium, Robinson published works on the dates of flowering of Arboretum plants and lectured on their ornamental attributes.

After planting and construction were under way, Sargent started another project aimed at fulfilling the institution's educational mission. Since there was no money in the Arboretum budget for a publication, he convinced a few interested individuals to back *Garden and Forest,* a journal of horticulture, landscape art, and forestry. Based in New York and managed by a capable journalist, William Stiles, *Garden and Forest* was a high-quality, well-illustrated journal, free of advertising. It debuted in February 1888 and appeared weekly for ten years. Its articles reflected the diversity of interests of its editor, Sargent, and his collaborators, including F. L. Olmsted and his partners. Week after week editorials on landscape design, preservation of urban park scenery, and conservation of forest resources were combined with informative articles on horticulture and botany. Sargent and Stiles contributed frequently, and Sargent called on his fellow Harvard botanists as well as experts from around the country for contributions. The result was a stimulating, wide-ranging discussion of plants and their value in human life.

The handling of the shrub collection exemplifies modification of Arboretum displays in response to visitor interest. A couple of years after its initial installation in 1885, the shrub collection became a highly visible and popular feature of the Arboretum; more than twelve hundred kinds of shrubs were arranged compactly according to the Bentham and Hooker sequence of families from Ranunculaceae to Liliaceae. Within each genus the plants were also arranged geographically.

Since shrubs generally grow more rapidly than trees, and the shrub garden had a one-year head start on the main tree collections, it was noted and praised in many articles published on the new Arboretum, even before all of the tree groups were in place. When the Committee on Gardens of the Massachusetts Horticultural Society

52. By the early 1890s the shrub collection had become such a popular feature that Sargent decided to retain it rather than plant its great variety of exotic shrubs out among the established tree collection (as had been the original plan). In the 1896 and 1897 seasons the shrub garden was completely renovated to provide slightly expanded planting space and more ample visitor walkways. The main building of the Bussey Institution can be seen in the background in this photograph.

visited in August 1885, members were impressed by the number and variety of shrubs, reporting: "Here anyone with even a limited knowledge of plants can take notes of these growing specimens, which will enable him to gain more knowledge than months or even years of study from books or catalogues could give" (Barker, 1885, p. 334). A living catalog it was, yet the beauty of individual specimens was equally admired. One garden writer raved about the "great group of systematically arranged shrubs two and a half miles of them, if you count the length of the walks between the different beds, where they are grouped in splendid luxuriance; such specimens! so vigorous, so hand-

some, so laden with blossom or fruit according to the season!" (Robbins, 1892, p. 28).

The shrub collection was originally intended as a temporary holding area until the implementation of the Arboretum plan reached the stage when the various shrubs would be incorporated. In the meantime, its value as a self-contained educational exhibit became apparent to Sargent; he also became more convinced of the appropriateness of using native shrubs almost exclusively along the roads and among the major tree groups. By the early 1890s he admitted that what the Arboretum really needed was ten acres upon which to plant a well-designed and com-

prehensive garden of shrubs. Once he knew that the only additional land to be made available was the steeply sloped Peters Hill, Sargent became resigned to improving the shrub garden within its existing borders. The completion of the Arborway and the Forest Hills gate rendered it possible to make better use of the level ground there.

The expansion and the rearrangement of the shrub collection took place over two years. To allow more spacious accommodation for the shrubs, the adjacent nursery, nearly equal in size to the shrub collection, was given up. Every shrub was lifted and, before replanting, the entire area was regraded. The layout was changed from

53. John George Jack (1861–1949), a native of Quebec, started working for the Arboretum in 1886 as keeper of plant records, maps, and labels. In 1891 Jack conducted the first field classes offered to the public, thereafter serving as chief interpreter of the living collections for some forty years.

thirty-seven short rows to fifteen long ones that extended some 550 feet from the edge of the North Woods to Forest Hills Road. The result was more room for shrubs and for visitors to stroll between the beds.

With the encouragement of the Harvard administration, the Arboretum began an experiment of furnishing popular instruction about trees and shrubs in 1891. Since the mid-1880s, when his own son desired to educate himself for a landscape designer's career, President Charles Eliot had been urging that instruction be given at the Arboretum: "public teaching, by lectures and by demonstrations, is an indispensable part of any proper and complete execution of the Arnold trust" (AAA, Eliot to Sargent, 21 January 1887). Although Sargent probably felt that the systematically arranged and carefully labeled collection would serve as an adequate demonstration in and of itself, he came around to the view that the Arboretum could more actively educate its visitors. Once construction

and planting of the original grounds had made sufficient progress, it was possible to pursue this goal.

Sargent, always a rather reticent personality, apparently did not consider himself a suitable lecturer; he preferred to communicate through the written word and rarely, if ever, spoke in public. Accordingly, he chose John George Jack from among his small staff to give outdoor instruction in the Arboretum. Sargent had engaged Jack, a lanky Canadian farmer and naturalist, as his first "working student" in the spring of 1886. In order to further his own education in natural history, the twenty-five-year-old Jack had been studying during winters with the paleontologist Alpheus Hyatt, curator of the Boston Society of Natural History Museum. Jack apparently desired more experience with plants than Hyatt could give him. When he came to Brookline with a letter of introduction and a keen desire to improve his knowledge of trees and shrubs, Sargent amiably told him he could do manual labor in the collections for a dollar a day. Jack agreed, but was soon set to work inventorying the nurseries, labeling plants, and recording information for each species on the card system. He remained at the Arboretum for fifty years.

When Sargent asked Jack to conduct the field class, he had been in charge of plant records and had assisted with specimen preparation and seed collecting for five years. His work required constant checking of the nurseries and grounds. Taking up a task that John Robinson had begun, he annually recorded the blooming and fruiting dates of plants on the grounds. He also continued his own education by attending Goodale's summer-school course in botany at Harvard.

With the first course of popular lectures in May and June 1891, Jack was appointed lecturer at the Arboretum. Every Saturday morning and Wednesday afternoon his students, mostly teachers, assembled at the Bussey Institution's lecture room for a half-hour talk, then toured the Arboretum collections with Jack for another two or three hours. Like the Boston Society of Natural History's Teachers School of Science, the Arboretum intended to contribute to public school instruction by making its resources known to teachers. Students of the Bussey Institution were also encouraged to attend. The following year the dendrological instruction was expanded to both spring and fall courses. The preliminary indoor lecture was dropped after a couple of years, and the "field classes" took place entirely among the emerging living collections. Jack apparently had an aptitude for teaching and an enthusiasm for trees, shrubs, and the Arboretum landscape that he readily shared with students. For the next twenty-five years, with rare exception, John George Jack taught the class both spring and fall to an average of thirty-two people, many of whom were teachers.

While Sargent declined to teach classes or lead tours, his efforts to educate and inform through the medium of publication were prodigious. Although *Garden and Forest* was financially distinct from the Arboretum, Sargent considered it "the organ of the Arboretum." Over the course of its existence Sargent authored nearly 370 articles and numerous unsigned editorials for its pages. The articles brought forth facts on "New or Little Known Plants," many of them trees and shrubs being cultivated for the first time at the Arboretum. Week after week Jack supplied a column, "Notes from the Arnold Arboretum," in which he discussed the ornamental attributes and cultural requirements of woody plants.

THE VISITOR EXPERIENCE

In the last decade of the century Sargent commented: "The number of people who visit the Arboretum from curiosity, to enjoy its sylvan beauties, or to avail themselves of the opportunity to study its collections, appears to be increasing rapidly and its educational value as a great object lesson must already be considerable" (C. S. Sargent, 1895a, p. 188).

The grounds had become accessible to the public little by little as each section of carriageway was completed, starting with Valley Road in 1886. Once most of the construction was accomplished, the Arboretum management welcomed reporters from newspapers and horticultural publications. These journalists universally praised the blending of the scientific with the aesthetic in its arrangement. It was admired not only as a demonstration of

woody plants but as an example of the best in naturalistic planting style. One *Garden and Forest* writer exclaimed that although the Arboretum must be considered a school of dendrology as well as a pleasure ground, "information is here administered so delightfully that one is hardly conscious of being at school" (Robbins, 1892, p. 27). Contemporary newspaper articles, many of them preserved in scrapbooks in the Arboretum archives, present a lively picture of the new tree garden and reveal how the place was utilized by the public.

Visitors who came by carriage while the Arborway was still under construction entered the main gate by a temporary road from Centre Street. Meadow Road apparently served so well for driving between Centre Street and Forest Hills that the parks department put off final surfacing of the parallel section of the Arborway in favor of other construction in 1893. This brought more people into the eastern end of the Arboretum just after the museum building and adjacent plantings were established. Like the carriages, horseback riders were restricted to the graveled drives that gently meandered among hill and vale.

Foot travelers arrived in Jamaica Plain by at least two forms of public transportation in the 1890s. For many years steam trains had been operating between Boston and Forest Hills. As soon as the entrance on South Street was finished, passengers who left Forest Hills Station walked down South Street to enter the Arboretum where Bussey Brook washed the base of Hemlock Hill. After 1889, when electric streetcars from Boston replaced horse-drawn trolleys and began service down Centre Street, the Civil War Monument in Jamaica Plain became a popular stopping-off place where a short walk led to the main gate on the Arborway. For pedestrians there were separate, hard-surfaced paths that accompanied the drives and diverged from time to time deeper into the collections. Additional miles of ways to explore the Arboretum's "wild wood beauty" had been made by mowing paths across its meadowy turf.

Whether riding or walking, Arboretum visitors at the turn of the century gained a wealth of sensory and mentally stimulating experiences. Inside the vine-clad granite gateposts of the Jamaica Plain entrance stood the museum,

54. *Soon after planting and construction of Hemlock Hill Road, visitors began to enjoy the mountain laurel's snowy masses of bloom each June.*

encircled by shrubbery. A mantle of climbing hydrangea clothed one end wall. The living handbook of trees began there, with the magnolias, dense and shrublike in their youth. Across the low, well-watered North Meadow, a broad view to the east disclosed graceful willows just starting to overtop the Arborway wall. A lofty American elm dating to Bussey's time beckoned the traveler onward past the dense shrubbery. To the right, young tulip trees and a twenty-year-old belt of native oaks screened out the Adams Nervine Asylum. Beyond the elm and the tulips, thrifty juvenile trees of the linden, rue, and maple families stood forth, generously separated to allow for their eventual dimensions. Massed plantings of fragrant sumac, yellowroot, and wild roses enveloped the drive's edge.

Beyond these, glacial knolls embraced the roadway; their wooded slopes allowed scarcely a glimpse of the collections beyond. The maturity and diversity of this section of native growth stood in pleasing contrast to the monocultural genus groups. Here, in the woods, strict scientific order relaxed a moment before resuming once again with the fine-textured trees of the legume family that airily deported themselves within the great curve of the drive. Along the side paths visitors were encouraged to dwell on Arboretum scenery while resting on benches made of wood and curved iron.

Once out of the North Woods the whole northeast slope of Bussey Hill loomed into view. Behind the rosaceous trees that stood around the ponds on the left was the shrub garden. With row upon row of sunny beds like an amphitheater, the fruticetum was an ever blooming outdoor lesson plan. Here a multitude of shrubs in their prime of life, each with a label standing in the ground before it, invited convenient comparison for those who chose to stroll the mown-grass lanes. It was more like a botanical "cabinet" or the "order" of a botanical garden than any other part of the Arboretum, but here the individual shrubs were free to grow to perfection unshaded by any trees.

As the drive turned to leave the ponded valley, it passed an assortment representing several families, each of which contained a relatively small number of hardy trees, that followed the rose family in the Bentham and Hooker sequence: witch hazel, tupelo, castor aralia, persimmon,

and cornelian cherry. Where Bussey Hill Road began its carefully graded rise, the side hill was grounded in forsythia, privet, and lilacs of the olive family for about 125 feet of its extent. Above these, some of the oldest of the generic tree collections flourished as a result of the capacious provisions with which they had been planted. The ashes, tree brethren of those "showy garden shrubs" just mentioned, had trunks already ten or twelve inches in diameter. Their high-thrown, straight-armed branching enhanced their illusion of stature in youth. The catalpas, too, were among the largest specimens noted by a contemporary garden writer; beyond them, the elm collection boasted some specimens nearly thirty feet in height. Next in the botanical sequence were the birches, advantageously finding their place on the northern, wet slope of "Overlook Hill," as Bussey Hill was termed during this era. As in the case of every tree genus, the native American species—golden birch, river birch, black and paper birches—wholly occupied the primary view from the drive. For each kind, a single individual stood boldly forth surrounded by ample space for its fullest future breadth and height, while near it clustered a small population to mock its habit when living under forest conditions. Here the roadside shrub masses included their kin, dwarf birches of the high mountaintops and northern zones.

From the overlook, visitors enjoyed distant views of the Blue Hills, Boston Harbor, and even the hills of southern New Hampshire. Closer to hand, the Arboretum's sylvan beauties could be admired comprehensively from the concourse. There were no genus collections located on the upper hill, since this area had been excluded when the original plan was made. The hilltop plantings still retained the mixed shrubs and native trees installed for parks department use as well as remains of Benjamin Bussey's landscape: the grand hedge of lilacs, stately white pines, beeches, and other trees.

The perceptive visitor of 1900 might have noticed that Valley Road, which carried forward the living catalog after the birches, was the longest-established section of the Arboretum. Many of the first descriptions of the planted Arboretum dwell on this section, since the early habit was to enter at the South Street gate. The valley from which the road took its name marked the lowest point between

55. *Among the natural features of the Arboretum that Olmsted argued were the most important to preserve within Boston's "sylvan system" of parks were its "eminences commanding distant prospects, in one direction seaward over the city, in the other across charming country-side to blue distant hills." Although the Arboretum took over care of the plantings on the heights of Bussey Hill in 1895, preservation of its inspiring views continued to be a priority.*

the Arboretum's Bussey Hill and neighboring Moss Hill. Here the nut-bearing and cupuliferal collections promised arboreal dignity for the valley floor. Walnuts, hickories, oaks, chestnuts, beeches, and hornbeams primarily made up these groups. Many visitors marveled at the scattered old oaks already renewed by Sargent's early plan of heavy pruning, soil aeration, and fertilization. These ancient oaks, and the native stand covering rocky Central Woods, lent a venerable character to a scene otherwise bursting with fresh young growth. Skirting the road near the chestnut tribe were shrubby chinquapins, while across the drive low-growing bushy types of oak footed the display of their taller relatives. The roadside luxuriance was supple-

mented by masses of natives such as sweet pepperbush, bayberry, sweet fern, azalea, dogwood, shadbush, and vines that were encouraged to trail over the ground and soften the verge.

The main drive plunged toward the next hill, dark in its hemlock raiment. At its base, Hemlock Hill Road formed a right-hand branch to cross the brook, offering visitors two alternatives. The continuation of Valley Road led tree watchers to the exterior of South Street. An ample grove of red pines and white pines had been planted on the north side of this road, mirroring the evergreen cover on the precipitous cliffs on the south. At the juncture of Valley and Hemlock Hill Roads, masses of young rhodo-

56. The South Street gate was the earliest-established visitor access to the Arboretum. A lofty American elm graced this entry. Beyond its welcoming portal, Valley Road led to wild woodlands and planted groves of beech, hornbeam, oak, hickory, walnut, and more.

dendron and mountain laurel contrasted their broad, sculptural leaves with the feathery hemlocks above them. The adventurous visitor could leave the road, and almost all signs of civilization, to enter the somber shades of what many likened to a primeval forest. Only footpaths traversed Hemlock Hill. Here were plenty of murmuring pines and hemlocks, although an occasional sweet birch or white oak brightened the canopy. A stroll on the ferny, needle-floored slopes presented a quieting experience unlike any other within the city limits. If the living botanical handbook was to be seen through to its conclusion, however, the turn onto Hemlock Hill Road afforded the route.

Along its length—opposite the indigenous hemlock forest—were planted the gymnosperm families: resinous, needle-leaved, cone-bearing woody plants. The carriageway's course followed Sawmill Brook upstream at enough of a distance to leave a delightful vista up its verdant floodplain, bordered on the opposite side by a bluff. Despite the recently installed junipers, yews, umbrella pine, and cupressaceous genera, the view was rural, punctuated by an occasional venerable white oak or American elm. A thicket of bright-barked beech, sprung from the roots of an aged parent tree, graced brookside just before the pine collection was encountered. Where these "swishing pines" had begun to cover the ground with a carpet of russet needles, the slope pulled away from the brook to form a wide bowl. On its sides squatted the pyramidal forms of youthful spruces and firs "wrapped like esquimaux in their furry . . . garments"; nearly uniform in shape, their different hues betrayed their diverse speciation (Robbins, 1892, p. 28). At the Walter Street gate, larches formed the final link in the botanical sequence.

Across Bussey Street, Peters Hill's ample slopes accommodated the supplement, or addendum, to the living handbook. Still under development in 1900, its circuit drive skirted recent plantations complemented by longstanding woods and scattered mature oaks. Those who left the graveled drives to ascend the summit were rewarded by a grand panorama, for no trees screened the view.

Although all was new (the ten- and twenty-year-old trees were only just beginning to hint of their eventual

stature), the place had certainly been transformed since the days in the 1870s when Sargent contemplated the "worn-out farm." With expanded boundaries encircled by stone, its family groups of trees threaded together by topographically harmonious drives—its soil drained, graded, enriched—the Arboretum was now a garden, yet unlike any other garden known to Bostonians or to Americans. "Has no equal in the world. . . . The Arboretum a collection of trees and shrubs arranged in botanical sequence––nothing like the one in Boston"—thus was headed an article for the *Rochester Democrat* by that city's superintendent of parks, John Dunbar, in 1897. In 1905 the horticultural publisher J. Horace McFarland called it "a tree garden to last a thousand years." Another contemporary article (for *Country Life in America* in 1904) was titled "The World's Greatest Tree Garden."

While the Arboretum was still in its infancy at the turn of the century, writers of the time comprehended the forethought shown by its design and management. How well the foundations had been laid for its future growth, in its careful installation and in its two-part institutional affiliation. Here the trees were "safely insured for centuries of growth" where they could always be studied and observed by amateur and professional alike. While the specimen plants in each of the tree collections did "seem needlessly remote," the lushness of native shrubs run riot along drives compensated in these early years. Although preserving the existing natural woods had made it more difficult to arrange the Arboretum, think how bald it would have looked for the first fifty years without this well-wooded backdrop.

Its comprehensiveness, taxonomic arrangement, record keeping, labeling, field classes, research, publication, and staff expertise were components of the Arboretum's contribution to science and education. The beauty and variety of its landscape composition enriched the lives of all who partook of it. Here were "sweet country restfulness and peace," a place where one could "lead for an hour the life of a woodlander." From bold outlook to secluded corner, there were myriad encounters with the arboreal world to be had, especially when seasonal change and annual growth were considered in the mix. The secret of its success was perhaps stated best by a writer for the *Critic and Literary World:* "The Arboretum does not tell all it knows on a first acquaintance" (Duncan, 1905, p. 117).

The presence of visitors was a further transformation from the days when Sargent and his assistants first surveyed the faded farmland. It was now "gay with carriages" in lilac time, and the attendance of students was frequently noted. Every spring and fall, John G. Jack could be seen leading a coterie of teachers and the horticulturally inclined from plant to plant. At times in between, Benjamin M. Watson's horticultural students from the Bussey Institution, or scholars of landscape gardening from Harvard's Lawrence Scientific School or the Massachusetts Institute of Technology were observed, notebook in hand, pacing up and down the shrub collection rows or scrutinizing a label on the trunk of a healthy specimen tree. "Special students"—professionals in training, often from foreign botanic gardens and institutions—stayed for months of intensive observation of the collections under Jack's wing. Sargent personally conducted such visiting dignitaries as France's premier nurseryman, Maurice de Vilmorin, and the assistant director of Kew, George Nicholson, through the Arboretum.

As President Eliot expressed it, there was satisfaction with the growing usefulness and influence of the Arboretum in the nineties:

The number of visitors to the Arboretum has been much increased by the completion of the approaches to [it] by the Boston Park Commission. The Arboretum now makes part of a drive which thousands of people follow on pleasant afternoons through all the milder weather of the year. The natural woods and the systematic collections attract the attention of the greater part of these visitors chiefly for their beauty, which varies with the succession of the seasons; but there is a considerable number of visitors on foot who visit the Arboretum for study combined with enjoyment. (Eliot, 1895, p. 30)

CHAPTER FOUR

DEVELOPMENT OF THE SCIENTIFIC PROGRAM

W hile the Arboretum was growing from an idea to a serene garden of trees arranged to display classification, its mission to teach the knowledge of trees also grew and changed, with research becoming an increasingly important goal. Although the Arboretum budget rarely allowed what Sargent considered adequate funds to pursue research, he had the knack of seizing opportunities as they arose and carrying projects through for the institution with perseverance. In Sargent's day, the research program was thoroughly entwined with the management of the living collections. For both, the emphasis turned from America's forests and trees as sources of timber to Asia's temperate woodlands and woody plants for ornament. Nonetheless, neither topic ever entirely replaced the other. As the science of forestry came under the province of government agencies and schools of forestry, the Arboretum specialized in dendrology (the study of tree biology and classification) and descriptive botany. From its start with North American trees, it extended to worldwide coverage of this subject. Essential to the Arboretum's investigative program was the formation of herbarium and library collections to supplement observation of living plants. Equally important was the means to communicate its scientific findings—its many publications, always aimed at the highest standards.

TENTH CENSUS AND RELATED PROJECTS

It was through the influence of Asa Gray and George B. Emerson that Sargent participated in the Tenth Census of the United States. As far back as 1848, Gray had agreed to report on the nation's forest trees for the Smithsonian Institution. But as other tasks took precedence, and he realized that the need was for practical information outside of his botanical interests, Gray relinquished this duty. In the 1870s, when the movement toward governmental management and protection of American forests gained momentum, Gray found himself drawn into the forestry question once again after a meeting of the American Association for the Advancement of Science in 1873. The AAAS formed a committee to urge Congress to action. Gray served on it, as did George B. Emerson, who had just returned from six months of observing forestry practices in Europe. The

57. Alfred Rehder (1863–1949) searches for information in the Arboretum library soon after his arrival at the turn of the century. Born and trained in Germany, Rehder did much for the Arboretum's research program in plant systematics and horticulture. In his work on the Bradley Bibliography, *Rehder contributed to the development of the library. Next to living plants, books about them were Charles Sargent's professional passion. Over a period of years, he donated the majority of volumes to the library, which was housed on the second floor of the museum building.*

committee composed a memorial on the need for fact-finding and action to ensure future timber supplies. Emerson took this to Washington, where he spent several days talking to senators, congressmen, and President Grant. Although nothing immediate came of their memo, Gray and Emerson had made contact with federal officials interested in the forest question.

When it was time to plan the census of 1880, Secretary of the Interior Carl Schurz realized that up-to-date documentation on forests was needed. He consulted with Gray, who suggested Sargent for the position of expert on forests and forest products for the census. Sargent had demonstrated his knowledge of the subject in his two tree-planting papers of 1876 and 1878 and a report on the forests of Nevada written after his journey to study the trees of the Great Basin in September 1875. Late in 1879, just after Sargent's responsibilities to Harvard's botanical garden had ended, he agreed to survey the nation's woodlands for federal authorities. Olmsted's plans for the Arboretum seemed to be well in hand, and Sargent thought his participation in the government-funded investigation would enhance the institution. He wrote to Hooker that although "the labor will be enormous, I think that in the interest of the Arboretum, I should accept it. It will give me great facilities for travel at Government expense to every part of the Continent and will enable me to enrich the Arboretum in every way, besides doing much needed work in showing the great wealth and value of our forests" (AAA, C. S. Sargent correspondence, 13 October 1879).

Sargent proposed to gather information firsthand, personally or through correspondents in the field, rather than rely on published literature, which tended to be out-of-date or incomplete. That was how Emerson had approached his Massachusetts trees report. In January 1880 Sargent called on George Engelmann in Saint Louis to discuss the project and meet some other Missouri botanists. He then headed south to look at pines, cypress, and other conifers until late February. Sargent traveled southeast once more that spring, to North Carolina with the Delaware botanist William Canby. When not in the field, Sargent drew up a preliminary catalog of all tree species known in the United States, which he published and

58. Charles Sargent, Francis Skinner, and George Engelmann traveled together in the summer of 1880 to examine trees and forests of the American West for the Tenth Census of the United States. Engelmann was an active physician as well as botanist, and he rarely had the opportunity to leave Saint Louis to work in the field. Nonetheless, Sargent found him an excellent traveling companion. Skinner was a Brookline neighbor who volunteered his assistance to the Arboretum on several projects.

distributed with a request for notes and corrections. He also wrote numerous letters recruiting botanists around the country to send him information, specimens, and wood samples.

That summer Sargent took a four-month excursion to the Far West accompanied by Francis Skinner, a neighbor and friend who often volunteered his support for the Arboretum. The two left Brookline in mid-June and botanized in Colorado for a few days before meeting George Engelmann and his colleague, Charles C. Parry, in Utah. The four then headed to San Francisco and took a coastwise steamer to British Columbia. The botanists then worked their way south through the Pacific states, moving up rivers, climbing mountains, looking at as many kinds of trees as possible, trying to solve questions about identity and extent of distribution, comparing forests of the Cascades with those of the Coast Range. After Parry left them in early August, the party traveled to southern California and crossed Arizona along the Mexican boundary, where many species occur that are endemic to the Sonoran desert. They covered a vast territory, and Engelmann (then seventy-one) chided that Sargent never allowed more than "five minutes for anything." Despite the four months of tramping and collecting over thousands of miles and diverse climatic zones, Sargent returned to Brookline in mid-October disappointed that he had not done more. His wide-ranging and fast-paced tour had shown him how much work was left to be done. It was paramount for the census to disseminate information quickly, however, rather than wait to resolve every question of range or nomenclature.

For the next two and a half years Sargent compiled the results of his observations and the information that came in from his agents and correspondents around the country. His brother-in-law, Andrew Robeson, helped with the correspondence, tabulated the quantitative data, and prepared maps to accompany the final report. Wood samples were obtained for every species when possible. To test them for various physical attributes, Sargent enlisted the services of the Cambridge chemist Stephen Sharples, and the army made its materials laboratory at Watertown available. In 1881 Sargent visited Engelmann once more, then

journeyed to look over the wooded regions of Texas and Louisiana.

Sargent finished writing the final document by mid-1883, although it was not published until 1884. *Report on the Forests of North America* consisted of three parts: an expanded catalog of trees, the results of experiments on the wood of nearly four hundred species, and an analysis of the economic aspects of the forests of each state and region. Nothing so comprehensive or current had ever been published on America's trees and forests. It combined succinct biological information with descriptions of human-use patterns to form a valuable reference for what Sargent hoped would be the basis for future decisions to manage and better protect the nation's forests.

The census work had its predicted positive effects on Sargent and the Arboretum, enhancing the reputations of both; it also greatly broadened the director's perception and knowledge of American trees. Influenced as well by Asa Gray's taxonomic achievements, Sargent had come to see the need for further systematic research, and his awareness of all American forest regions had been heightened. He apprehended their diversity, their relationship to topography and climate, and the present and potential value of their products to humanity, as well as humankind's threat to the forests. Sargent continued to campaign for conservation and management of forests for many more years, usually as an adjunct to his administration of the Arboretum. The special agent on forests to the tenth census had made contacts with botanists around the country, and many of these collaborators added to the collections of the Arboretum for years afterward.

Even before the census investigations were published, Sargent was drawn into related projects. For the American Museum of Natural History in New York, he amassed logs of every tree species obtainable for the Jesup Collection of North American Woods. This exhibit, which ultimately contained some five hundred carefully prepared samples, was financed by the multimillionaire Morris K. Jesup, one of the founders and presidents of the museum. Mary Sargent painted life-size depictions of foliage, flowers, and fruits to accompany the specimens of wood. The display was opened to the public in 1885, although additional logs and watercolors were added for several years afterward. Sargent simultaneously obtained smaller samples of wood for an exhibit he hoped to display at the Arboretum.

In view of his knowledge of western forests, Sargent was asked to join the Northern Transcontinental Survey, funded by an affiliation of northwestern railroad interests. His part in the survey was to participate in a field trip to the Montana Rockies in the summer of 1883. While Sargent admired the noble forest covering this region, it was not only the beauty of its natural features that impressed him. The greater importance of the region was the snow and ice its peaks captured and how its vast forests held the meltwater in the soil while it made its way gradually to three great river systems. Because of the water-retaining capacity of these forests, Sargent called the area the dome of the continent and urged that it be set aside as a forest preserve. (The area later became Glacier National Park and is now Waterton-Glacier International Peace Park.) Whenever Sargent campaigned for forest conservation he stressed the role of wooded regions in the protection of watersheds. During the 1880s, he took his campaign for forest protection closer to home, to New York's Adirondacks, where what was left of that region's woodlands protected the waters of the Hudson River.

SILVA OF NORTH AMERICA

Another consequence of Sargent's growing reputation was a request from Spencer Baird, secretary of the Smithsonian Institution, to produce a silva of North America, a work to treat extensively the botany of trees. The census was necessarily a broad-based gathering of practical information, with the descriptive and comparative biology of trees being only briefly summarized in its first part. What the overseers of the Smithsonian wanted was an expansion of that catalog to include more complete descriptions and biological information for each American tree. Having done much of the groundwork, Sargent was ready for the job. He and Baird agreed that accurate illustrations to facilitate recognition of the hundreds of North American trees were essential to supplement the silva's text. And

Sargent knew just the person to do the drawings: Charles Edward Faxon.

Faxon was a lifelong Jamaica Plain resident. He had been raised by his brother, Edwin, an active amateur botanist who shared this enthusiasm with Charles. Before reaching the age of fifteen, C. E. Faxon had taught himself to draw by using lesson books and by making copies of such works as Audubon's. He graduated from Jamaica Plain public schools, then studied civil engineering at the Lawrence Scientific School, where he perfected his skill

59. Charles Edward Faxon (1846–1918), a lifelong Jamaica Plain resident, illustrated biological subjects for Harvard's Museum of Comparative Zoology and taught at the Bussey Institution before joining the Arboretum staff in 1882. He produced 744 beautiful drawings of trees for The Silva of North America. *Faxon, who also managed both library and herbarium, is seen here with some of the original wooden cabinets that contained the pressed-plant collections.*

in mechanical drawing. Although he clerked for a few years in the family leather business, natural history was a stronger calling. Edwin Faxon had a large personal herbarium in their home, which his brother undoubtedly helped to curate. In the late 1870s Charles Faxon produced color plates for the Yale professor Daniel C. Eaton's *Ferns of North America.* Professor Alexander Agassiz, who had taken over direction of Harvard's Museum of Comparative Zoology from his father, hired Faxon to delineate organisms for his scientific publications. Subsequently, Faxon taught basic biology courses for the museum, as well as for the Bussey Institution, where Sargent came to know him.

Faxon joined the Arboretum staff in May 1882 to take over care of the library and the dried-plant collections. In addition to managing the Arboretum's incipient herbarium, he was set to illustrating trees for the government-funded silva project. As Faxon's work progressed, Sargent realized that (at the rate the Smithsonian was able to pay for the services of an artist) it would be seventy-five years before illustrations of all the North American trees could be completed. Involvement with the Smithsonian was amicably terminated, and Sargent made private arrangements for publication. Even so, the work could not be completed overnight. Sargent aspired to the kind of originality that had gone into Bentham and Hooker's *Genera Plantarum,* the final volume of which had just been published. Despite its outside funding, the *Silva* became part of the ongoing research-and-publication program of the Arboretum for the next twenty years.

Sargent worked on the project whenever he found time, accumulating information, weighing it against his own observations, and then writing up concise treatments of each genus and species. Despite all the activities on the Arboretum grounds and the founding of the Garden and Forest Publishing Company, he managed to take at least one field trip a year for the *Silva* during the 1880s. Faxon often accompanied him; his original observations enlivened the drawings he made to complement Sargent's text. In the summer of 1887 the Arboretum's director

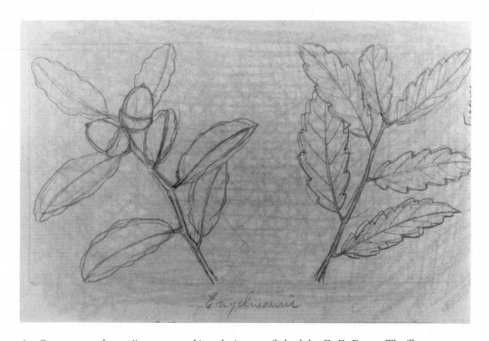

60. Quercus engelmannii, *evergreen white oak, in a pencil sketch by C. E. Faxon. The illustrator worked with living material whenever possible to create botanically accurate, lively depictions of flowering and fruiting branchlets of every species to be included in* The Silva of North America. *Many sketches were made preliminarily to rendering a final drawing in ink.*

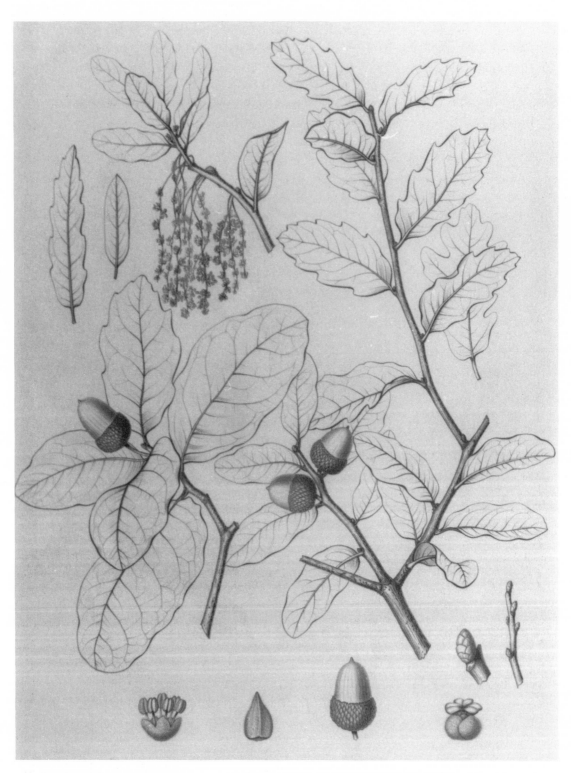

61. Quercus engelmannii, *plate 387 by C. E. Faxon for* The Silva of North America. *The Engel-mann oak is an evergreen species native to southern California and northern Baja California, Mexico. Faxon's drawings were sent to France to be engraved for publication.*

went to Europe to discuss the projected book with botanists there and to examine specimens of North American plants at Kew and the British Museum.

Sargent used the pages of *Garden and Forest* as a trial ground for the findings of his research. In a fifteen-part series of 1889, "Notes upon Some North American Trees," he discussed revised opinions on nomenclature, taxonomic rank, tree versus shrub status, extent of natural range, and more, all "for the purpose of calling out criticism" preliminary to inclusion in the *Silva*. As his research continued in the 1890s, subsequent notes on North American trees were released in the periodical.

As the first volume of the *Silva* neared completion, Sargent paused to note a sad event for American botany—Asa Gray died in January 1888. He expressed personal and universal loss:

The whole civilized world is mourning the death of Asa Gray with a depth of feeling and appreciation perhaps never accorded before to a scholar and man of science. . . . He was something more than a mere systematist. He showed himself capable of drawing broad philosophical conclusions from the dry facts he collected and elaborated with such untiring industry and zeal. . . . His long list of educational works [has] had a remarkable influence upon the study of botany in this country during the half century which has elapsed since the first of the series appeared. . . .

One of Asa Gray's chief claims to distinction is the prominent and commanding position he took in the great intellectual and scientific struggle of modern times, in which, almost alone and single handed he bore in America the brunt of the disbelief in the Darwinian theory shared by most of the leading naturalists of the time. . . .

But his botanical writings and his scientific fame are not the most valuable legacy which Asa Gray has left to the American people. More precious to us is the example of his life in this age of grasping materialism. . . . Great as were his intellectual gifts, Asa Gray was greatest in the simplicity of his character and in the beauty of his pure and stainless life.
(C. S. Sargent, 1888, p. 1)

Sargent's unabashed admiration for Gray stirred him to join with others to memorialize the botanist by publishing his collected works and raising a permanent endowment for the herbarium. Gray's exacting scholarship remained the model to which Arboretum research aspired.

When the first volume of *The Silva of North America* appeared in October 1890 (copyright date), it was dedicated to the memory of Asa Gray, "friend and master." This initial volume was just the start, however. At that time Sargent estimated that 422 species of trees native to North America (exclusive of Mexico) would be described and illustrated in twelve volumes. Volume 1 covered the first thirteen families of the Bentham and Hooker sequence that had woody genera, some thirty-three North American species. For each one, Sargent produced a complete list of synonyms, a full description, and a discussion of its geographical distribution, history, and economic importance. Faxon usually executed at least two illustrations for each species, one a branch in flower and another a branch with fruit.

When it came to producing the book, Sargent and his publishers spared no expense. On the advice of British and French colleagues, he had copper plates of Faxon's beautiful illustrations engraved under the direction of A. Riocreux by the Parisian artists Eugène and Philibert Picart, famous for their excellent plant portraits. The text was produced and the work published by Houghton, Mifflin and Company. The result was a sumptuous book, printed on high-quality paper, with generous margins. With the issue of the first volume, *The Silva of North America* was hailed as a "great work," a "noble volume," a "remarkable undertaking." Reviewers lauded Sargent's careful research and encyclopedic text, while Faxon's illustrations were admired for their lifelike and precise rendering. Of course this was only the beginning; Sargent and Faxon toiled steadily for another decade until the final volume came out.

Although much of his energy was focused on oversight of the Arboretum, Sargent rarely let up investigating and writing the *Silva*. Volume 2 appeared shortly after its predecessor in May 1891; two more volumes were published before Sargent left for an excursion to Japan in 1892. The next two years saw only one *Silva* volume each while Sargent was simultaneously evaluating the results of his Japan field trip. Starting in 1894 he resumed yearly excursions to view trees on which he felt his knowledge was rusty in order to present as lively a picture for the species in remaining volumes as he had in the early issues.

62. Among the richest forest areas in North America, in terms of numbers of species, are those of the southeastern United States. Sargent traveled to this region several times while investigating for the census and the Silva. *This photograph, taken by the son of architect H. H. Richardson, shows a woodland thick with rhododendrons in North Carolina.*

A welcome companion on several of these trips was John Muir. Sargent and Muir had become acquainted years earlier through George Engelmann and found they had a common determination to preserve the forest. The two arboriphiles were often joined by William Marriott Canby, a banker and railroad director from Wilmington, Delaware, who botanized every chance he could. Canby and Sargent had explored the magnificent Montana Rockies together as members of the Northern Transcontinental Survey in 1883. In the fall of 1897 Sargent, Muir, and Canby went to Alaska and the Canadian Rockies. They met again a year later to peruse the southern Alleghenies during fruiting season.

From Muir's perspective, Sargent's pursuit of knowledge among the trees was single-minded and untiring:

All his studies were bent toward this book and with unflagging industry . . . he has labored to make it complete, traveling, studying, writing, determined to see every tree on the continent, known or unknown, growing with its companions in its own native home. And with few exceptions, he has seen them all, most of them in the different seasons of the year, in leaf and flower, and fruit, or disrobed at rest in winter . . . flitting from side to side of the continent . . . traveling thousands of miles every year, mostly by rail of course, but long distances by canoe or sailboat . . . , through swamps, along lagoons, and from one palmy island to another, jolting in wagons or on horseback over the plains and deserts and mountain chains of the West, . . . enduring all things—weather, hunger, squalor, hardships, the extent and variety of which only those who from time to time were his companions can begin to appreciate. (Muir, 1903, pp. 9–10)

The *Silva*, like the Arboretum, was completed within a couple of years of the turn of the century, with the last of the twelve planned volumes being released in January 1898. The intervening years of research and observation necessitated two supplemental volumes for additions and revisions, and with their publication on 15 December 1902 the work concluded.

As volume after volume appeared, notices in botanical and horticultural periodicals sustained praise for the work. The aspect that piqued the most criticism was Sargent's rigid application of the rule of priority in nomenclature, a departure from the preferences of Asa Gray, yet in line with practices favored by a new generation of American botanists. A reviewer representing the viewpoint of the layperson rather than that of Sargent's fellow botanists was John Muir, who exalted the book and the forests he thought it so aptly brought to the light. Claiming to have "read it through twice, as if it were a novel, and wished it were longer," Muir thought the work was worthy of the vast continent of woodlands it divulged. "Of all the nature-books I have ever read, the *Silva* is the largest and best, everywhere breathing the peace of the wilderness, restful, yet inciting to action, infinitely suggestive and picturesque" (Muir, 1903, p. 21). Some eighty-five years later a science writer for the *Boston Globe* called it a "tree book as Yankee as a cod," insofar as it represents the pinnacle of Victorian science, a compendium, tastefully and durably rendered into magnificent volumes. In writing the *Silva* and fostering a role in forest conservation for the Arboretum, Sargent took up tasks left to him by George B. Emerson, Asa Gray, George Engelmann, and other early students of the American forest.

AMERICAN TREES AND ARBORETUM POLICY

His intimate knowledge of American trees in their natural habitats and his witnessing of wasteful and illegal harvest practices fired Sargent's attempts to promote forest conservation. After a disappointing attempt to save the Adirondacks from waste in the late 1880s, he limited his forest advocacy to editorials in *Garden and Forest* for many years. Sargent was drawn once more into active participation in 1896 when he chaired a commission of the National Academy of Sciences to investigate and make recommendations for forestlands of the United States. The commission took a field trip to the West in the summer of 1896. Gifford Pinchot, a young German-trained American forester, was invited to participate; John Muir joined the group in an unofficial capacity.

Sargent was enthusiastic about the trip from a botanical standpoint, calling it the "longest, hardest and probably the most interesting and instructive journey" he had made in western America. He gained valuable information for the *Silva,* especially on the range of various species of conifers, but he was not optimistic about the commission's results in terms of obtaining substantial protective measures from Congress. Although some new forestland was set aside in federal reserves as a result of its report, many of the commission's recommendations were passed over in the political change from the Cleveland to the McKinley presidency. The ultimate fate of America's forest policy was to fall into the hands of Pinchot, who received an appointment as special forest agent under the Department of the Interior, an action unrelated to the National Academy's recommendations.

After 1897 Sargent rarely took an active role in the promotion of forest legislation. Instead, he concentrated on research that was useful to forest science and education that would open the public's eye to the value of trees. As the science and profession of forestry began finally to develop in this country, it became clear that the Arboretum could include only some facets of this specialty in its programs. Not long after Yale University established the first school of forestry in the United States, Harvard commenced its forestry offering with undergraduate instruction through the Lawrence Scientific School in 1903. Once the twenty-one-hundred-acre Harvard Forest in Petersham, Massachusetts, was acquired to serve as a practical demonstration and experimental ground in 1907, the university established it as an institute for research and graduate-level study. The Arboretum's resources, however, were best suited for studies of dendrology and of arboriculture, the knowledge of how to propagate and grow trees as individuals; scientific forestry required a greater scale for its investigations. Still, the basic knowledge of trees was the starting point for this work, and here the Arboretum had its place.

Distinct from the value of the *Silva* itself is the influence that investigation of American trees had on the Arboretum. It established native American woody plants as a priority in early planting policy and in presentation of the collections. American trees were given first place and allotted more space in the planting scheme of 1885. Native American shrubs were lined out among the major tree collections, while introduced shrubs were for the most part kept separate in the special shrub collection. Great thought and energy were given to the proper care and management of the native stands on the Arboretum grounds. The *Silva* established the institution's reputation for careful research and meticulous publication. Although Sargent was several times disappointed with the results of his efforts to secure protective legislation for the country's forests, his publications and expertise contributed greatly to this process—one that still needs his kind of knowledge and advocacy.

After only a short break from work on the *Silva,* Sargent made up a compact version suitable for field use. Not content to rest on the larger work's laurels (its first volume was nearly fifteen years old), Sargent brought American tree classification and nomenclature up to date for the abridgment. *A Manual of Trees of North America Exclusive of Mexico,* which came out in 1905, presented 826 pages, with 642 illustrations adapted from Faxon's drawings, in one volume. Sargent hoped this book would be a convenient, affordable means for students to use the information assembled for the fourteen-volume work. A noteworthy deviation from his previous compendium was that Sargent adopted the arrangement of families and genera recently published by the German botanists Adolf Engler and Karl von Prantl. During the remainder of his career Sargent kept abreast of developments in the study of North American trees, even while the Arboretum's scientific program diverged into eastern Asia's woody flora. The director's correspondence with fellow tree biologists around the United States was prodigious and unrelenting. He constantly encouraged collectors to send specimens of American plants, and he worked continually toward perfecting the *Manual,* bringing out a second edition in 1922.

When Sargent gave up working in Cambridge and took up the census survey, he began both an herbarium and a library for the Arboretum in temporary quarters in a large vacant house on his father's estate. Sargent's acquaintance with Francis Parkman and Asa Gray instilled in him a high regard for books not only as sources of information but as objects of artistic and historical interest. Gray also demonstrated the value of pressed-plant samples as permanent references to observations made in the field and as documentation for published research. For twelve years the "Dwight house" on Ignatius Sargent's estate served as administrative offices and museum for the Arboretum. John Robinson, the first botanist to take charge of the dried-plant collection, had studied with Gray and had experience conducting the herbarium of the Peabody Museum of Salem. He organized the specimens that Sargent and his correspondents collected to document the census work. Robinson established procedures for mounting and had cabinets of cherry wood built to house the herbarium. He also started a program of collecting specimens to voucher woody plants at the Arboretum and other locations in eastern Massachusetts. After Robinson left to rejoin the Peabody Museum, Faxon took charge of these collections, which grew considerably with the research on American trees.

A permanent building to house the library, herbarium, and administrative offices of the Arboretum was constructed in 1891 and 1892, largely with funds provided by H. H. Hunnewell. Its site was chosen early during planning by Sargent and Olmsted to be near the principal gateway without encroaching on the view that greets visitors to the grounds. Designed by the architectural firm of Longfellow, Alden and Harlow, the building's proportions and subtly detailed brick façade were influenced by the style of H. H. Richardson, with whom Olmsted and the architect Alexander W. Longfellow had worked. Its materials and construction were intended to be as fireproof as possible, since it was to house valuable scientific collections.

Late in 1892, Faxon and Sargent oversaw the move of herbarium and offices from the Sargent estate to "the museum," as the building was called for several years. Now that there was adequate housing on the grounds, Sargent donated a library of more than six thousand volumes on dendrology, silviculture, forestry, systematic botany, and geography that he had assembled over some twenty years. Although less obvious to visitors, books and pressed specimens were considered as essential to the institution's mission as were living plants on the grounds. The library, especially, continued to be the object of Sargent's personal generosity, since the endowment income rarely covered all the Arboretum's needs. In its development also originated the Arboretum's specialty in bibliographic research.

ALFRED REHDER PERFECTS BIBLIOGRAPHY AND NOMENCLATURE

His correspondence with leading botanists and his efforts to develop the Arboretum collections as valuable research tools brought Sargent to the realization that a simplified means to access the literature on woody plants was needed. Thus, he determined to compile a guide to the literature of the ligneous plants of the world. His experience with the *Silva* gave him perspective on what it would take to bring such a long-term work to completion and confidence to tackle another encyclopedic project. Fortunately, the Arboretum received a grant that would fund the work.

Just before the century's end, Abby A. Bradley desired to make a substantial gift to the Arnold Arboretum to serve as a memorial to her father, William Lambert Bradley, a manufacturer of agricultural fertilizers. He had been interested in all aspects of farming, and the cultivation of trees on his Hingham estate had been a particular source of pleasure. In order to strengthen the institutional base for research, the director accepted Bradley's generosity.

Sargent hired a trained bibliographer to start the work in 1899 and cautiously predicted the project would take ten years to be readied for printing. Obviously he had in mind impressive tomes that would aptly memorialize Mr. Bradley, for whom the work would be named. For reasons unknown the trained bibliographer did not remain with the Arboretum for more than a year. As luck would have it, the next man to assume the task completed it

63. *The museum building, which came to be referred to as the administration building after the turn of the century, served as headquarters for Arboretum research activities. Sometime before the mid-teens most of the thick belt of shrubs placed along Meadow Road in the 1890s was removed, revealing the structure's Boston-ivy-covered walls to visitors entering the Arborway gate.*

admirably, nearly within the ten years Sargent had predicted, and continued a career of prolific scientific publication under Arboretum auspices that spanned nearly fifty years. In fact Alfred Rehder's work on the taxonomy and nomenclature of woody plants would do much to guide the institution's scientific programs.

It was not quite chance that put Rehder at the Arboretum at just the time when Sargent needed someone with his training and inclinations to compile the bibliography, but neither had Sargent called him there for that purpose. Rehder came to Boston in 1898 in early spring, intending to stay in the United States for six months, on assignments from two employers. For *Möller's Deutsche Gärtner-Zeitung,* an illustrated weekly, he was to report on American horticulture and woody plants. The thirty-four-year-old Rehder also had a modest allowance from the German government to investigate pomological practices and viniculture. These journalistic intents belie his broad experience and grasp of the scientific aspects of botany, however. When Rehder came to the Arboretum he already had some 140 publications (many on woody plants) to his credit, and he had held positions in some of his country's prestigious botanical gardens.

Rehder had grown up in the surroundings of Germany's finest naturalistically styled estate gardens, both his father and grandfather having been managers of such private collections. He was born in Waldenburg, Saxony, where his father landscaped and managed the estate of Prince Otto Friedrich. The prince's well-planted park on the river Mulde, and his large greenhouse filled with rare exotic plants, were known throughout the region.

Rehder attended school in Waldenburg and apparently showed enough promise to be sent to live with an uncle for two years of studying Latin, French, and ancient history in preparation to enter a gymnasium, or high school. Although he graduated from the gymnasium in 1881 fully qualified to attend a university, his father's profession did not yield an income sufficient to pay for such an education. Instead, Alfred Rehder was apprenticed to his father for three years and then held a series of gardening positions.

During a twenty-month stint at the botanic garden in Berlin, he took the opportunity to attend lectures, including those of Augustus Wilhelm Eichler and Paul Ascherson, leading plant taxonomists of the day. After spending a year in Darmstadt, from 1884 to 1895 Rehder was head gardener at the botanic garden in Göttingen, site of the most important university in Hanover. There he masterminded a reorganization of the gardens, carried on research and publication, and began to correspond with his country's many top-rank botanists. When it became apparent that Rehder's initiative and scientific ambitions were at odds with the expectations of Göttingen's director, he left the position to write for *Möller's Deutsche Gärtner-Zeitung,* moving to Erfurt. For three years he produced numerous articles, including a regular column on woody ornamentals. Since he had mastered English, the magazine thought the trip to America a good investment.

With his well-developed interest in woody plants and botanical gardens, it is not surprising that Rehder chose Boston and the new Arnold Arboretum for his initial stop. One of the first articles he sent back to Germany was "Das Arnold Arboretum," giving high praise not only to the beauty of its landscape and the richness of its plant collection but to its ample library and herbarium resources. The exact circumstances that led Rehder to prolong his stay are not known, but apparently the expense of living and traveling in America was greater than he had planned. On application to Sargent he obtained grounds work in the nursery and shrub collection for a dollar a day, the same rate offered John George Jack ten years earlier. (This was Sargent's modus operandi for testing the mettle of prospective assistants.)

Crucially, Rehder also made contact with Liberty Hyde Bailey, professor of horticulture at Cornell University, who was enlisting collaborators for his monumental *Cyclopedia of American Horticulture.* At Bailey's request Rehder began composing concise treatments of the woody plants cultivated in the United States. Over the next two or three years, Rehder contributed articles on two hundred genera, from *Abelia* to *Zizyphus.* At about the time the *Cyclopedia* went to press in early 1900, Sargent asked Rehder to take up the *Bradley Bibliography* where it had recently been left off. For a few years Rehder continued his association with *Möller's Deutsche Gärtner-Zeitung,* sending numerous articles on such topics as North American plants, greenhouse practices, the cut-flower industry, fruit growing, rural cemeteries, street trees, and parkway plantings.

During the initial work on the bibliography, Rehder returned to Germany on two separate trips. While there for the second, extended stay from 1904 to 1906, primarily to examine libraries throughout Europe for the bibliog-

raphy, he took time out to marry Anneliese Schrefeld, the daughter of the park director at Muskau, for whom Rehder had worked briefly.

In Europe he made steady progress on the *Bradley Bibliography,* sending back typewritten cards and manuscript as he went along. Rehder's bibliographic investigations also pointed up shortcomings of the Arboretum's library resources. Determined to maintain the precious book collection at the same high standards he had for the living plant collections, Sargent assigned Rehder to obtain books; he collected specimens for the herbarium and plants for the grounds as well. When Rehder came back, bringing his bride to live in the United States, he continued to search out and organize his citations to the woody-plant literature for several years. Since he was handling tens of thousands of entries, the task was a formidable one.

Rehder saw the first volume of the *Bradley Bibliography* in print in July 1911. Aside from all the effort that had gone into gathering the data, reading proof must have been an astonishing feat. Six years later the final volume came out. The *Bibliography* was organized by subjects and ultimately comprised five volumes. It covered four centuries of literature on woody plants in most languages. The first volume included all books and articles of a general botanical nature. The second was a family-by-family, genus-by-genus listing of taxonomic works on woody plants. Volume 3 covered arboriculture and economic uses, while volume 4 dealt with forestry and silviculture. The final volume, published in January 1918, contained the index to authors and titles. The five volumes comprised 3,895 double-column pages, with more than one hundred thousand entries. Sargent was satisfied with Rehder's execution of the project: "The preparation of the *Bradley Bibliography* has been entrusted to Mr. Alfred Rehder of the Arboretum, who has devoted himself to it . . . with intelligence, industry and enthusiasm. He has explored the principal libraries in the United States for works on these subjects, and has examined all the great botanical libraries in every country of Europe" (C. S. Sargent, 1911b, p. iv).

The result was an invaluable reference to "the knowledge of trees." Using the *Bradley Bibliography* could be likened to being able to examine the shelves of all the Western world's libraries where trees are the topic. Get-

64. Spruces and firs in the conifer collection, seen from under the pines. The Arboretum's living catalog of plants was a reference constantly consulted by Sargent, Faxon, Rehder, and other botanists who joined them.

ting the information from cards to the final printed format was a drawn-out process. Rehder spent some sixteen years on it, but not without interruption for many other Arboretum projects.

Besides his knowledge of woody plants, painstaking attention to bibliographic details, and facility with languages, including Latin, Rehder had a genius for the intricacies of plant nomenclature with which Sargent was particularly pleased. Ever since the first attempt at devising a universal code for the naming of plants in 1867, presided over by Alphonse de Candolle, European and American botanists had not been consistent in applying these rules. In the 1890s more conferences were held in an attempt to bring uniformity, but met with no real success. In general, most American and British botanists disagreed with their professional colleagues in the rest of the Western world on several points. As a relative newcomer to plant systematics and eager to get through his *Silva*, Sargent had little interest in debating the fine points of the nomenclatural process. He was more concerned that some uniform system be accepted by all workers than in arguing for one particular set of rules over another.

Luckily, Rehder was well acquainted with the debaters, the issues, and their ramifications for published botany, and he played an active role toward developing unanimity. As a result he guided Arboretum policies in this regard. Rehder attended the 1905 International Botanical Congress in Vienna, at which the "Paris Code" of 1867 was debated, revised, and voted on by the international community of botanists. Its general acceptance notwithstanding, many American botanists, who had a few fundamental and several minor differences with the "Vienna Code," adopted their own set of rules, which came to be known as the "Philadelphia Code" or the "American Code" by about 1910. After 1905 the British botanists generally adopted the Vienna Code, as did the Arnold Arboretum, nearly alone among American institutions. Eventually, over several decades of international congresses the "opposing" factions became more unified, adapting the best from the American and Vienna Codes. Throughout his career, Alfred Rehder worked toward the goal of a universal code for naming plants, contributing

greatly to the prestige of Harvard's Arboretum by his effort.

It was also Rehder's influence that led to the adoption of the Engler and Prantl sequence of plant families for Arboretum publications and the organization of its herbarium. Their work, *Die Natürlichen Pflanzenfamilien* (1887–99), was the first widely accepted classification of the world's genera of plants that attempted to use evolution as its basis. Engler's system was influenced by that of Augustus Wilhelm Eichler, whose lectures Rehder had attended.

USING THE COLLECTIONS, COMMUNICATING THE RESULTS

The living catalog of plants, as well as the Arboretum's library and herbarium, became the basis for taxonomic study by its staff and collaborators starting at the turn of the century. Research in taxonomic botany can take one of two basic approaches. It may be floristic, aimed at delineating all the species of a geographic region, or it may be monographic, attempting to treat extensively one taxon (any taxonomic element regardless of its classification level, e.g., a species, genus, or family). A monograph usually includes analysis and synthesis of existing knowledge and results of original research on the taxonomy of a genus or, more rarely, a family. A monograph is generally not limited to representatives of a geographic area and accounts for all elements in the group under study: all species of a genus or all genera of a family. Monographic work that is more limited in scope is termed a "taxonomic revision." Monographs and revisions of many woody-plant groups were produced by botanists at the Arboretum.

As Sargent wrapped up his work on the *Silva*, he welcomed George Russell Shaw's keen interest in pines, especially the species of western America and Mexico that puzzled many students of the genus. Shaw (1848–1937), a native of Parkman, Maine, graduated from Harvard in 1869, studied architecture, and went into business with one of the Hunnewell family of Wellesley. Just as he passed fifty years of age, Shaw started a thorough study of the pines, using the Arboretum as his headquarters. So important did his research become that when Shaw wrote up his biographical notes for his fiftieth college reunion, he said his principal occupation had been the study of pines.

Shaw journeyed to Cuba, Mexico, and Europe to see and collect living pines as well as consult libraries and herbaria. In Mexico he traveled with Cyrus Gurnsey Pringle, who had collected in the American West for the census. Shaw's first major publication, *The Pines of Mexico,* came out in March 1909. In this revision, he recognized eighteen species of pines native to Mexico out of a large number of names that had been confusing botanists since the mid-nineteenth century. To round out his systematic arrangement, he described each species and variety, and included a key, notes on distribution, and discussion of previous botanical treatments. It was a precise and invaluable resource for every conifer enthusiast, but what gave it appeal beyond its technical information were the illustrations, rendered by Shaw himself. They depicted all the characters of needles and cones frequently used to distinguish species. More endearing are occasional habit sketches showing an entire tree in its typical surroundings. Shaw's *Pinus* plates have a spontaneous quality, revealing a commitment to teach based on personal experience. George R. Shaw published a second book, *The Genus Pinus,* in 1914. Similar to the work on Mexican pines, this monograph treated the genus on a worldwide basis.

While Alfred Rehder undertook many projects during these years, he continued a monographic study of the genus *Lonicera* started before leaving Germany. The group includes many shrubs and vines of the Northern Hemisphere commonly known as honeysuckles. Rehder also brought his taxonomic expertise to bear on recent Arboretum acquisitions, especially trees and shrubs from overseas, including the results of Sargent's Japan seed of 1892, many of which reached flowering age at the turn of the century. Some of these plants were being grown for the first time in the United States. In the Arboretum collections and nurseries Rehder even found species and hybrids never before described by botanists.

65. *George R. Shaw (1848–1937) started studying pines at the Arboretum at the turn of the century. In addition to publishing works on these trees, he assembled an exhaustive exhibit that was displayed in what became known as the conifer room of the administration building. Shaw once wrote to Rehder about a plant identification inquiry from a fellow botanist: "It is a relief to me to learn that it is not another pine for I have more than I can handle already" (AAA, 18 July 1933).*

Sargent also concentrated on a pet genus, *Crataegus,* a member of the rose family that includes trees and a few shrubby species, collectively termed hawthorns (or thorns, as Sargent often alluded to them). Sargent's involvement with *Crataegus* was an outgrowth of *Silva* research during which he and his many nationwide correspondents kept finding what they thought were new kinds, never described before. Between the time of publication of the last

66. *One of Alfred Rehder's genera of specialty was* Lonicera, *a group of shrubs and vines that occur throughout the Northern Hemisphere and are known popularly as honeysuckles. Many of them, such as* Lonicera chrysantha *variety* longipes *from western China pictured here, are admired for their translucent, brilliantly colored berries. Rehder became a leader in the movement toward an internationally accepted code of botanical nomenclature.*

supplement to the *Silva* in 1902 and the issue of Sargent's *Manual* in 1905, the number of hawthorn species named increased considerably.

Simultaneously, an exhaustive collection of hawthorns, both living and pressed specimens, was assembled at the Arboretum. Starting in about 1899, Sargent gathered seed of *Crataegus* on his own field excursions and through his many botanical correspondents. Jackson Dawson handled some two thousand lots of hawthorn seed over a five-year period. Many of the resultant plants were large enough to be planted out on the grounds by 1905. The hundreds of species could not be accommodated in the main rose-

family area that surrounded the shrub collection, however. Peters Hill, where supplementary groups of crabapples and conifers had already been placed, was designated the site for a new North American thorn collection.

This planting as it subsequently developed was a prime example of "a collection for investigation which need not necessarily be permanent," as outlined by Sargent in his policy statements of 1885. The plantation of *Crataegus,* laid out in squares, was of an experimental nature. The plants were on trial for evaluation of hardiness, behavior under cultivation, suitability for park and home planting, and confirmation of botanical distinctness. The bare mound of

7

67. *Detail from a plate illustrating* Pinus gregii *from George R. Shaw's* Pines of Mexico. *His habit sketches of some of the thirty-five species and varieties found there capture the essence of the Mexican countryside as well as of the trees. Mexico is one of the world centers of diversity for the genus* Pinus.

Peters Hill was the best the director could do with the acreage available for a "trial ground." Here the plants interfered little with the original arrangement of the collections.

Ironically, there could hardly have been a worse choice than *Crataegus* for an experimental plantation. The comparison of progeny with their wild parents only confounded confusion in this genus. The question of botanical distinctness was the one that most troubled Sargent. Collecting, identifying, and describing species and varieties of *Crataegus* became a scientific obsession with the director, as his biographer, S. B. Sutton, recounts in a chapter appropriately titled "Crataegus: A Thorny Problem." From a modest seventeen or so species that such nineteenth-century botanists as Pursh, Michaux, Nuttall, Torrey, and Gray (and even Sargent in the fourth volume of *Silva*) had observed, the number of hawthorn kinds had risen to more than three hundred by 1905. Sargent was not the only one naming new species; his colleagues W. W. Ashe of the United States Forest Service and Chauncy Beadle of the Biltmore Estate Arboretum in North Carolina both described many new ones. In the end, Sargent elucidated what he believed were seven hundred distinct species of thorn and grew them all on the side of Peters Hill.

In a defense of this species naming, Sargent stated that it resulted from the observation, made under cultivation, that plants from different localities that were supposed to be the same species differed from each other in such characters as time of flowering, number of stamens, color of anthers, and shape, color, and time of ripening of fruit. The differences among plants that were thought to be one species were observed once the plants were grown at the Arboretum. Plants raised from seed collected in the wild were identical to the parent plant from which they were obtained. Seedlings raised from Arboretum plants, in turn, were identical to their parents. Each slightly differing type apparently "bred true," confirming in Sargent's mind that they were distinct species.

What Sargent, and most of his colleagues at the time, did not know is that *Crataegus* commonly displays a reproductive quirk known as apomixis. During flowering and embryo development, the plants produce viable seed

*68. Thicket hawthorn (*Crataegus punctata*), planted near the summit of Bussey Hill, shows the horizontal branching habit typical of many hawthorns. Sargent conducted an exhaustive scrutiny of Crataegus, ultimately naming and describing hundreds of new kinds. Although the majority were planted on Peters Hill, Sargent placed a few individuals that he deemed the best in other sites on the Arboretum grounds.*

without the process of cross-fertilization, thus rendering the seed genetically identical to the tree producing it. In essence, when this occurs the seedling is a replicate of the parent, as though the plant reproduced vegetatively rather than "sexually," the more common process during which there is always a mixing or recombination of genes. Unknown to them, each wild population of hawthorn that Sargent and his collectors encountered was a colony of identical individuals, a clone; and it was the distinctness of each clone that Sargent came to recognize as species.

The director's fast-paced proliferation of *Crataegus* was carried on until 1913, but even then he pursued the topic doggedly for years as he continued to revise his *Manual*. It was as if commitment to completeness entirely outweighed all other factors concerning *Crataegus*. Many times Sargent despaired of the problem, remarking once to John Dunbar of the Rochester Parks Department that it would take a century to complete a study of the genus. To Muir Sargent wrote: "I cannot shake off this confounded *Crataegus* business which I fear is going to shorten my life, or at any rate deprive me of a good deal of pleasure in other directions" (AAA, C. S. Sargent correspondence, 18 August 1907).

In order to quickly disseminate observations that did not merit lengthy treatises, Arboretum scholars needed a scientific journal. With Charles E. Faxon on hand to render excellent drawings, the Arboretum ventured a new serial, *Trees and Shrubs,* in which botanists could broadcast their findings on woody plants. Between 1902 and 1913 eight parts of *Trees and Shrubs* were issued, with twenty-five plants depicted in each. Taxonomic and nomenclatural notes, diagnoses, and descriptions were provided by Sargent, Rehder, Shaw, and their fellow botanists at the Gray Herbarium. A few plants represented by herbarium collections sent back from E. H. Wilson's first trip to China for the Veitch nursery firm occasioned note in *Trees and Shrubs*.

Although the journal was discontinued, there was still the need for a periodical publication through which scientific information could promptly reach the public. Once such compendiums as the *Bradley Bibliography,* a two-volume catalog of the library, and a three-volume account of the results of exploration in China were out of the way,

the Arboretum scientific staff turned to a project it had long had in mind. The first number of the *Journal of the Arnold Arboretum* appeared in July 1919. Initially, its pages included notes on trees and shrubs, with descriptions of new species and their relationships, letters from correspondents, and notes on vegetation of countries visited by officers and agents of the Arboretum, but it soon extended to many botanical topics.

ASIATIC SILVA

Without a doubt the most significant program in the second half of Sargent's administration was exploration and interpretation of the ligneous flora of temperate eastern Asia. Taking a cue from Asa Gray's groundbreaking hypothesis of the floristic relationship between eastern North America and eastern Asia, Sargent inaugurated investigation of this vegetatively and climatically similar region. Aside from the knowledge gained, the results of several expeditions to Japan, Korea, and China had practical implications for the display of hardy plants in Jamaica Plain and for the Arboretum's educational role.

Although woody plants introduced from the Orient before the establishment of the Arboretum thrived at Jamaica Plain and the tree museum accessioned seed sent by botanists in China and Japan, the director warmed only gradually to the idea of bringing in more. Before 1900 Sargent's attention was on the American silva, and on trees as subjects for reforestation as much as for ornamentation. Nonetheless, from July to December 1892, he traveled with his nephew, Philip Codman, to northern Japan to study firsthand the resemblance of its forest flora to that of eastern North America, which had so fascinated Asa Gray. This tour of another north-temperate forest system extended, by comparison, Sargent's understanding of the trees of North America.

The director wrote up his observations for a series of articles, "Notes on the Forest Flora of Japan," in *Garden and Forest* in 1893; these were reprinted in book form as *Forest Flora of Japan* (1894). Sargent introduced the book with a reiteration of Gray's theory, focusing the discussion with his own views on the woody flora. Despite an oc-

69. *The Arboretum's active study of the woody plants of eastern Asia began when Sargent traveled to Japan in 1892. There he and his nephew Philip Codman collected specimens and seed on Hokkaido and northern Honshu. One of the plants brought into cultivation in North America for the first time as a result of the trip was anise magnolia* (Magnolia salicifolia), *an early-blooming variety that rivals the beauty of the more commonly cultivated star magnolia* (M. stellata).

casional favorable comparison of a Japanese plant with its American relative, Sargent's preoccupation with the tree flora of North America greatly influenced this assessment of the "foreign" plants. Right at the start of *Forest Flora of Japan,* he stated that "while Japan is extremely rich in the number of its tree species, . . . the claim . . . that the forests of eastern America contain the noblest deciduous trees of all temperate regions can, so far as Japan is concerned, be substantiated, for, with few exceptions, the deciduous trees of eastern America surpass their Asiatic relatives in size and beauty" (C. S. Sargent, 1894, p. 8). This statement notwithstanding, Sargent returned from his quite extensive travels in Japan with seeds of some 200 kinds of plants for Dawson to sow at the Arboretum greenhouses and 1,225 specimens for the herbarium.

Several factors account for the Arboretum's apparent change of emphasis from strictly "American trees for America" toward Far Eastern trees and shrubs. Once it became clear that forestry would not be an important part of the Arboretum's mission, horticulture and the introduction of new ornamental plants took on a greater significance. This role for the institution appealed greatly to its supporters, many of whom were plant collectors in their own right. The lack of scientific knowledge of vast forested areas like those in China's interior presented a challenge to a garden mandated to cultivate all the hardy trees. With Alfred Rehder on the staff there was greater taxonomic expertise to evaluate the flora of regions largely unknown to Western botanists. Sargent's firsthand glimpse of Asia's arboreal wealth in Japan must have left a greater impression than his summary of the trip revealed. And it was while he was there that the seeds of the Arboretum's involvement with China were sown.

In Japan, Sargent and his nephew traveled for a time with James Herbert Veitch. He was in the Far East to obtain plants for the Veitch nursery firm of Chelsea, England, then in the charge of his uncle, Harry J. Veitch. Young Veitch was eager to explore for plants in China, and he and Sargent discussed the possibilities at length. Sargent knew that the flora of much of China's interior was imperfectly known and that relatively few seeds or plants had been successfully retrieved from its vast mountainous terrain. James Veitch wished to leave Japan for China then and there, but his uncle had other plans. Instead, James completed his assignments in the Far East and returned to assist in the business at home. Sargent kept up a correspondence with the senior Veitch on the topic of Chinese plants, however.

Eventually, Harry J. Veitch was convinced of the potential profitability of sending an agent to China. Sargent's prodding, among other things, led to his decision. So did specimens of plants arriving at Kew from Augustine

Henry, a physician with the imperial customs service stationed at Yijing (Ichang) on the upper Ch'ang Chiang (Yangtze River). On the recommendation of the Kew administration, Veitch selected Ernest Henry Wilson, a recent graduate of the Royal Botanic Gardens' diploma course, for the mission. After some six months of preparation during which Wilson worked at Veitch's Coombe Wood nursery and studied herbarium specimens at Kew, he left in April 1899 for China by way of Boston. The twenty-three-year-old Wilson spent five days at Sargent's place, Holm Lea, going daily to the Arboretum. Here he studied the collections carefully and received valuable instruction from Jackson Dawson on how to prepare and pack plant material for shipment. Sargent, too, shared advice and information on what to expect of the Chinese flora. Wilson left a good impression, and the director was confident about the prospects of Veitch's expedition.

One of the goals of the trip was to collect seed of the dove tree (*Davidia involucrata*), a rare and beautiful tree with flower clusters surrounded by two large white bracts said to resemble the wings of a dove. Wilson's two-and-a-half-year trip to Hubei (Hupeh) and Sichuan (Szechuan) was successful, so much so that Veitch sent him once more to western Sichuan from 1903 to 1905, this time to obtain *Meconopsis integrifolia,* the yellow poppywort, near the Tibetan border. The poppywort was an herbaceous perennial Veitch was intent on offering to Britain's gardening public. Sargent and Veitch kept up a close correspondence on Wilson's activities and results, the Arboretum director urging the nurseryman to see that the plants be accurately described and identified.

Meanwhile, Sargent took an excursion of his own in 1903, traveling around the world with his son, Robeson, who, after graduating from Harvard in 1900, had decided to become a landscape architect. The year before, the publication of the final volumes of the *Silva* having brought an end to this project, Charles S. Sargent received an honorary doctor of laws degree from Harvard University. He already had in mind the production of the *Manual,* but Sargent needed a break and he was eager to see something of China. The director convinced John Muir to accompany them by promising visits to the great Siberian wilderness. Since the trip was largely for the education of Robeson, considerable time was spent in Britain and on the Continent visiting the best of gardens and museums. After stops in Berlin, Saint Petersburg, and Moscow the travelers finally saw some wild nature in the Caucasus Mountains before boarding the Trans-Siberian Railroad.

For nearly a month they journeyed across Russia's "vast forests and broad steppes," leaving the train to botanize and collect specimens wherever possible. Such stops were all too infrequent for Muir, however. Once they reached Harbin in Manchuria, Sargent, who had once been accused by George Engelmann of never allowing more than five minutes for anything, led the threesome south to Peking, across the Yellow Sea to Korea, on to Japan, and thence to Shanghai. Muir, by this time tired of Sargent's pace, struck out on his own. The two Sargents continued south, calling at as many of China's coastal cities as possible, then went on to Java and Singapore before crossing the Pacific for home six months after having started.

While Charles S. Sargent mulled over what he had seen and heard of China and kept up with the news of Wilson's exploits, John George Jack acted on his desire to see the flora of Asia. Although Jack relished botanizing in the field and collecting specimens, he never published technical papers. He once stated to the New York Botanical Garden's N. L. Britton, "I have no particular desire to get involved in any of these identification or nomenclature troubles . . . because I am fond of a peaceful warless life" (AAA, Jack correspondence, February 1895). Taken with wanderlust, and perhaps intrigued by reports on the Far Eastern flora, Jack journeyed to Japan and Korea in 1905. Sargent did not send him; Jack financed the trip himself, making stops along the coast of northern China as well. Nevertheless, the Arboretum assistant brought back seeds and specimens of woody plants to the institution, several of them new to cultivation in the West. Dawson, propagation wizard that he was, worked on the 650-odd seed lots obtained by Jack in the gardens and wilds of the Orient. Although Jack did not penetrate China's politically tumultuous interior, his results proved welcome additions to the Arboretum collections.

WILSON TRAVELS TO CHINA FOR THE ARBORETUM

After an excursion to botanize in Peru and Chile in the winter of 1905–6, Sargent traveled through the Strait of Magellan and embarked for England before heading home in April. The trans-Atlantic detour was made with a singular purpose: Sargent was determined to see Wilson and sound him out about another trip to China.

Desiring to stay home after his second trip for the Veitch nursery firm, Wilson took a position at London's Imperial Institute to work on plant collections from Hong Kong. When Sargent turned up, the two men visited Veitch's Coombe Wood nursery to look over the young Chinese plants, and Sargent questioned him at length on his trips. Having received duplicate herbarium specimens from the Veitch-funded expeditions, Sargent was by now very enthusiastic about China's potential and Wilson's abilities as a collector.

While Harry J. Veitch was satisfied with what had been accomplished, Sargent wanted more. Veitch's understandably commercial motives had differed from the Arboretum's needs. Whereas herbarium specimens were less important to Veitch than his proprietary interest in the plant material obtained, Sargent's interests were scientific: herbarium specimens, seeds, and propagation material of as many previously unknown woody plants as possible to be widely exchanged with colleagues and sister institutions. From Sargent's point of view there was much left to be done in China, and Wilson had demonstrated exceptional ability to accomplish things botanical in the "Flowery Kingdom." He knew his plants, and he was resourceful, energetic, and enjoyed working with the Chinese people. Since Wilson was now a free agent, at least with regard to Veitch's interests, Sargent asked him to make a trip under Arboretum auspices. The younger man did not readily accept the proposition. He had just started a government job that promised long-term security and good chances of promotion. Moreover, he had lived with his wife for only eighteen months of their four-year marriage, and now they were expecting a child.

On returning to the Arboretum, however, Sargent continued his pursuit of Wilson by mail and began the process of sounding out sponsors for the expedition he hoped to launch; by mid-September, Wilson agreed to go. Leaving his family in England, he came to Boston in December 1906 to work out details with Sargent before signing a contract to spend two full winter and summer seasons in China. The instructions for his mission were summarized in a six-page letter from Sargent written shortly before Wilson's departure (overland to San Francisco, thence sailing to Shanghai). Herbarium specimens in sets of six, seeds enough to raise four or five hundred plants of all woody species likely to prove hardy (except those already well known in cultivation), orchid specimens in sets of a dozen for Professor Oakes Ames of Harvard, and one hundred fifty bulbs of each kind of lily encountered for certain sponsors were among the many items on Wilson's list, not the least of them being photographs of "as many trees as possible provided the tree [in the] photograph can be named." Wilson took with him a large "field camera" requiring three boxes to hold it and the attendant bellows and tripod; with it were transported cases of glass-plate negatives. Wilson's photographic results alone could be considered worth the cost of the journey—on his first trip to China for the Arboretum he exposed 720 glass plates.

While Sargent counseled Wilson extensively on the methods and goals of the expedition, he also wisely gave his explorer leeway to make judgments as circumstances warranted. The two kept up a constant correspondence, Wilson reporting on his finds and questioning plant identities, Sargent answering those queries and sending feedback on the condition of material received and, eventually, progress of the resultant progeny. After sending the Arboretum some 2,262 lots of seed, cuttings representing 1,473 plants, and 30,000 herbarium specimens during this sojourn in the Ch'ang (Yangtze) Valley in Hubei and Sichuan, Wilson returned to England in the spring of 1909 via Siberia. Stopping at the botanic garden in Saint Petersburg, he was pleased to see living offspring of seed from his 1907 collections that Sargent had already distributed to the Russian garden officials.

After a four-month stay at home, during which he oversaw the processing of his photographic negatives at Kew, E. H. Wilson traveled with his wife and daughter across the Atlantic to Boston in September 1908 to work on his collections for the Arboretum. As Wilson unpacked and organized his herbarium specimens, he and Alfred Rehder started the task of identifying and cataloging them. Their work would eventually result in a series published as *Plantae Wilsonianae*.

Wilson's revealing photographs of China's vegetation and scenery were a sensation. Pleased with these images of the journey, he and Sargent offered sets of them to the expedition's sponsors and to the institutions that were to receive corresponding herbarium specimens and plants. A selection of the photographs was processed in quantity and an accompanying index was published. The index, entitled *Vegetation of Western China,* together with five hundred beautiful photographic prints, sold for $250. Sargent wrote a concise, informative explanation of the geography and ecology of the forests Wilson investigated, giving the reasons for the area's floristic diversity:

In central and western China narrow valleys are separated by high ridges. These in Hupeh rise gradually, but in Szech'uan the slopes are steeper and finally extend up to the region of perpetual snow on summits equalling in height some of the highest peaks of the Himalayas. In each valley, in every glade, and in each 500 feet of altitude, species of plants occur which are not found in adjacent valleys or at lower or higher altitudes. The region therefore explored by Mr. Wilson is probably richer in its number of species of woody plants than any other part of the world beyond the tropics. (C. S. Sargent, 1912b, p. 3)

Funds realized from the sale of photographs aided somewhat in covering the costs of further exploration. Wilson took the camera on subsequent trips with equally valuable results, and sets of pictures were again offered to subscribers. His detailed images were used to illustrate the many articles and books he wrote speading the fame of the Arboretum's work on eastern Asia. The photographs remain a valuable and frequently consulted archive today.

Despite the successes of the Arboretum's first Chinese venture, the fall of 1908 was a poor season for seed production on conifers in Hubei and Sichuan. Mindful of the need to complete the Arboretum collection of cone bearers, Sargent persuaded Wilson to make another journey to get the spruces, firs, and hemlocks that grew in

70. *Travel permit issued to Ernest H. Wilson in Yijing (Ichang), western China, when exploring for the Arboretum in 1908. During this trip he was accompanied by Walter Zappey, who collected birds for Harvard's Museum of Comparative Zoology; the document includes permission to carry guns and cartridges.*

abundance and great variety on the Tibetan plateau. The China veteran left Boston in April 1910 and, after escorting his wife and child to their English home, boarded the Trans-Siberian Railroad. In China he headed straight to Yijing (Ichang) in western Hubei, his customary headquarters. By this time Wilson was known in the city, and several residents who had accompanied him on his three previous explorations joined Wilson once more. Some of these men were well versed in techniques of collecting, processing, and packing seed, as well as in preparation of dried botanical specimens, and Wilson welcomed their assistance once again.

After a productive summer locating trees and pressing specimens in the tributaries of the Ch'ang (Yangtze), the fall promised a good conifer harvest. Unfortunately Wilson's caravan was caught in a landslide in early September, the leader suffering a severe compound fracture of the leg. His men carried Wilson for three days to Chengdu (Chengtu), the capital of Sichuan Province, where residents of the Friends' Mission took him in and gave medical aid. Because of the unavoidable delay in getting proper medical attention the leg was slow to heal, and at one point the necessity of amputation was feared. Wilson was forced to stay in Chengdu for five months of frustrating recuperation. Now his genial relations with his men over the years paid off. In the fall and early winter, these assistants revisited previously marked trees and collected ample seeds of many of the plants Wilson and Sargent desired.

Wilson returned to the Arboretum with his collections in the spring of 1911, but he soon headed to England for more rest. Even though the trip had ended in disaster for Wilson, it resulted in some 1,285 packets of seeds and 462 herbarium collections in multiple sets. With the tasks of working up his collections and collaborating with Rehder to produce *Plantae Wilsonianae* ahead of him, Wilson brought his family to Jamaica Plain and settled into life at the Arboretum in late August. Although he was much improved, he still felt some strain from the aftermath of his ordeal.

For the next six years Rehder and Wilson scrutinized the pressed specimens, pored over Wilson's field notes, and followed the progress of the living plants Dawson

71. *Cyclamen cherry* (Prunus cyclamina) *is one of many fine plants brought into cultivation in the West for the first time by E. H. Wilson. Native to Hubei, Sichuan, and Sikang Provinces, this cherry has proved cold hardy and disease resistant over the eighty-odd years it has been grown at the Arboretum.*

raised from seed and cuttings. They sent duplicates of selected specimens to specialists on certain genera or families, enlisting their aid to correctly identify and classify the Chinese plants. This collaborative interpretation of the ligneous vegetation of central China was published in nine parts, collected into three volumes. The first part was issued 31 July 1911 and the ninth was issued 31 January 1917. The series constituted "an enumeration of the woody plants collected in Western China for the Arnold Arboretum of Harvard University during the years 1907, 1908 and 1910 by E. H. Wilson." Charles S. Sargent was listed as editor, since so many botanists had contributed their expertise to assess Wilson's finds. The book gave the results of the examination of some forty-five hundred collections made by Wilson for the Arboretum. Wilson, Rehder, and their collaborators cited another sixty-five hundred specimens they observed to aid in identifying the Wilson plants. Many of these were Wilson collections for Veitch, as well as the Augustine Henry specimens that had inspired the Wilson expeditions in the first place. The Arboretum had also received materials from other collectors in China, notably Joseph Hers, a Belgian railroad official who botanized in northern China, and William Purdom, who was sent jointly by Veitch and the Arboretum to the northern provinces in 1909.

Rehder masterminded the bulk of the work and authored about one-third of the articles in *Plantae Wilsonianae;* he and Wilson together did another half; the remaining articles were contributed by eighteen botanists at institutions around the world. Sargent contributed to the treatments of *Crataegus, Phellodendron,* and *Carya,* and George R. Shaw worked on the pines. The collaborators, or their respective institutions, retained duplicate herbarium material for their collections. In classifying and identifying the Wilson collections, Rehder and his collaborators were often covering new ground. Their problems were similar to those that Asa Gray had faced when collections from western North America poured in to him sixty years earlier. There existed no one book or catalog where they could look things up; there were no keys to Chinese plants in the Western literature, nor in the Chinese literature. For every species they found, the authors provided references to pertinent publications.

Many of the trees and shrubs were species or varieties that had never before been named or described in botanical literature. In these cases, a full description in Latin was prepared, a name designated, and additional discussion in English provided. Sometimes the scrutiny of the specimens from central China shed light on questions of identity of related plants known from other places. This led to the preparation of "mini-monographs" or elaborations of wider groups than occurred in Hubei and Sichuan alone. In all, Rehder, Wilson, and their collaborators enumerated 2,716 species and 640 varieties and forms of woody plants representing 429 genera in 100 natural families. Among these, 4 genera and 382 species were described and named for the first time. *Plantae Wilsonianae* was akin to the first step toward a "Silva" for China (or at least western central China), an undertaking that is yet to be completed.

In the early teens, Wilson was kept busy by the work on *Plantae,* as he termed it to Rehder. For more than two years he remained at the Arboretum, working on the Chinese specimens and preparing manuscript for publication. Wilson had another literary work under way, as well. In 1913 his first book for a general readership, *A Naturalist in Western China, with Vasculum, Camera, and Gun, Being an Account of Eleven Years' Travel, Exploration, and Observation in the More Remote Parts of the Flowery Kingdom,* was published in two volumes. The book thoroughly elaborated his adventures and exploits in vivid and entertaining style. If truth be known, Wilson had a much greater affinity for this kind of writing than for exacting taxonomic work; and, even more, he preferred the travel itself.

EXPLORATION OF THE JAPANESE EMPIRE

Although Wilson's injured leg precluded strenuous journeying, he and Sargent agreed on another field trip, to start in January 1914. This time Wilson traveled to Japan, where comfortable accommodations and transportation were more readily obtained. In addition, the country's botanists and horticulturists had their flora well under control. His wife and daughter accompanied him, adding home comforts and support to life abroad.

Once he established his family in Tokyo, Wilson visited botanic gardens and nurseries and took short trips to many wild locales throughout the island nation. During the spring he made an intense study of flowering cherries, especially the numerous named varieties planted in gardens. Wilson documented these with herbarium collections and photographs, and compared what he saw in cultivation with the species that could be found in the wild. Similar observations and gatherings were made for other woody plants, with conifers coming in for special scrutiny.

While Wilson kept up an informative correspondence with Sargent as usual, he and Rehder also continued a written dialogue on identifications and taxonomic problems of Japanese species not unlike what their *Plantae* work must have been when they were together in Jamaica Plain. Rehder even sent the just-issued fourth part of *Plantae Wilsonianae* to Wilson in Kamakura, which prompted his colleague to write: "I have looked over the pages with feelings of much pleasure and think that all concerned in the production of this work may feel pardonable pride" (AAA, Wilson correspondence, 3 May 1914). When Rehder apparently voiced frustration at the enormity of their tasks, Wilson attempted to encourage him with a rare, but pithy, comment on Sargent's administrative style: "I can understand the interruptions you suffer from but you must console yourself with the thought that the Professor is the mainspring of the particular machine we, for the time being at any rate, form a part" (AAA, Wilson correspondence, 14 April 1914).

Before Wilson left for Japan, the hope was that some of its outlying islands and Korea (then a Japanese territory) could be included in his itinerary, if his health and other factors allowed. Korea, especially, was of interest since so many of the plants Jack brought back had proved hardy at the Arboretum. Sargent directed Wilson to be on the lookout for opportunities to join with a Japanese institution for a Korean expedition. When war broke out in Europe, Sargent wrote to Wilson to cut his trip short. Although trouble was not expected in eastern Asia, Sargent feared that war—even if only Europe remained involved—would mean financial difficulties for the Arbore-

tum. Wilson completed seed collecting planned for the autumn season and sailed across the Pacific, reaching the Arboretum in January 1915.

On this comparatively relaxed but productive trip, Wilson exposed some six hundred photographic plates, collected two thousand herbarium numbers in multiple sets, and harvested abundant seed to be grown at the Arboretum and distributed. He brought back living plants or cuttings of numerous varieties of cultivated cherries in addition to the usual complement of propagating material. Once back at the Arboretum, Wilson set his Japanese collections in order, preparing herbarium material for identification and distribution and checking these against the seed-lot numbers in the propagation department. In 1916 he also published monographs on two groups that had been given special study during the trip: *The Cherries of Japan* and *The Conifers and Taxads of Japan*.

Before the final part of *Plantae Wilsonianae* was issued, Wilson was on his way to Asia once again. His assignment this time was to explore the extremes of the Japanese Empire: Korea, the Bonin and Ryukyu Islands, and Formosa (Taiwan). A patriot, yet unable to fight for England because of his lame leg, the news of World War I distracted Wilson. Sargent thought he would be better off on a botanical excursion than at home, where Americans were still debating whether to take a stand in the conflict. Despite his infirmity, however, Wilson set a whirlwind pace. He spent the first few months hopping between the Ryukyus and the Bonins, subtropical isles that would yield no hardy plants for the Arboretum but whose floras were little known and contained peculiar trees that grew nowhere else.

In mid-May Wilson traveled north to Seoul and spent the remainder of 1917 exploring the "virgin forests of northern Korea," penetrating regions Sargent believed had never been visited by Westerners. Late winter and early spring of 1918 found Wilson a thousand miles south on Formosa, where the Japanese Forest Service aided his attempts to observe and photograph every species of conifer in the mountain forests. He returned to Korea by way of Japan and spent the summer seeking specimens and seeds in both countries. Wilson wandered through For-

72. *Ernest H. Wilson (1876–1930) at an inn in the Tumen-Yalu divide in northern Kankyo Province, Korea, August 1917. Although he much preferred botanizing in the field, Wilson spent many an hour working with Rehder to place his discoveries in scholarly context for publication.*

mosa once again from October to December, making the climb to the thirteen-thousand-foot summit of Mount Morrison (Yü Shan). After twenty-six months away, Wilson arrived in Boston with the usual wealth of dried specimens, propagation material, and photographs. Among the more celebrated "finds" of this trip were the Kurume azaleas, Japanese cultivated forms of *Rhododendron obtusum* variety *japonicum;* although not hardy in New England, these became popular greenhouse and florist subjects. But for outdoor cultivation the Arboretum obtained many fine Korean plants, such as Korean mountain ash (*Sorbus alnifolia*) and Korean boxwood (*Buxus microphylla* var. *koreana*), one of the hardiest forms of boxwood for northern gardens. Wilson and Rehder published their analysis of the vegetation of these regions and enumeration of species encountered in the *Journal of the Arnold Arboretum.*

Wilson's 1917 to 1919 expedition in the Japanese Empire was his last to temperate regions of Asia for the Arboretum. When the first federal plant-import legislation, passed in 1912, gained stronger teeth as the "Quarantine 37" law in 1917, Sargent took it as a serious blow to the Arboretum's exploration program. While the scientific results had been an important component of the program, horticultural interest in plant introduction was its main source of funding. Without the promise of a share in new plants for gardens, the director feared support for exploration would not be forthcoming. Eventually American botanical gardens learned to work with the regulations, which were aimed at curbing the introduction of pests, but for a time the Arboretum's activity in the Far East was suspended.

SARGENT EXPANDS SCOPE OF RESEARCH

Once the commitment to library and herbarium collections to round out the work of the Arboretum was made, housing them became an intermittent concern. Before all the material from Wilson's second trip for the Arboretum had arrived, an addition to the museum building was needed. In 1909 both herbarium and library had grown

73. Beauty bush (Kolkwitzia amabilis) *was one of the most immediately successful of Wilson's introductions from China. This large shrub produces abundant, delicate pink flowers each spring and sports interesting light tan exfoliating bark. The specimen pictured here grows on Bussey Hill Road near the birch collection.*

to the point that they required room for expansion. With two generous donations, a four-story brick wing, fifty by thirty-six feet, was planned, using the layout of the Gray Herbarium and that of Kew as models. On three of the floors, steel cases designed to protect herbarium specimens from dust, humidity, insects, and fire were built around a central clerestory, which allowed light to penetrate and materials to be lifted up and down by a pulley. This is believed to be the first use of steel for herbarium cabinets in the United States. Earlier cabinets were made of wood, which was sometimes permeable by destructive beetles. Without even using the fourth floor, which was reserved for later expansion, Sargent thought the new arrangement would hold eight times as many specimens as were then in the Arboretum herbarium.

The new wing also had space for twenty-eight worktables. Here, Wilson and Rehder spread out the Chinese specimens and made comparative observations with those acquired on previous expeditions and from other collections. Here, too, Shaw worked on his pines and Sargent on *Crataegus* and the revision of the *Manual*.

With the removal of the herbarium from the original wing, there was space on the upper floor for the growing library. In 1904 Sargent had hired an assistant under Charles E. Faxon for the library, Ethelyn M. Tucker. With more assistance and the impetus of Rehder's bibliographic work, the Arboretum sought acquisitions with renewed vigor. As soon as she joined the Arboretum staff, Tucker was set to cataloging the library. The building addition allowed Faxon and Tucker to rearrange the books completely, and the catalog progressed similarly. The published version came out in two volumes in 1914 and 1918.

A turning point in the management of the research collections came with Faxon's death in February 1918. Sargent greatly felt the loss of his longtime associate.

Faxon, assistant director when he died, had been with the Arboretum since 1882, quietly managing both herbarium and library, skillfully delineating portraits of plants for its published research. Although, like J. G. Jack, he never wrote scientific papers, Sargent, Rehder, Wilson, and Shaw relied on Faxon's advice and assistance when writing theirs. His work as a botanical illustrator of the first rank brought him and the institution lifelong distinction. During his connection with the Arboretum, nearly two thousand of his drawings were published.

With Faxon gone, Alfred Rehder stepped into the position of curator of the herbarium and Ethelyn Tucker became librarian. Now that the *Bradley Bibliography* and *Plantae Wilsonianae* were finished, Rehder was free to guide the herbarium and to inaugurate production of the *Journal*. Tucker, too, had recently completed her large published work, the *Catalogue of the Library*. The reorganization of the books themselves into subjects, which she and Faxon worked out based on the catalog, had recently necessitated expansion of library shelving into the fourth

74. In the late teens and early twenties, Sargent increasingly stressed the importance of extending the Arboretum's role in dendrological research to the whole world. He predicted that botanical inventory and evaluation of tropical forest regions would become as important as the study of north-temperate trees. He also knew that expansion of facilities for the library and herbarium would be needed.

floor of the herbarium wing. When Tucker took over its management, the library contained some 31,525 bound volumes and 8,629 pamphlets.

As most American educational institutions reconsidered their role in the world's scholarly affairs at the close of World War I, Sargent contemplated new directions for the Arboretum. The director wished his institution to advance and expand its research program. With the aim to have the Arboretum maintain first rank as a center for gathering and publishing information about trees, he developed a four-part scheme. Although they were not publicly delineated as such until 1922 (in a report in the *Journal of the Arnold Arboretum,* "The First Fifty Years of the Arnold Arboretum"), he worked on the ideas for years, often in conjunction with efforts to raise funds.

First and foremost was a continuation of the exploratory, systematic, nomenclatural, floristic, and bibliographic work that he, Rehder, Wilson, and other colleagues had carried on. With it went the development of herbarium, library, and photograph collections. He envisioned expansion in this field to woody plants of the world, not just those of the north temperate zone. He knew dendrology would become as important in tropical regions as it was in northern forests: "The exploration of the tropical forests of the world will require perhaps a century and a large expenditure of money to accomplish. It is work that this Arboretum should begin and steadily push forward" (C. S. Sargent, 1922, p. 170). Its administrator was fond of saying that the Arboretum should develop the facilities and expertise to answer any question asked about any tree growing in any part of the world!

The second and third elements of the plan involved two fields suitable to be pursued at the Arboretum, studies of diseases of trees and of insects injurious to trees. If funding were forthcoming, these two departments could be properly equipped and staffed. The Arboretum's collection would provide ample documented material with which to work, and its location near the Bussey Institution would allow access to experts in related fields in the Harvard community. Undoubtedly the phenomena that had inspired the tightening of plant-importation regulations in the teens—chestnut blight, Dutch elm disease,

gypsy moth, and Japanese beetle—prompted the idea that the Arboretum join the battle against such ailments and pests.

The fourth component of Sargent's hoped-for research program was experimental hybridization of woody plants. Genetics was the "hot topic" in biological science throughout the teens and twenties in America. The neighboring Bussey Institution had an extensive research and teaching program in theoretical genetics. While Sargent thought of this proposed department in terms of the practical results of breeding superior trees for ornament, timber, or fruit and nut production, he acknowledged that observations made in this field would add to the understanding of basic biological processes. An advantage of the Arboretum for this work was its permanent location, fixed as it was by the contract between the city and the college for a thousand years. This longevity in site and in stewardship meant the breeding of some kinds of trees, for which one might wait fifty years for results, could well be undertaken at the Arboretum. In many discussions, letters, and reports he argued the need for funds to develop these research programs, and he several times attempted to interest the new charitable foundations (like Carnegie and Rockefeller) that were coming into existence, or private donors, in endowing one or another of them. While the first of these programs continued admirably and the fourth influenced Sargent's planting policy for Peters Hill, their major implementation did not take place until after his death in 1927.

The early 1920s were years of unprecedented growth for the Arboretum herbarium through institutional exchange. Once living-plant introduction was restricted by quarantine regulations, Sargent's annual reports gave increasing emphasis to this department. In 1920 the herbarium collection was completely rearranged, but additions numbering between ten thousand and twenty thousand sheets per year meant that managing the collection was more than Rehder could do while still pursuing his taxonomic studies and editing the *Journal.* In 1921, Sargent offered Ernest J. Palmer a position on the Arboretum staff to assist in the herbarium. A native of Missouri, Palmer had been subsidized jointly by the Arboretum and the

Missouri Botanical Garden to collect specimens since 1913. For the Arboretum, he specialized in hawthorns, oaks, and other woody plants for Sargent's *Manual,* but Palmer was widely versed in the floras of the central and southern states from Illinois to Missouri, Arkansas, and Texas. He began sending botanical papers for publication in the *Journal of the Arnold Arboretum* soon after its establishment. After moving to Jamaica Plain, Palmer aided in the curation of the herbarium and continued his field-work, usually taking one excursion per year to his region of specialty to obtain specimens for the Arboretum. Palmer remained for twenty-seven years until his retirement in 1948.

To reinforce the Arboretum's strength in systematic botany, Sargent actively extended the herbarium toward worldwide coverage. Wilson journeyed for the Arboretum once more from July 1920 to August 1922, concentrating on gathering pressed specimens. The lands he visited—Australia and Tasmania, New Zealand, the Malay Archipelago, India, and southern and eastern Africa—would not provide living plants for the Arboretum grounds, but they did yield a harvest of scientific data on the ligneous vegetation of the Southern Hemisphere. Sargent tried to enlist the support of California institutions and donors for the expedition, since plants from the south of the world would thrive with them. The assistant director's mission on this trip included building institutional relationships with botanical gardens, scholars, and forest establishments as the basis for future exchange of herbarium material. Wilson returned to Boston with 5,000 herbarium collections, 1,830 photographs, and 500 publications, having paved the way for exchange of literature and scientific specimens and for future exploration programs.

Although Sargent fretted over the impairment of the Arboretum's plant-exploration activities by federal import regulations, he rarely passed up a good opportunity to obtain material from the Far East. In 1924, the Arboretum made arrangements for Joseph F. Rock to explore the northwestern province of China, Gansu (Kansu), and nearby Tibet. The area was quite unknown botanically, consisting of forest-covered mountains and high grasslands on the outskirts of the Himalayan Range. Rock was a veteran of expeditions to Southeast Asia and China for the United States Department of Agriculture and the National Geographic Society. The first twenty years of his career had been spent teaching botany and Chinese at the College of Hawaii, where the Viennese-born Rock had intensively surveyed the islands' flora.

As proficient in several languages as he was in botany, Rock spoke some of the Chinese dialects fluently. This skill may have saved him more than once, for he endured frequent encounters with bandits in areas remote from governmental authority where he journeyed for the Arboretum. Harvard's Museum of Comparative Zoology jointly sponsored the Kansu-Tibet expedition in order to procure bird specimens for its collection. The trip was planned for at least three years, and each year Sargent reported that Rock's explorations were the "most valuable work accomplished by the Arboretum." By June 1926 Rock had already sent back 402 seed lots, cuttings from 42 plants, and 200 photographs. Evaluation of the complete results of his journey would be left to the administration succeeding Sargent's, however.

THE SCIENCE OF CULTIVATED PLANTS

Just as the Arboretum's exploration of eastern Asia had the twofold purpose of increasing scientific knowledge and bringing home plants potentially worthy for gardens, its publications bridged the gap between botany and horticulture. The institution's mandate to contain a comprehensive collection of hardy woody plants required constant interplay between the science of plant classification and the science of cultivation. Knowledge of identification and classification was essential to the decision of what should be grown, and the experience of growing such a wide variety of material yielded information valuable to cultivators. Rehder, undoubtedly influenced by Liberty Hyde Bailey, who pioneered this field, was especially instrumental in bringing intense scrutiny and exhaustive bibliographic research to the enumeration of cultivated plants. Wilson exercised his eye for good garden subjects while he scoured the Asian forests for anything new to science. The monographs they coauthored covered the botany of their subjects, but they were not without information of interest to growers.

Rehder and Wilson collaborated once more in the late teens to produce *A Monograph of Azaleas* (1921). For this volume, the two divided the field geographically, Rehder enumerating the North American species and Wilson those of the Old World. Utilizing Wilson's experience with eastern Asian species in the wild, and Rehder's ability to evaluate the specimens from many field collectors in North America, the two produced a much needed resource on a group of shrubs important in American gardens. Each of the authors acknowledged Sargent's helpful understanding of garden forms and the value of observing living specimens in the Arboretum.

Rehder also worked for some time on another compilation that he and Sargent plotted. They wished to bring together in one compact volume an account of all the ligneous plants cultivated in the United States. This would include not only many native American plants but also the multitude of trees, shrubs, and vines that had been introduced from foreign countries and selected by the work of hybridizers. Thus, work for this manual drew from Sargent's *Silva,* from Rehder and Wilson's Asiatic plant studies, and from Rehder's knowledge of European cultivated plants and his work for L. H. Bailey's *Cyclopedia.* With this project in mind, Sargent requested his collectors (such as Alice Eastwood of the California Academy of Sciences, and Ernest J. Palmer) to press specimens of trees and shrubs cultivated in their regions. On his trips to Japan, Wilson collected many garden forms of woody shrubs and trees in order to authenticate what was being imported into the United States.

While Rock explored the Tibetan hinterlands for new woody plants, Alfred Rehder made final preparations for what was to be his most widely utilized book, *Manual of Cultivated Trees and Shrubs Hardy in North America Exclusive of the Subtropical and Warmer Temperate Regions,* more briefly known as *Rehder's Manual.* Published in January 1927 by the Macmillan Company, Rehder's book was designed to complement Sargent's and Gray's manuals, which enumerated the native vegetation of the same re-

gion. Rehder's goal of a concise compilation of all the woody plants known to be cultivated in eastern North America was achieved in the 930-page volume. In the introduction he stated his expectation that introduced ligneous plants would become increasingly prevalent and would greatly modify the original vegetation wherever man made his home.

The book was arranged in the Engler and Prantl sequence of families, as were most other American manuals. For each of the 2,300 species of trees and shrubs, Rehder included a concise physical description, as well as time of flowering and fruiting, references to published illustrations, region of native habitat, date of introduction into cultivation, hardiness zone, and comments on horticultural merit or special cultural requirements. He gave distinguishing features and pertinent information for 2,465 varieties. Brief mention was included of another 1,265 species and 507 hybrids that were either less well known or rarely cultivated.

Although information on the climatic zone in which each cultivated plant can survive is now quite commonly included in horticultural books and commercial plant catalogs, it is not often appreciated that Rehder pioneered this concept with the first edition of his *Manual.* He divided North America into eight zones, each characterized by five-degree increments in the lowest mean temperature of the coldest month, and included a map of the zones in the book. Using information gathered from many sources—such as the climatic conditions of each species' native habitat, records of its performance at the Arboretum and sister institutions round the world, and correspondence with cultivators—Rehder indicated as accurately as possible the northernmost reliable hardiness zone for each species and for many varieties.

To aid growers and horticulturists in identifying unknown woody plants, Rehder included detailed keys to the families, genera, and species. As he explained in his introduction, Rehder followed international rules of botanical nomenclature so that the names in his manual agreed with those used by Bailey, Sargent, and Merritt L. Fernald and Benjamin L. Robinson, authors of the most recent edition of *Gray's Manual.* Exception to the international rules was made for the numerous horticultural

75. *In compiling his* Manual of Cultivated Trees and Shrubs Hardy in North America, *Alfred Rehder included not only all the new species and varieties brought into cultivation from the wilds of both hemispheres but also many forms that were selected and named by gardeners and nursery owners. One of the twenty-eight garden forms of Norway spruce included in the second edition of Rehder's* Manual *is this pendant form, growing in the Arboretum conifer collection* (Picea abies *forma* pendula).

forms for which there was as yet no generally accepted code for naming. For these "trinomials" Rehder used a system for giving rank and author citation that he and Bailey worked out as a practical solution for the gardening public.

The book was not light reading; rather, it was chock-full of dense, encoded data that had never before been gathered in one small volume. Its value to nurserymen, gardeners, arborists, botanical garden curators, and anyone else needing to identify trees and shrubs became readily apparent. Years later, Rehder's understudy Clarence Kobuski said that the *Manual* "became almost overnight a bible for . . . those interested in gardening and cultivated plants" (Kobuski, 1950, p. 6). Of the sixty-five hundred kinds of plants mentioned in the *Manual,* most were grown at the Arnold Arboretum. It was these living examples, as well as the institution's library, herbarium, and human resources, that enabled Rehder to produce the compendium.

Another person who contributed to its program in descriptive botany of cultivated plants initiated her association with the world-famed tree museum as a volunteer in 1921. Susan Delano McKelvey moved to Boston in 1920, thinking she might pursue the field of landscape architecture. She had graduated from Bryn Mawr College in 1907 but had received no horticultural or botanical training. Long a supporter of the Arboretum, in 1921 McKelvey volunteered her services to Sargent in return for instruction in the identification of plant materials. To test her commitment, as he was wont to do, the director assigned her to the greenhouse to wash pots. Her keenness readily became apparent, and the new propagator, William Judd, soon taught her techniques of plant propagation. Fascinated by the wealth of resources in the Arboretum library, McKelvey began to spend time studying there. In the summer she accompanied J. G. Jack and others on a field trip to collect specimens in Glacier National Park, and the following year she botanized with him in New Hampshire's White Mountains. She accompanied E. H. Wilson on his tours of inspection to document the Arboretum collections on film. With her appreciation of the institution's tripartite resources thus

established, McKelvey began a scholarly study of *Syringa,* the lilac, an important group for the Arboretum and one for which a monograph was needed. Her interest shifted from landscape design to botanical research. McKelvey's association with the Arboretum, during which she completed three major botanical works, lasted until her death in 1964.

Although McKelvey's book *The Lilac* (1928) appeared after Sargent's death, it was directly influenced by his encyclopedic vision. Published by the Macmillan Company, the quarto volume included black-and-white photographs and four-color charts pocketed inside the cover. McKelvey enlisted others to contribute to the exhaustive work. Wilson wrote a chapter on the history and geographic distribution of the genus; Rehder provided a taxonomic description of the genus and its sections, with a key to the species. Other experts added chapters on cultivation and on diseases and pests. The greater part of the text, however, consisted of McKelvey's descriptions and informative discussions of the species and many of the garden forms derived from them. The majority of the photographs, which very clearly depicted identifying characters, were of Arboretum plants documented by accession numbers. As the author stated in the preface, the most significant contribution of the volume was the provision of information on garden forms. Like Rehder's *Manual,* the attempt was to bring the rigors of botanical science to garden subjects, a difficult task that the Arboretum continued to address under subsequent administrations.

The taxonomic work of the Arboretum staff was pursued with methods and goals similar to those of their cohorts in the nineteenth century. Despite acceptance of Darwin's evolutionary hypothesis, its impact on the science of diversity was only gradually felt. This was true not just at the Arboretum; taxonomy remained in a descriptive mode for much of the early twentieth century. The first attempts to formulate classifications based on evolutionary principles were geared at the family level—systems such as that of Engler, which ordered families from the primitive to the advanced. At the level of species there was still fundamental cataloging to be done, and the pre-Darwinian concept of species as well-defined physical entities lingered well into the twentieth century. It would be for the next generation of Arboretum scientists to participate in its replacement by a "biological species" concept that recognized species as living populations of organisms.

The very act of assembling the living collection of the Arboretum affected the topics investigated by the institution's botanists. Their observation of growing specimens on the grounds greatly augmented the ability of Rehder, Sargent, Wilson, and others to communicate botanical findings. On the other hand, the pursuit of knowledge in Asia's forests greatly added to the living collections, resulting in many adjustments and changes to the original layout of the grounds. How the plantings and the public programs were adapted to the twentieth-century explorations explains much about the landscape as it exists today.

The great growth of the herbarium, however, provided the first inkling that the science of taxonomy was becoming the science of diversity. The postage-stamp model no longer applied, though no one quite realized it. Collecting "one of every kind" was outmoded for investigation of evolutionary relationships in the plant kingdom. This was the beginning of a gap between making a living collection and undertaking modern scientific scrutiny.

76. *After her monograph* The Lilac *was published, Susan D. McKelvey turned to the study of* Yucca, *spiny-leaved members of the lily family native in the southwestern United States. From 1928 to 1932, she made eight trips to collect and photograph these plants. Here McKelvey and her driver stand before the auto that took them to Arizona and neighboring states in 1932. The trailer is full of specimens and equipment. Her* Yuccas of the Southwestern United States *was published in two parts (1938 and 1947); she then wrote a history,* Botanical Exploration of the Trans-Mississippi West, 1790–1850.

CHAPTER FIVE

LANDSCAPE AND
PUBLIC PROGRAMS ADAPT

Twenty-five years of planning notwithstanding, the Arnold Arboretum grounds were changed substantially by its managers in the early twentieth century. This is not unusual in gardens, where growth and the elements make unpredicted alterations to original visions. In a garden with a diversity of users for science and pleasure, such as Harvard's Arboretum gathered to it, flexibility was even more necessary because of human activities, not the least of which was discovery of more plants. The forces of nature occasioned minor action in the final stages of construction. As the emphasis of research changed from American forestry to descriptive botany of the world's north-temperate woody plants, an audience of horticulturists, both amateur and professional, emerged to support the Arboretum's plant introduction program.

Once the influx of plants from the Far East began, the not-so-simple botanical arrangement of trees was scrambled by the need to display far greater numbers of kinds, especially of shrubs, than had been planned. While clarity of the botany lesson to be derived was sacrificed, the integrity of the overall design and the naturalistic beauty survived the rearrangements and insertions surprisingly well. Ironically, the one thing Sargent thought could save the simplicity of the original scheme, additional land, came too late in his administration for effective incorporation in the arrangements. As it happened, those who taught and wanted to learn botanical relationships were perceptive enough to continue to observe them, while the still apparent generic grouping was sufficient to serve horticultural and aesthetic ends.

In its second quarter century the Arboretum gained a widespread constituency as diverse as its collections and landscape. Its small staff responded by providing guidance and information in ways that were economical and limited to each one's area of expertise; none were trained educators or experienced publicists. As surrounding Jamaica Plain became a more densely populated part of the city, no longer the site of country homes, and modern transportation made the place accessible, Boston citizens took to its springtime glory, making it their own. Its most critical audience was made up of the growing ranks of botanists and horticulturists at institutions of like purpose.

77. Oak Path, 1903. After traversing the planted collection of oaks, this mown-grass path went by some ancient white oaks from colonial days as it approached the western plateau of Bussey Hill. The grass paths were initiated by Sargent to expand access to the collections once the planted trees began to mature.

FINISHING TOUCHES

Perfection of the landscape after 1900 included finishing the surface and ground cover. While many areas under the new tree collections had a well-developed grass flora derived from former orchards and pastures, other sections needed improvement, especially those disturbed during road construction. Most of the north side of Bussey Hill, including ground under the arborescent legumes, lilacs, ashes, catalpas, elms, hackberries, and birches, was plowed, graded, and seeded as early as 1893. Between 1900 and 1902, Arboretum laborers prepared the surface and "laid down permanently in grass" the remainder of the north side of Bussey Hill (under the birch and hackberry collections) and much of the North Meadow on the west side of Meadow Road (the land under the tulip trees, lindens, and horse chestnuts). Across Meadow Road from the last, the area of the maple collection was similarly treated, as was the conifer slope.

In spite of this careful preparation, only the grass paths and a few other key areas were regularly mown all season during Sargent's administration. The turf-covered paths, initiated in 1896, were so welcomed by visitors who wished to study the trees intimately that at least another mile of them was added in 1903. Closely following the land's contour, they were inexpensive to create and maintain. The routes could be easily changed to pass a particular plant or collection in its season of interest, or to allow overused ground to rest. Throughout most of the tree collections, native and naturalized herbaceous plants, such as buttercups, daisies, asters, and goldenrods, were allowed to grow up with the grass all spring and summer. Like the natural woods areas, the high grass and herbaceous vegetation prevented soil compaction and absorbed and retained more moisture than short turf. The mown paths provided an orderly means to enjoy the meadowy luxuriance and approach the planted trees. The wooded areas still had a rich herbaceous layer, with many wildflowers. Very occasionally this was enhanced, as when masses of ferns were installed on the lower slopes of Hemlock Hill soon after the turn of the century.

In a few densely planted areas, a lawn was maintained by mowing throughout the growing season. The photo-graphic record shows regularly cut expanses around the administration building, between the shrub-collection rows, and immediately adjacent to sections of the drive where there were no shrub masses. While the design was to look naturalistic, the Arboretum was never allowed to go wild or be unkempt. Mown areas provided an important transition between man-made drives and lush vegetation. Where vines and shrubs were permitted to grow over the edges of paths and drives, they were annually pruned back to prevent rank obstruction of passages. At the base of young trees and specimen shrubs a neatly edged circle was kept free of grass and weeds through the season by repeated tilling. This was done as much for the health of the plants as for appearance. In the shrub collection the margins of lawn were similarly maintained.

The management of watercourses was a case in which natural conditions were often at odds with the care of tree collections. After 1900 Sargent still objected to inadequate drainage in the valley of Bussey Brook and in the North Meadow. The problem of spring freshets that eroded the banks of Bussey Brook below Hemlock Hill was corrected in 1903 when the city enlarged the culvert that carried the brook out of the Arboretum under South Street. Efforts to control excess soil moisture in the North Meadow were more complicated. Prized by colonial farmers as a natural source of hay, and apportioned and leased for haying rights by neighboring landowners in the seventeenth and eighteenth centuries, the meadow had historically been wetland. Two branches of Goldsmith Brook shed waters from Moss Hill into the North Meadow and merged just prior to exiting the Arboretum through a culvert under the Arborway. The channels dug through the meadow by Arboretum workers, and the improved outlet made by the City of Boston in 1892, did not have the hoped-for drying effect. Since Meadow Road and the Arborway were both constructed at nearly the same time, effectively damming parts of the meadowland, they may have negated any improvement.

Two more attempts were made to rectify the situation. In 1900 city officials let a contract to improve the conduit that took Goldsmith Brook out of the Arboretum, and the following winter the Arboretum crew cut a canal to a depth of five feet and laid tile drains across the meadow.

78. Bussey Brook valley, looking east. The area just beyond the stone bridge was a swampy alder thicket before Arboretum construction began. Although the brook channel was improved during original construction, better control of erosion in this area was made possible by installation of a larger culvert under South Street by the city in 1903.

79. *This aerial view looking north, with Peters Hill in the foreground, shows the orchardlike planting of hawthorns on the north slope. Most of the hawthorns were planted there between 1905 and 1910. The photograph was taken in 1927, soon after mass plantings of oak and larch were added on the southeast side, where their planting circles are evident.*

The director expected a fourteen-acre increase in land available for the growth of trees, but four years later he complained once more: "The brook that enters the Arboretum from the grounds of the Adams Nervine Asylum flows irregularly. For a few weeks of the year it is a torrent, tearing away its banks and choking its bed with stones and gravel; for the remainder of the year it is a dry and unsightly ditch. The annual cost of repairing the damage done by the floods of this brook has been considerable" (C. S. Sargent, 1906, p. 246). This problem was with the southern branch, which crossed the linden collection and passed under Meadow Road. To remedy the situation, the Arboretum installed 1,237 feet of culvert to carry the brook across its property. After this, trouble with North Meadow drainage was not mentioned again by Sargent. Nevertheless, many acres remained too wet for the growth of trees and, in times of extreme weather, the area has betrayed its original wetland status by flooding.

SUPPLEMENTARY COLLECTIONS

Accessions to the living collection increased after the turn of the century as a result of exploration and scientific discovery, as well as exchange with other botanic gardens and the nursery trade. To a certain extent, visitor interest caused the expansion of some groups. When finding the space to include new plants with their relatives became difficult, the problem was handled by disrupting the classificatory sequence, although this was not admitted at the time. There were generally two ways new plantings were made on the grounds. One was to form supplementary collections of one genus in a new location; the Peters Hill extension largely comprised this type of innovation. The other was to form what might be termed subsidiary shrub collections, mixed plantings of species and varieties considered to be on trial.

When supplementary collections were created, the original planting of the genus was left in place. Taking a cue from the increasing popularity of the Arboretum lilac display, Sargent decided to extend the planting of another May-blooming group in the hopes of widening the visitor experience. In 1901 an additional collection of flowering crabapples, plants of the genus *Malus,* was planted on the eastern base of Peters Hill between the circuit drive and the woods that screened Bussey Street. The original area designated for crabapples, on the south side of Forest Hills Road, was by this time too limited. The plants there were growing well, but there was no room for more. These were left in their place at the foot of the Bussey Institution bank, while recently obtained *Malus* species and varieties were arrayed on Peters Hill, where there was still plenty of space unplanted. Blooming usually a week before the lilacs, the new crabapple display would extend the time and place of spring floral attraction. The main collection of willows and poplars and a supplementary conifer grouping, as mentioned previously, were the other recent plantings on Peters Hill.

The collection of hawthorns (*Crataegus*) assembled as a result of Sargent's investigation was the most extensive supplementary group placed on Peters Hill. In 1905, there were hundreds of species in the nursery in need of a site on the grounds. The earlier-known American types and those of the Old World already lined the Arborway wall at the margin of the main rose-family collection. With the decision to locate the recently accessioned "thorns" (as Sargent called them) on the north slope, on the uphill side of Peters Hill Road, an initial planting of three hundred species was made in the spring of 1905; this was followed by several major additions until about 1910. Up to five individuals of each species or seed lot were planted.

Aside from its ramifications for Sargent's work as a taxonomist, the living *Crataegus* planting long remained a dominant feature of the landscape on Peters Hill. Its existence, and Sargent's perseverance, precluded other developments that could have taken place on the twenty-five acres the collection occupied. Hawthorns are small trees, most with horizontally spreading branches; the few that are shrubby were grown together at the western end of the group. Even the tallest kinds have branches that reach to the ground and crowns spreading as wide as high. With the crabapples across the road being also relatively low, wide-shaped trees, the whole side of Peters Hill took on a rather uniform aspect. There is no evidence that shrubs were used to ornament the roadsides in this area. To relieve the monotony, there was the backdrop of native woods on the rocky ledges behind the crabs and near the hilltop above the hawthorns, along with a few large old white oaks that stood among the thorns. Yet, most of this side of the hill retained an open, pastoral character.

If the Arboretum had amassed oaks, pines, hickories, or some other large-tree genus, the area might have returned to the look of its prehistoric forest cover, and travelers on the drive would not be able to perceive the size of the hill. The thorn collection was, instead, reminiscent of colonial days, when many a New England hillside was studded with orchard trees. It was bright and spacious in contrast to the shades of Valley Road or the columned dark of Hemlock Hill. While botanical scrutiny rather than public display was the original basis for the collection, its numerical superiority within one genus became a fixture that outlasted Sargent's tenure. While always noted, it rarely caused as great a stir among chroniclers and admirers of the Arboretum as the azaleas, lilacs, or crabapples.

The shrub collection was completely overhauled in 1906 and 1907. After its last renovation in 1895, the Arboretum administrator became resigned to its existence, although he still thought it "not entirely satisfactory." It was overcrowded; many shrubs had outgrown their allotted space. Others lacked vigor because they were marginally hardy or required better soil drainage or were dominated by healthier neighbors. An unusually dry summer in 1904 followed by two severe winters took a toll on shrubs and trees as well. The relatively narrow walks between beds and the lack of approaches were further drawbacks. In late 1905 Sargent and his staff worked out plans for a new arrangement and readied the ground. All the plants were either moved to temporary quarters or repropagated in Dawson's teeming workshop.

The area to be occupied by the new shrub garden was enlarged, the grade was changed to slope gently and uniformly toward the ponds, and tile drains were laid across the area. To help with the drainage, Sargent convinced park officials to reset a culvert carrying the outlet from the ponds that had been installed during original construction. Replanting of shrubs and reseeding of grass paths began in the fall of 1906. Nineteen beds, ten feet in width,

that curved slightly with the terrain were spaced at five-foot intervals, thereby increasing the width of walking area between the rows. To improve access, one grass path was cut through at right angles to the rows, allowing greater options among routes when studying the shrubs. In a further departure from previous practices, Sargent positioned a trellis, constructed of ten-foot-tall concrete posts strung with courses of heavy galvanized wire, along the northern and eastern perimeter. Its dual function would be to display vines in generic groups and to protect the shrub garden from cold winds.

Many of the shrubs were returned to the systematic arrangement in single lines down the center of neatly edged beds. Once replanted, the total 7,765 feet accommodated some 1,200 shrubs, but this was far fewer than the wilds and gardens of Asia and Europe yielded to the Arboretum. Correspondingly, the decision was made to move several shrub groups elsewhere on the grounds. The stated intention was to transfer shrubs of the genera that contained trees to sites close to their tree allies. In practice, the genera thus moved were not necessarily ones that had tree relatives! Often, a few representatives of these groups were left in the shrub collection for the sake of illustrating their place in the botanical order. In the long term, this solution to overcrowding in the shrub collection rendered the botanical sequence in the main collection less apparent.

On the recently claimed land around the North Meadow, shrubby willows were appropriately set opposite a grass path from their tree brethren that lined the Arborway wall. On the west side of the meadow, just across the road from the museum building, a long bed for the subsidiary collection of *Ribes,* the genus comprising gooseberries and currants, was prepared. These are members of the Saxifragaceae, a chiefly north-temperate family that contains many ornamental shrubs but no trees hardy in the New England climate. The family would have come between the rose and witch hazel families in the Bentham and Hooker sequence (in the vicinity of Rehder Pond), but no room had been allotted it. The location of the new *Ribes* collection was behind a border of American maples and native shrubs that lined Meadow Road, accessible by a grass path that came to be known as Maple Path. Why this spot was chosen can only be guessed. Since most *Ribes* are low growing, rarely exceeding three feet, they could be placed there to utilize recently drained ground without interfering with the view across the meadow. At about this time, beds of Saint-John's-worts (*Hypericum*), honeysuckles (*Lonicera*), and tamarisks (*Tamarix*) were installed on the west side of Linden Path, a grassy alley that meandered between the *Tilia* collection and the border plantation of native oaks along the Adams Nervine Asylum property.

Another genus of the saxifrage family moved from the shrub collection was a numerous group, the mock oranges (*Philadelphus*). They were placed within the curve of the main drive around Rehder Pond, just past the junction of Meadow Road and Bussey Hill Road, a site in order within the Bentham and Hooker sequence. Included were hybrids developed in France by the Lemoine Nursery as

well as their parental western American species. Once established, these large, white-flowered shrubs spread their arching branches across the grass under the canopy of the tupelos and the tree legumes.

While the numerous garden forms of lilac, varieties of *Syringa vulgaris,* lined the south side of Bussey Hill Road, the species of wild origin from eastern Europe and western Asia had been placed in the shrub collection. During the 1906 alterations, all the species were moved to the eastern end of Bussey Hill Road between the forsythias and the garden-form lilacs.

Two genera of the Caprifoliaceae, another family with no tree representatives, were transplanted to the opposite side of Bussey Hill Road from the lilacs; their new location was roughly correct in terms of the botanical arrangement. Between the sidewalk and road several of Rehder's

80. After the shrub collection was overhauled in 1906 and 1907, several genera were moved to new sites throughout the Arboretum. The mock oranges (Philadelphus) *were placed on the north side of Bussey Hill Road, above Rehder Pond. Pictured here is the fragrant-flowered 'Avalanche,' a variety originated at the nursery of Victor Lemoine at Nancy, France, in 1896.*

specialty, honeysuckles (*Lonicera*), were placed to supplement those that remained in the shrub-garden rows and the ones along Linden Path. Further up Bussey Hill Road, the second caprifoliaceous genus, *Viburnum*, was placed, a few of the kinds across from the upper end of lilacs and many more in front of the plane trees, near the junction with Valley Road.

As part of the rearrangements of 1906–7, a large addition to the rhododendron collection at the base of Hem-lock Hill was made, extending it further along Bussey Brook toward the South Street gate. Additionally, masses of azalea species were shifted to a new habitat on the "western slope of Bussey Hill," on the shoulder of the plateau that spread below the summit overlook. The area was part of the former city reservation and had not yet been developed with Arboretum collections. Azalea Path, as the flowery way came to be known, struck a southward course roughly following the old city reservation bound-

81. Azalea Path was created along the western edge of the Bussey Hill plateau to relieve overcrowding in the original shrub collection after 1907. Masses of North American and Asiatic azaleas were placed under a canopy of oaks, white pine, and hemlock that remained from Bussey's hilltop concourse.

ary that in turn had been defined by the western leg of Benjamin Bussey's tree-lined hilltop concourse.

When the azalea collection was placed there, many trees from Bussey's time remained, notably, large white oaks and sentinel white pines, giving high shade and an arboreal framework to the new path. In early photos the pines look a bit scrawny or scant of foliage, as though they had been recently released from a crowding of brush and other trees. Over the years they filled out to form the high, bushy crowns typical of open-grown, mature white pines. Both North American and Asian azaleas were sited here. Within a few years the royal azalea (*Rhododendron schlippenbachii*), brought from Korea by J. G. Jack, burst forth its large, delightfully clear pink flowers in early May, where the path left Bussey Hill Road. Further down the path another of Jack's finds, the Poukhan azalea (*Rhododendron yedoense* var. *poukhanense*) was massed, its mauve tone rendering it more difficult to use in combination with other spring bloom, but an admirable, dependable plant nevertheless. At the far end of the path was a group of *Rhododendron obtusum* var. *kaempferi,* the torch (or Kaempfer) azalea. Carried back from Japan by Sargent in 1892, this plant came to be favored by Wilson even more than by his director. Sargent observed that their salmon-orange corollas tended to fade quickly if grown in full sun; thus, he planted the Azalea Path group where the pines and a remnant grove of hemlocks provided shade. The native American azalea species blooming generally later and for longer periods than these showy orientals extended the season of interest here, even though these were well represented in the roadside shrubberies throughout the Arboretum. A few years later many more rare and interesting trees and shrubs were interspersed with the azaleas.

In the early 1920s, Sargent and Wilson greatly expanded the Bussey Hill plantings when many azaleas from seed collected on later Arboretum expeditions were finally mature enough to put on display. New material was added extending the beds further along the southern end of Azalea Path and closer to Oak Path below it. Another change to the Azalea Path plantings in the early 1920s was the relocation of the collection of Scotch brooms, plants of the genera *Cytisus* and *Genista,* to the upper side of Azalea Path. The reason given for this move was that these

82. *Sargent obtained a hardy form of cedar of Lebanon* (Cedrus libani *variety* stenocoma) *by commissioning a botanist to travel high in the Taurus Mountains of southern Turkey. The foliage and cones of this cedar are exceedingly handsome.*

ADDITIONAL GENERIC COLLECTIONS

After the many relocations that resulted from overhaul of the shrub collection, other genera were repositioned or added to sites from time to time without following the botanical sequence. The reasons for such changes varied, and often were not explained.

On the opposite side of Bussey Hill's western plateau, a new, experimental plantation was added before 1910. For years, such conifer fanciers as H. H. Hunnewell and Henry Winthrop Sargent had tried to grow the cedar of Lebanon (*Cedrus libani*) in the Northeast with little success. Prized for its stature and historic associations, it had been raised to maturity in Pennsylvania but no further north. Sargent, in his characteristically thorough investigations into the native ranges of all remotely potential trees, realized that the species occurred at twelve-thousand-foot elevations in the Taurus Mountains of southern Turkey. In 1902 he commissioned Walter Siehe, a botanist

living in Smyrna, to obtain seed from this high-mountain population whose progeny were more likely to prove hardy at the Arboretum. The resultant seedlings did well in Dawson's nursery, and, to give them a trial, a belt of young *Cedrus libani* was planted on the east side of the meadowy plateau just above Bussey's old lilac hedge. At that time the location was along the Arboretum's boundary, since the grounds around the mansion were then part of the Bussey Institution. As the cedars continued to thrive despite the cold New England winters, and displayed a more upright growth than the plants of southern origin, the new strain was later named variety *stenocoma*.

At about this time, a new planting of shrubby kinds of *Amelanchier,* or shadbushes, was made in response to the findings of a recent taxonomic investigation of the eastern American species by a Cornell University professor, K. M. Wiegand. They were placed just beyond the new *Ribes* bed, further down Maple Path.

In the 1911 and 1912 seasons, two mixed groups of conifers were given special sites on the grounds. Dwarf and slow-growing conifers were placed on Conifer Path above its junction with Valley Road, not far from their tree relatives. The second planting was made in a clearing on the ridge of Hemlock Hill, where the surrounding evergreens would provide cover. Here some cone-bearing species thought to be less winter hardy were planted. Among them were cedar of Lebanon, Japanese cedar (*Cryptomeria japonica*), incense cedar (*Calocedrus decurrens*), and Lawson cypress (*Chamaecyparis lawsoniana*), the latter two native to the western United States. This was one of very few times accessioned plantings were made on the upper slopes of Hemlock Hill.

By 1915, a supplementary collection of *Pyrus,* flowering pears, was established in the Peters Hill tract near the Bussey Street Gate. Shortly afterward, a supplementary plum collection was located on the southwest side of the hill. Both American and eastern Asian species of plums were eventually included in the Peters Hill group. A large proportion of the great mound was now devoted to tree genera of the rose family, although the original collection remained in the vicinity of the ponds.

If strict adherence to the Bentham and Hooker order were kept, the genus *Cornus* would have to fit in the tightly planted area around Rehder Pond, within the curve of the drive as Meadow Road turned into Bussey Hill Road. The modest representation of this genus of shrubs and small trees that was located in sequence had recently been joined by new plantings of *Philadelphus* and *Lonicera,* among others. As a result, an overflow group of *Cornus* was placed behind the cork trees that lined Meadow Road. Here the Arboretum crew planted the early-spring-blooming cornelian cherry (*Cornus mas*) and its close relative, *C. officinalis,* as well as other small trees in the genus.

As Wilson became more active in management of the grounds in the final years of Sargent's administration, finding room for the diverse and numerous species in the rose family was a persistent problem. Asiatic explorations had yielded greater numbers of species of *Sorbus* than could be placed in the originally designated area along the Arborway wall above the shrub collection. *Sorbus,* or mountain ashes, are trees of cool habitats in the wild. Sargent observed that the relatively hot, dry conditions of the southwest-facing slope along the Arborway caused poor growth and performance among the mountain ashes that had been planted there. He and Wilson chose what they thought would be a colder site, along Valley Road

in the foreground and as an understory of the American oak collection, to develop a subsidiary *Sorbus* collection in 1920. Eventually, more than forty accessions of *Sorbus* representing species from three continents were placed in this location.

After his travels in Japan, Wilson became enthusiastic about the Japanese garden forms of cherries, especially the double-flowered ones such as the now widely planted Kwanzan cherry. Although some varieties were added to the original *Prunus* collection along Forest Hills Road in 1915 and many more were lined out in the Arboretum nurseries for trial, he really wished the Arboretum could have a display to rival that of Washington, D.C., or the Rochester, New York, parks. Once the young trees were evaluated in the nursery for several years, in 1922 Wilson lined out a collection of some fifty garden forms of flowering cherries on Bussey Hill, completely filling the lawn on the plateau. While the limited space precluded creating the kind of show that he admired in Washington and Rochester, the collection made up in diversity for what it lacked in size.

Additional subsidiary collection beds were established in 1925. Euonymus and sumac (*Rhus*) collections were placed in the foreground of the birch collection on the east side of the summit section of Bussey Hill Road. Little room for expansion remained on the original areas designated for these two genera on opposite sides of Meadow Road between the major tree groups of horse chestnuts and maples. In the conifer collection a new bed of several yew (*Taxus*) varieties was laid out between Bussey Street and Hemlock Hill Road; this bed subsequently grew to screen the street from the conifer-clad brook valley.

ASIAN INTRODUCTIONS FIND SPECIAL SITES ON GROUNDS

When Wilson arrived to take up scientific work with Rehder in the fall of 1911, he saw other changes on the grounds besides the relocation of generic groups. Many of the latest additions were plants with which he had greater familiarity than anyone on the staff, having seen them full-grown in their native habitat. The plants from Jack's

83. *E. H. Wilson became very enthusiastic about the many double-flowered forms of flowering cherries during his expeditions to Japan. He sited a large collection of them on the Bussey Hill plateau, completely filling in the area encircled by the Chinese plant bed. In this photograph, taken from the summit of the hill looking south, can be seen (from left to right): a row of Japanese double cherries, a row of Chinese crabapples, the Chinese shrub bed, and the upper bed of Azalea Path.*

84. *The Chinese bed, or "collection of Chinese shrubs," was first laid out in 1911 to test some of Wilson's introductions. Here, Louis Victor Schmitt, the grounds superintendent, examines several cotoneasters Wilson obtained in China.*

expedition of 1905, a collection of Wilson's introductions for Veitch that Sargent obtained on a trip to England in 1907, and the acquisitions from the two recent Wilson expeditions had all outgrown their space in the Arboretum's modest nursery. As far as possible, Sargent attempted to place new kinds from the Far East in the vicinity of already existing groups. While trees could often be used "to extend existing collections," the multitude of Korean, Japanese, and Chinese shrubs posed another problem; even among trees, there were many whose ability to withstand the Arboretum climate was unknown. The solution to the quandary of placing shrubs arriving from the Orient whose hardiness was untested was to open shrub beds "in more sheltered and less conspicuous positions." It was thought better to restrict them to certain areas where their novelty and trial status would be understood.

Already, some of Jack's and Wilson's finds had been placed among the azaleas on the recently created Azalea Path on Bussey Hill. One was *Eucommia ulmoides*, a latex-producing tree that might provide a source of rubber hardy in temperate regions, a source of fascination to Sargent and his dendrological colleagues. The dove tree

(*Davidia involucrata*), the prize of Wilson's first trip for Veitch, was planted among the Asiatic azaleas. It had yet to bloom at the Arboretum, however. Other small trees and shrubs that were the only one of their genus or family known, such as *Idesia polycarpa, Poliothyrsis sinensis,* and *Euptelia polyandra,* found their place in the beds that lined the new path. While azaleas continued to dominate for some time, the introduction of monotypic and rare plants eventually changed the character of the planting from a generic collection to another mixed shrub collection.

Between 1909 and 1911 two new areas were opened up for the trial and demonstration of lesser-known plants. Adjacent to Azalea Path, on the plateau below Bussey Hill Overlook, the Arboretum planters dug a broad bed encircling the level meadow. Sargent explained that the spot was chosen because its exposure would give the new

85. Additional planting spaces, opened up between 1910 and 1920 to test Asiatic plants, flanked Hickory Path. The neatly edged beds can just be seen through the trees in this photograph; the area subsequently became known as the Centre Street beds. The foreground shows how the native herbaceous vegetation was mown through to create grass paths.

plants a real test of hardiness. The shrubs installed here were largely members of the rose family: cotoneasters, spiraeas, roses, flowering pears, *Neillia sinensis,* and *Exochorda,* among others. Plants from other families set out for evaluation in what came to be known as the "collection of Chinese shrubs" were *Kolkwitzia amabilis, Dipelta floribunda,* and some *Lonicera* species—all members of the Caprifoliaceae—Chinese witch hazel, Kousa dogwood, and many more. The *Kolkwitzia,* or beauty bush, was one of the first of Wilson's introductions to gain wide popularity with American gardeners. Since the area had only been planted that year, the experimental collection from western China was mentioned only briefly in the new guide to the Arboretum published in the spring of 1911. However, as the young shrubs grew year by year, filling in their bed, blooming and fruiting in their turn, their performance was repeatedly lauded in Arboretum publications.

A second planting space to hold specimens that could not be accommodated in the shrub collection or within an existing systematic group was created along Hickory Path. It had the advantage of being a relatively protected site on a south-facing slope below the stone boundary wall. Many species were planted both here and on Bussey Hill to ensure survival in at least one area should conditions prove unfavorable in the other. The path itself was a mown-grass strip flanked on either side by cultivated beds. It paralleled Centre Street from near the gate of that name almost to the border of Central Woods, where it curved to meet Conifer Path. It skirted the far side of the collection of hickories, hence its name; subsequently, the area was alternately referred to as Centre Street Path. More recently, it has been called the Centre Street beds. Besides what was termed a supplementary collection of barberries, shrubs of small stature like *Daphne genkwa,* several *Mahonia* species, and *Indigofera kirilowii* (a J. G. Jack introduction) were planted in the beds. Larger-growing rarities to be found in the vicinity were *Styrax japonica, Parrotia persica,* and a group of *Pterocaryas,* including a hybrid between a Caucasian and a Chinese species that originated at the Arboretum. Additionally, since it was thought best to try the new oriental plants in more than

one place, a few, such as *Eucommia ulmoides* and *Malus theifera,* were placed among the shrubs of Linden Path, which diverged from Meadow Road. Since such groups as *Hypericum* and *Tamarix* were already growing there, Linden Path, like Azalea Path, became an area of mixed shrubbery.

Photographs taken in the teens and twenties show that the plantings of Azalea Path and the Chinese shrub collection on Bussey Hill, as well as Hickory Path and Linden Path, were maintained similarly to the original shrub collection. The grass was kept closely mown, and the edges of the beds were neatly trimmed. No mulch was used, the soil being lightly tilled to keep down the growth of weeds all season. Metal labels with raised letters on stands stuck into the ground in front of the plants named each shrub or small tree for visitors' information. Sometimes wooden stakes were similarly placed until metal labels were obtained. Throughout the Arboretum an additional zinc record label was also attached to every accessioned plant.

In a sense, Azalea Path, the Chinese shrub collection, and Hickory and Linden Paths were subsidiary shrub collections. They were maintained in a more formal state mainly because it was easier to keep track of numerous small plants that way. They were closer to the idea of experimental plantings, the secondary, more temporary kind of collection that Sargent had envisioned in the 1880s, and he would have preferred to locate them on additional land rather than within the original Arboretum planting. He continually stressed the Arboretum's need for more land from 1900 onward. The areas he was obliged to use for subsidiary shrub groups were ones that Sargent thought were "inconspicuous." Bussey Hill had not been part of the original grounds, and the Centre Street and Adams Nervine boundaries were peripheral and more or less screened from the carriage drive by the trees of the hickory and linden collections.

Once initiated, the plantings were expanded or rearranged from time to time, especially those on the Bussey Hill plateau. Even though there was ample space for them on Peters Hill, selected pears and crabapples from China were installed in the Chinese shrub collection on Bussey Hill in 1915. Wilson apparently rearranged much of this collection before going off to Japan for the first time. The addition of a diverse group of flowering cherries to fill in the lawn encircled by the Chinese shrubs in the 1920s greatly changed the landscape. The Chinese collection was eventually confused with Azalea Path, and today what was originally the latter is referred to as Chinese Path.

FURTHER ALTERATIONS TO COLLECTIONS

Not long after the influx of oriental plants began, many of the new arrivals were held in a nursery on the west side of Peters Hill, a place thought to be sheltered from winter winds. Here examples were often kept even after sister seedlings were tried in other places on the grounds. Nonetheless, even this area suffered serious damage in the winter of 1913–14, when almost every Asiatic oak there was killed. In the 1920s the location came to be referred to as "the mixed plantation of trees," rather than the Peters Hill nursery. Many of these trees were of great interest and had reached a size worthy of notice. With little room to accommodate them with others of their genus in the original layout, many of the trees were left in place. Once other land was developed for growing young plants, the southwest slope was no longer needed as nursery. Turf was allowed to grow up under the mixed plantation, where visitors were encouraged to observe such worthy Arboretum introductions as Amur chokecherry (*Prunus maackii*), toon tree (*Cedrela sinensis*), and some Asiatic birches and oaks.

While shrubs of naturally small stature and those from high alpine regions were placed in the shrub collection or the beds along Azalea and Hickory Paths, many failed in these locations. Such plants have special drainage requirements and are less tolerant of extreme summer heat, and their size makes them more vulnerable to various threats among a general collection: encroaching weeds, vandalism, or being overlooked. In 1919 Sargent announced that a new alpine shrub bed had been constructed at the Orchard and Prince Street nursery complex. He designed a six-by-fifty-foot raised bed divided by rocks into numerous small planting pockets. Each plant was provided its preferred soil type, and the whole garden was shaded in summer by lath screens on an overarching pergola.

Under the watchful eye and meticulous care of the propagation staff, the miniature shrubs were watered and protected according to their individual requirements. Included was an assortment of natives and exotics of northern and high-mountain climates, such as twinflower (*Linnaea borealis*), dwarf birches (*Salix herbacea* and *S. uva-ursi*), and Labrador tea (*Ledum groenlandicum*). In all, more than three hundred kinds of dwarf shrubs were cultivated; even plants from the mountains of New Zealand were tried in the specialized habitat later on. Once the bed was established, Sargent alerted his collectors to obtain seed from alpine woody plants when possible.

While the hawthorn plantation still dominated the north and east sides of Peters Hill, in the 1920s Sargent developed the south side with what could be interpreted as miniature reforestation experiments. A band of native woodland separated the *Crataegus* collection from the south slope. Starting first in about 1920 with installation of *Larix leptolepis*, Japanese larch, a series of mass plantings of a few forest tree species were made during subsequent years. The Japanese larches were obtained from an American nursery company. More modest clusters of several other larch species, European and Asian, were located there as well. Sargent and Wilson also set out masses of oaks on this slope. Extensively planted were *Quercus bicolor*, the native American swamp white oak, and *Quercus mongolica*, Mongolian oak, but many other species of oaks of diverse origin were set out between 1920 and 1927. There is little on record to explain why these two genera were massed on the far slope of Peters Hill. Most likely, Sargent still wished to display as many important timber trees as possible in the group-planting format and Peters Hill was the only area remaining to accommodate them. Since he was also quite seriously trying to raise funds for an Arnold Arboretum timber tree breeding program, he could well have been intending their use in such experiments.

The planting rearrangements during this phase of Arboretum existence were a few of many steps toward the collections and landscape that are seen today. There was probably a greater change in appearance from the time of Sargent's death to the present than during the second half of his administration; nonetheless, knowing what transpired as a result of early-twentieth-century programs sheds much light on subsequent management of the collections and the present landscape. For the second twenty-five years these changes are difficult to trace, since the detailed records of the plantings, kept on "sheets of a large scale map" on which every plant was carefully plotted, have not survived. Sargent's annual reports, the 1911 *Guide,* the *Bulletin of Popular Information,* and the photograph collection are the only institutional record of the changes to the arrangement of the collections since the original plantings of 1886–92.

Because he was constantly at a loss for space to pleasingly grow all the woody plants he and his skilled agents had found, Sargent became fond of recommending one thousand acres as the minimum needed to start a new arboretum. The arrival of never-before-grown trees and shrubs from Asia was driving the desire for additional land, but even from regions better known there was material to be given trial. Sargent now interpreted the original mandate "to grow all plants hardy in said West Roxbury" rather broadly to include at least some of what are now called cultivars.

The growth of the United States nursery industry as well as those of Europe and Japan was the source of this new material. The formation of new kinds by hybridization, and the selection and naming of "sports," or forms, added to the plants that Sargent felt bound to represent in the Arboretum's collection. Not long after the turn of the century a deliberate attempt was begun to expand the collection of lilacs to include all garden varieties known. There was plenty of interest in this group shown by Arboretum visitors. The comprehensive collection became the basis for S. D. McKelvey's monograph. Likewise, cultivated forms in such genera as *Philadelphus, Deutzia, Malus, Prunus, Rhododendron, Viburnum,* and *Lonicera* were added to the collection. At the Arboretum, comparison and evaluation of their performance for gardens in the Northeast were made by the staff and by horticulturists from around the country. Many of them were treated, albeit briefly, in Rehder's *Manual.*

One category of woody plants that the Arboretum continued to exclude from its collections was domestic fruit. Although the wild progenitors of such comestibles as apples, plums, and peaches were accessioned, no attempt was made to include the numerous varieties selected by growers for quality of fruit production. Successful cultivation of such plants would have required maintenance procedures not possible within the Arboretum's means, and its acreage would never have begun to accommodate all of them. Studying and exhibiting pomological specimens was appropriately left to agricultural schools and experiment stations, which were taking rapid strides in this specialty.

Although the removal of plants was never mentioned in annual reports and the popular *Bulletin,* elimination of some sections of roadside shrubberies took place during the teens. This is apparent from photographs in at least two areas. The dense thicket that lined the west side of Meadow Road from in front of the administration building to the *Liriodendron* collection was gone by 1917. Out of this group, several large shrubs, among them Canby's viburnum (*Viburnum canbyi*) and vernal witch hazel (*Hamamelis vernalis*), were left to line the service drive and frame the entrance gate, however. Once the magnolias around the building reached the age to bloom reliably and the stature to form a display on their own, pulling out many of the shrubs probably improved the view of this early-spring-flowering group. Two cucumber trees (*Magnolia acuminata*) were planted in the lawn astride the entry walk in 1918, creating a more formal style for the front of the building than was usual for the Arboretum. Photographs also reveal that a line of low-growing roses that occupied the narrow strip between Forest Hills Road and the sidewalk on both sides was replaced by mown turf as of the late teens. Although it is probable that other areas were thus thinned, the roadside plantings of native shrubs

86. *One of several areas originally preserved as natural woods was the slope above the junction of Meadow, Forest Hills, and Bussey Hill Roads, as photographed in 1898 by Alfred Rehder. Shadbushes bloomed in this area each spring. Unfortunately, most of the native trees here were devastated in the hurricane of 1938.*

A·996

continued to be the subject of frequent remark in the *Bulletin*s during the remainder of Sargent's tenure.

The natural woods were never encroached upon to make room for Arboretum accessions. Although Sargent continued to have the areas maintained with thinning and pruning, they were otherwise left alone. The native woods were considered essential to the beauty of the place, and the pleasing landscape as important as the specimens that composed it. Their preservation was further argument in favor of obtaining more land for collections.

THE PROPAGATION DEPARTMENT

Crucial to the introduction of new and unknown woody plants from the far corners of the temperate world, and to their care once placed in the collections, was the activity of the Arboretum's propagator. For nearly forty years this role was combined with supervision of planting and maintenance in the person of Jackson Dawson. The glasshouse on the Bussey Institution grounds was where Dawson began his Arboretum operations in the 1870s. Part and parcel of the Arboretum's struggle to work with limited land and budget was the necessity to relocate its modest propagation, nursery, and maintenance facilities from time to time.

The initial planting of great numbers of trees and shrubs onto the grounds in 1886, which depleted the nursery, provided an opportunity to move Dawson's headquarters away from the Bussey Institution. From the trustees of the Adams Nervine Asylum, the Arboretum took a long lease on a "dwelling house and an acre of ground adjoining the Centre Street entrance." There, a fifty-by-twenty-foot glasshouse was built, and a frame yard and nursery areas for small plants were relocated. Dawson, his wife, and their seven children moved into the clapboard farmhouse at 1090 Centre Street. From 1886 on, he superintended the Arboretum grounds from this spot. In addition to the propagation of specimens destined for the Arboretum collections and supervision of landscape maintenance, Dawson's responsibilities included the distribution of plant material.

"The interchange of plants and seeds with other botanical establishments" was an item that Sargent never failed

87. *Jackson Dawson, propagator from 1873 to 1916, was known for his humor, wit, and uncanny ability to coax seed or cuttings of untried plants into growth. Several of his children entered the nursery profession, and one taught at a school of landscape design.*

to mention in his annual report to the president of the university. Through the distribution of seeds and plants to organizations and individuals, the Arboretum not only received new material for its collections in exchange but also kept a high profile among fellow institutions. Once Arboretum-sponsored exploration was under way, sharing material with other growers assured successful introduction should Arboretum plants fail and provided data on their hardiness range. Dawson's own reputation as a master plant propagator grew along with that of the Arboretum, and the science of propagating woody plants became a specialty of the institution.

In 1905, a step toward improvement of grounds operations was the purchase of a house and forty-two thousand feet of land at the corner of Prince and Orchard Streets, just across Centre Street from the museum building. The plan was to use it as nursery area and dwelling for the grounds superintendent if and when the lease on the facility at 1090 Centre Street was rescinded. Nonetheless,

Jackson Dawson stayed at 1090 for another eleven years; the Orchard Street property was used to house an assistant and the labeling department. Having spent more than twenty years customizing his operations to the "tiny" wood-beamed greenhouse and raising his large and happy family in the adjoining century-old farmhouse, it is no wonder the propagator did not wish to move.

Visiting horticulturists and journalists always received a warm welcome at Dawson's compact workshop, "the cradle of the Arnold Arboretum," as Wilson termed it. There the talk was of plants and more plants, and if one could induce Dawson to take a walk on the grounds the cheery man would tell the story of origin of nearly every plant encountered. In later years he was fond of pointing to a seventy-foot pin oak not far from 1090 Centre Street in the original boundary plantation, explaining that he started it from an acorn he collected in 1873; the tree still stands. Typical of his mischievous humor was to point out a flat of what looked like green grass, tell an unsuspecting guest to lift it up, and say, "There, now you hold a Chinese spruce forest in your hands."

After the turn of the century, Sargent hired assistants to work under Dawson. From 1908 to 1912, Ralph W. Curtis took on much of the burden of the grounds management. Having obtained a master's degree from Cornell under Liberty Hyde Bailey, he came to Boston in 1906 to supervise the tree-pruning program for the park commission. At the Arboretum, Curtis introduced some new methods, encouraging the use of machine-driven equipment for spraying, for example. Under his oversight extra nursery areas were added at the Orchard and Prince Streets property. When the herbarium wing was built onto the administration building, Sargent took Curtis's recommendation that space for the storage of equipment and supplies be planned for the basement. Years later Curtis remarked that his time at the Arboretum had been a great learning experience. Although his tenure was brief, when Curtis left to return to his alma mater to teach ornamental horticulture in 1912 he took with him a commitment to the value of arboreta that would ultimately bear fruit in the formation of a tree garden at Cornell.

At about the time Curtis left, a young Dutchman named Christian Van der Voet, who had been in the nursery business and participated in the two-year horticul-tural training course at Kew, turned up at the Arboretum looking for employment. He had come to the States and traveled about in hopes of finding a position. Sargent hired him to assist Dawson and soon found him best suited to supervise the outdoor work.

Still lacking "an understudy" in the propagation department, Sargent wrote W. J. Bean, curator of the arboretum at Kew, in the spring of 1913, asking him to recommend someone. Bean suggested another Kew graduate who had stayed on to work in the Royal Botanic Gardens' arboretum. The promising young man was William H. Judd, twenty-five-year-old son of a Cheshire estate superintendent. Judd, who had just turned down an offer of a position on a rubber plantation in Uganda, did not hesitate to accept Sargent's proposal. What he knew of the Arnold Arboretum and Dawson's reputation as a propagator encouraged him. He booked passage and arrived in Boston on 7 July, barely more than two months after first hearing of Sargent's interest.

Judd got along well with Dawson and learned quickly from the master propagator. Aside from his own knowledge of horticultural practices, Judd's interests ranged to many aspects of natural history. Charles Faxon found him an enthusiastic bird-watcher and took pains to let the Englishman know of unusual sightings in the Arboretum. Judd soon made American friends and looked up former Kewites in the Boston area with whom he took walking excursions to the Middlesex Fells, Blue Hills, and other wooded areas. He joined the New England Botanical Club. On an assignment with the labor crew to collect sphagnum moss from bogs in the country for packaging plants, Judd botanized during his breaks while the others sipped cold cider. For years he assembled his own herbarium as a means to become familiar with the American flora.

Ernest Wilson and young Judd had a common bond in their early training at Kew. When Wilson was not overseas, he took the new propagation assistant along on his trips in New England to give lectures or take photographs. Judd also accompanied Sargent and Wilson from time to time to the well-planted estates of Arboretum patrons. Like Dawson, Judd participated in the Massachusetts Horticultural Society and the Boston Gardeners and Florists Club. After three years in the States, he married an Ameri-

88. *William Judd (1887–1946) came from England to the Arboretum in 1913 to assist Dawson in the green-houses and nurseries. Since both were graduates of Kew's professional gardening program, Wilson took Judd under his wing and showed him the Arboretum grounds. Here, Judd poses next to a Japanese white pine (Pinus parviflora) for Wilson's ever-busy camera.*

can in early March of 1916. Later that month Judd gave his first public lecture, "Flowering Shrubs," which he illustrated by showing some of his dried specimens. Shrubs were becoming a specialty of Judd's, and he often spoke and wrote on this topic for the Arboretum.

During that year Jackson Dawson died, bringing to a close forty-three years of nurturing the Arboretum plants. From his start in the 1870s collecting native shrubs in eastern Massachusetts to handling all the seeds and propagating material from Asian expeditions at the end of his career, Dawson raised more than a million plants. His avocation had been the hybridization of roses, in which he undoubtedly was inspired by his early relationship with Francis Parkman. He successfully created some much admired hybrids involving *Rosa multiflora* and was one of the innovators of the class known as ramblers by rose fanciers.

Next to Sargent and Wilson, Dawson was the best known of the Arboretum staff throughout the horticultural world. He was often likened to a wizard or an elf for his seemingly magical ability to coax things to life and for his endearingly friendly nature. Dawson was equally respected for his oversight of the safe distribution of Arboretum plants and seeds to the far corners of the globe. Grand totals of 47,993 packets of seeds and 450,718 plants (including cuttings and grafts) were sent out under his supervision. While the horticultural world mourned Dawson's death and paid tribute to his well-spent life and generous character, no one of his colleagues would miss him more than Charles Sargent; Dawson had been with him from the start.

Although there was no replacing a particular individual, Sargent had prepared the Arboretum for the loss by grooming William Judd for the position of propagator of trees and shrubs. Christian Van der Voet was officially made superintendent of grounds. Two years later, when Wilson was away on his second Japanese journey, Sargent could write him: "Judd has certainly done remarkably well and most of your Korean seeds have germinated. . . . I do not think any one could have done better with them than he has" (AAA, C. S. Sargent correspondence, 19 March 1918).

The Arboretum administrator took advantage of the change in propagation staff to make an innovation he

89. *After the propagation unit moved to a site at the corner of Prince and Orchard Streets, across the Arborway from the administration building, there was a protected location for a modest rock garden. The long, raised bed was created for naturally diminutive shrubs of northern and alpine regions.*

hoped would improve efficiency in this department. The lease on the 1090 Centre Street property from the Adams Nervine Asylum was given up, and in 1917 new greenhouse and propagation facilities were constructed on the Orchard and Prince Streets parcel. (This was the site of the house that had been used by the grounds superintendent and where Jack's assistant, Louis Victor Schmitt, prepared labels for plants.) Sargent hoped time and labor would be saved with this consolidation of the nurseries, despite what might seem a remote location. In the teens there was no traffic circle at the junction of Centre Street and the Arborway, and neither road was as heavily traveled as it is now. Once the move took place, Sargent thought the nursery work was better "systematized" under Judd.

After Wilson's return from Korea, Formosa, and Japan in March 1919, Sargent made another staffing decision. With Faxon also gone, Wilson was given the position of assistant director. Sargent had come to rely on his aid in choosing places on the overcrowded grounds to site the abundance of new Asian plants, and the title gave greater authority to Wilson's role as promoter of support for the Arboretum among amateur and professional horticulturists. And, too, the war years had been hard on the Arboretum. For the first time, Sargent had trouble finding laborers who met his high standards for quality, and he relied on Wilson to mitigate difficult labor relations.

BOUNDARIES EXPAND, PERIMETER BUFFERED

While Sargent never slackened his efforts to make the Arboretum living collection as complete as possible, he also needed funds to enlarge its territory and to carry on dendrological research. The backing for botanical explorations that yielded exciting horticultural results could always be found, but increasing the endowment and obtaining land were tougher problems. Although gifts came in and the permanent fund grew steadily after the turn of the century, the Arboretum's income was never sufficient to pay for all its programs. Sargent often acted as a one-man development office. He did this on top of pursuing his own botanical research, as well as managing that of the Arboretum staff, all at a time when he was past the retirement age of our day. He did have the support of the Harvard overseers' Committee to Visit the Arboretum, who often suggested that more of the acreage that had been retained for the Bussey Institution be turned over for use by the Arboretum. The idea gained greater credence after 1908, when the Bussey was reorganized from an undergraduate school of husbandry to a graduate school of applied biology, and thus no longer used its land for farm operations.

Some of the requested acreage finally came to the Arboretum in 1919, when the property between South Street and the Dedham branch of the New York, New Haven and Hartford Railroad, formerly utilized by the Bussey, was purchased by friends of the Arboretum from the college. The visiting committee and the trustees of the Massachusetts Society for Promoting Agriculture were largely responsible for this measure. The piece was sixteen acres of mostly low, fertile land that had been the pasture of farmer Ezra Davis when Benjamin Bussey obtained it. The northern end of the property consisted of wooded knolls, an extension of Bussey Hill's lower slopes. Below these, Bussey Brook flowed from under South Street, skirting the wet meadow to join Stony Brook just outside the new Arboretum territory. The poplar and willow collections were slated for relocation to the tract, and work soon began on the preparation of the ground.

Since about half the acreage was marsh, underlaid by deep peaty soil, and since it was abruptly bounded by the railroad embankment, the South Street property presented a challenge not unlike the North Meadow. A plan of existing conditions, made in 1920, includes evidence that the course of Bussey Brook was deliberately altered in an earlier attempt to improve drainage for farming purposes, with several ditches having been dug to let water flow into the brook. Nonetheless, the peaty substrate spanning the area between the brook and the railroad embankment contained a large area of standing water, sometimes referred to as a pond, that Sargent considered a pestiferous mosquito-breeding area.

He hired an engineer to study the drainage and recommend ways to lower the level of groundwater. From among the suggestions, Sargent chose to change the line of Bussey Brook to cross the swamp more directly and flow through the ponded area as an affordable solution. Directing the brook through the pond would render it less stagnant and perhaps more attractive. The course of Bussey Brook was changed in the fall of 1921. At the same time a good deal of peat was dug out adjacent to the existing pond, deepening and enlarging it and moving its center away from the railroad. The peat was used to raise and enrich the soil of surrounding land so that it could be planted.

A nursery was opened on the drier ground on the south side of the new brook channel, providing an alternate area to raise young plants once the Peters Hill nursery was designated a permanent plantation. Expansion of the poplar and willow collections into what became known as the South Street tract (now Stony Brook Marsh) took place after drainage improvements and fencing along the railroad were completed. These two genera had never been satisfactorily placed in the original layout. A dense line of tree species of willows was planted to define the north and east boundaries. More willows were planted on the west side of the pond in recently drained ground and along the site of the former brook channel. The southern end of the tract, from the nursery to the arch that allowed South Street to pass under the railway, was given over entirely to poplars of every kind and provenance. At the southern end the poplars were contiguous (except for the intruding South Street) with the original *Populus* collection that occupied the lower slope of the Peters Hill tract.

The City of Boston added a modest contribution to Arboretum acreage in its fiftieth year, 1922. An ancient and long-unused cemetery of about 3,650 square feet was turned over to the institution's use and care. The plot was on the southwestern boundary of the Peters Hill tract,

adjacent to Walter Street, a treeless area that contained memorials to the Weld family, members of the Roxbury Second Congregational parish, and some British soldiers who died during the American Revolution. The headstones were not to be interfered with, but the stone wall that separated the cemetery from the Arboretum was removed and a mass planting of tulip trees was installed to shade the graves. While the addition did not expand collections space, the continuity of the boundary in this portion could be visually improved by extending the native tree belt. Removal of the stone wall allowed efficiency in mowing the plot, but, in the long run, it may have contributed to a loss of appreciation for the historic significance of the old burying ground.

As the groundwork for expansion of the collections

90. Carolina hemlock (Tsuga caroliniana), one of Sargent's and Wilson's favorite trees, was thickly planted along Walter Street to form a boundary plantation on the Weld-Walter tract soon after it was added to the Arboretum in 1923.

onto the South Street land was under way, another parcel on the opposite side of the Arboretum was added for future development. In 1923 the Arboretum acquired an additional fifteen acres of land to the north of Peters Hill. This piece was bounded by Walter, Weld, and Centre Streets and abutted land purchased many years earlier by the city to extend the corridor of West Roxbury Parkway to the junction with Centre Street. From a low, fertile area bordering on Bussey Brook, the land rose southwestward to a dry, open hillside. For many years it was referred to as the Weld-Walter tract (now Puddingstone Hill).

Initial planting of the area followed Sargent's usual pattern: mass plantings to screen the boundaries were established first. Sargent and one of the Arboretum's most enthusiastic supporters in the nursery trade, Harlan P. Kelsey, had been corresponding for some time on the desirability of establishing a plantation of Carolina hemlock (*Tsuga caroliniana*) somewhere at the Arboretum. Wilson joined the two in partiality to this conifer from the mountains of the southeastern United States, which they agreed had potential for northern gardens. Sargent dubbed it "the handsomest conifer we can grow in New England," and Wilson included it when once asked for a list of his ten favorite trees of the world. Kelsey pledged to donate a sufficient number of trees to cover a hillside, if a site could be found. In 1926, numerous young Carolina hemlocks were planted in a wide belt that extended the length of the southern boundary of the Weld-Walter tract. The hemlocks, together with the white pine boundary plantation established on the Peters Hill tract some twenty years earlier, formed an evergreen allée along the public Walter Street. Sargent defined the Centre Street boundary of the parcel by planting a belt of Japanese larch (*Larix leptolepis*), one of the species recently massed on the far side of Peters Hill. There is little on record to indicate what further use he envisioned for the interior acreage of the tract, however.

A much smaller land addition in 1924 pleased Sargent as much as the Weld-Walter annex, if not more. A mere two and nine-tenths acres acquired from the Adams Nervine Asylum caused more comment in the annual report than the Weld-Walter parcel:

During the year the northern boundary of the Arboretum has been extended by the purchase of 125,060 feet of land from the Adams Nervine Asylum, including a heavily wooded ridge with an approach to it from Centre Street. There has always been danger that the trees on this ridge, which form a beautiful border to an important part of the Arboretum, might be cut down and the ridge levelled and covered with buildings. This would have been a great injury to the beauty of the Arboretum, and although the area purchased is not as large as some of the recent additions of land acquired by the Arboretum this is the most important it could make anywhere for the beauty of the Arboretum as a garden. (C. S. Sargent, 1926, p. 221)

The area was left undisturbed, to grow as natural woods like the other North Woods knolls. In the valley between the ridges a planting of deciduous hollies terminated the Linden Path shrubberies. Plant records indicate that a subsidiary collection of Asiatic maples was started behind the hollies once the new land was acquired.

Subsequently, two more adjacent parcels of land beyond the same ridge were obtained from the Adams institution. Sargent concluded negotiations for the first, a low meadow of some two and three-quarters acres that included an aboveground section of Goldsmith Brook, in February 1926. Purchase of the second piece, while planned by Sargent and the trustees of the Adams Nervine Asylum, was finalized by Wilson in June 1927, shortly after Sargent's death. Its two and a half acres extended from the boundary of the 1926 acquisition up a gentle slope behind the viburnum collection. The tract included the asylum's former nurses' quarters and the clapboard house at 1090 Centre Street, for years leased by the Arboretum for its propagation facility. After the Arboretum assumed ownership of the "Dawson Cottage," it was apparently leased back to the asylum for a time. The nurses' house was immediately torn down. The land was purchased because the Arboretum administrators wished to extend the institution's boundary to Centre Street and to prevent any other development that might mar its beauty if the asylum trustees were to find other buyers. The Arboretum made no systematic use of the land until the 1960s, except that the botanist E. J. Palmer occupied the house at 1090 Centre Street.

INTERPRETATION AND PUBLIC RELATIONS

Like its scientific program and its collections, the Arnold Arboretum's relationship with the public evolved many layers, or dimensions. The prevailing assumption about its educational role was that the multitude of plants, each labeled with its name and place of origin and set out in a pleasing arrangement, was educational in and of itself. Indeed, the collections gave pleasure to many; they won over the interest of a few; they educated those with a well-developed interest in plants.

What worked for the institution in its interaction with the public was the enthusiasm of its staff for the plants and the place. Staff members were frequently interviewed by the press, and they supplied many articles to horticultural and general magazines. The Arboretum's own interpretive publications were the educational tool that reached the greatest number of amateur and professional plantspeople. The institution responded to an increased interest in woody plants for ornamental planting by providing information and the exchange of plant material. With its emerging sister institutions, especially, the trade in plants and know-how became a two-way street.

In the early phases of Arboretum development, when American forests were uppermost in his mind, Sargent believed it should show not only living trees but also the valuable products obtained from them. Toward this goal he had two large halls constructed in the building to house educational exhibits. Soon after the turn of the century, the creation of exhibits was made possible in part by the generosity of two of Sargent's colleagues, Morris K. Jesup and George Russell Shaw. In 1902, Jesup donated the cost of furnishing one of the first-floor exhibition halls with glass cases to house a display of wood samples of nearly every North American tree. The samples, each a fourteen-inch-long half section of trunk showing bark and wood, polished and unpolished, were smaller pieces from the great collection in the American Museum of Natural History, assembled by Jesup with Sargent's assistance in the 1880s.

While Shaw used the Arboretum as his headquarters for the study of pines, he perceived needs for the museum building that could not be met by the institution's limited budget and offered his own aid. First, Shaw provided for fireproofing the attic's ceiling and installing a hardwood floor, thus converting that area into a tolerable workspace for herbarium specimen preparation and a useful storage place. Not to be outdone by Jesup, perhaps, the next year Shaw "fitted up [the other] of the halls in the Museum Building with handsome and convenient cases for the preservation and display of the collection of conifers." He took charge of this exhibit, mounting and organizing specimens of cones, foliage, and wood with carefully scribed labels for the public. For the next twenty years the two first-floor halls stood open to visitors daily, their ranks of polished wood and glass cases ordering the products of the forest for those who ventured in. The one hall became known as the Jesup room, the other the conifer room. More than one chronicler remarked on the hushed and somber air of the place.

From the press coverage, however, it is fair to say that the living exhibition was of far greater interest to the majority of Boston's public than the displays in the museum. As of 1900 visitors could obtain the aid and stimulation of a guide map printed by George Walker Company, "compliments of Macullar Parker Company Clothiers." This is the earliest known map of the grounds intended for public use. It was attractively printed in three colors, depicting the main roads and paths in gold and the grass paths, native woodland cover, and border plantings in green. The main tree genera were listed in a key to location on a grid and shown in place on the map with both common and scientific names. Roads, paths, brooks, and ponds were not named, nor were the gates, although their existence was indicated by symbols. The only geographic features highlighted were the North Meadow, called "swampland," and the three hills.

The Peters Hill tract, where only the poplars and some conifers had been planted so far, was shown to be almost bare of tree collections. Remarkably, this 1900 depiction included the "proposed new Bussey Street," the alternate route through the quarry site on the Peters Hill tract, suggested by the Olmsted firm in the late 1890s. Either this was an error or the hope of depressing Bussey Street, and thus more completely unifying the two sections for visitors, had not yet been given up. The publisher most likely used an Olmsted plan for the Arboretum as the basis for the guide map. The large sheet also included a smaller map that indicated transportation routes to the Arboretum and its position within Boston's park system.

It is not generally appreciated how much of a role everyone on the staff played in interpreting and publicizing the Arboretum for its diverse population of users. Alfred Rehder is usually remembered for his botanical contributions, yet early on he shaped its role as provider of horticultural information. Before 1911, Rehder found time to author a noteworthy series of articles for *Horticulture,* a new Boston-based weekly "devoted to the forest, plantsman, landscape gardener and kindred interests." During two growing seasons, starting in March 1908, his "Notes from the Arnold Arboretum" appeared nearly every week inside the front cover. Rehder's column informed readers what was in bloom, fruit, or fall color in Boston's tree garden, highlighted new introductions, and remarked on distinguishing characteristics and historic origins.

Rehder also gave carefully considered opinions on the ornamental or decorative aspects of the trees or shrubs he discussed. When the lilacs bloomed he suggested the best varieties of *Syringa vulgaris* in each color group and discussed the attributes of some of the wild types as well. He once remarked, for example, that "the Lilacs of the Persica-group should never be planted together with varieties of the Vulgaris-group, as the two are too different in habit" (A. Rehder, 1908, p. 717). Despite the enthusiasm for Alfred Rehder's up-to-date coverage of "new and useful trees and shrubs as tested at the Arnold Arboretum," the series stopped in September 1910. Most likely, the return of Wilson with more herbarium collections from a second trip to China for the Arboretum, coupled with the impending publication of the first volume of the *Bradley Bibliography,* left the taxonomist too busy to contribute to the horticultural press. His column set the stage for the institution's own published news briefs, however.

The following spring, just as the star magnolias in front of the administration building unfolded their pinwheels of white, the Arboretum's communication with the public resumed under a new guise. In early May 1911, Sargent was pleased to announce the commencement of the *Bulletin of Popular Information* in response to frequent inquiries about what to see and when certain things bloomed. Sargent described the intent of the new publication as being "to give, from time to time, in popular language, authentic information about the plants in bloom or otherwise worthy of special visits" (C. S. Sargent, 1912a, p. 188).

The *Bulletin* was available free of charge at the Arboretum's administration building; copies could also be obtained from the secretary of the Massachusetts Horticultural Society, from Houghton Mifflin Company, and from the Old Corner Bookstore in downtown Boston, as well as by subscription. It was published weekly from April to November usually omitting August and September. Sargent's announcement was published in *Horticulture,* and for a time that weekly included excerpts until the *Bulletin* came to be widely circulated on its own. Like Rehder's notes, the three- to four-page publication served both to prepare readers for a visit to the collections and to provide botanical and horticultural information on woody plants of interest to homeowners and growers.

That spring, Sargent also announced the publication of *A Guide to the Arnold Arboretum.* Written by the director and printed at the Riverside Press, Cambridge, this compact volume contained thirty pages of description of the grounds and plantings, an essay on birds in the Arboretum by Charles E. Faxon, and several fine black-and-white photographs by T. E. Marr. Included inside the covers were two maps, one of the vicinity around the Arboretum depicting how to reach it from nearby transportation terminals and a larger map of the grounds showing the position of all walks, roads, and groups of trees.

With the *Guide*'s thorough description of the grounds, Sargent inaugurated names for the roads and main paths to aid in orientation and location of plant groups. After giving a brief account of the roles of Harvard University, James Arnold, George B. Emerson, and Benjamin Bussey in the Arboretum's origin, as well as noting the participation of Frederick Law Olmsted and the City of Boston, Sargent explained the topography, systematic arrangement, use of native shrubs along roadways, preservation of natural woods, and the records and labels used to document the collections. In a whirlwind tour up and down the major named grass paths, as well as all the roads from the administration building to the height of Peters Hill and return, each and every collection was located and discussed briefly with comments on origins, outstanding ornamental qualities, and more.

To those who did not know the place like the back of their hands, the *Guide* could have been difficult reading if they simultaneously tried to keep oriented. For example, to cover everything the director wanted people to see between the administration building and the end of Meadow Road, Sargent had them go from the Jamaica Plain gate down Willow Path, then back to the Jamaica Plain gate; he then described everything along Meadow Road to the natural woods, then the trees that could be observed if one took Linden Path, then Maple Path (both paths roughly parallel Meadow Road). This would have been enough for one day for most enthusiastic tree students, but Sargent continued this pace and detail for the entire grounds. It shows that despite the well-planned single road following the sequenced planting, there was something worth seeing, to Sargent at least, on every part of the grounds. Fortunately, the map accompanying the *Guide* could be used for orientation, since it incorporated names for the roads and paths in addition to locating the major tree and shrub groups.

The *Bulletin* and the *Guide* opened a new era of public relations for the Arboretum. They were the first institutional attempts to interpret the tree garden fully for the public. Although many articles had been published in the papers and magazines, the *Guide* gave Sargent a chance to explain the whole layout, and the *Bulletin*s would routinely inform visitors, nurserymen, estate owners, groundskeepers—anyone who fancied or worked with woody plants.

VISITORS RESPOND WITH INTEREST IN ORNAMENTALS

The first edition of *A Guide to the Arnold Arboretum* was apparently such a success that a second printing was made in 1912. From the interest shown in the *Bulletin of Popular*

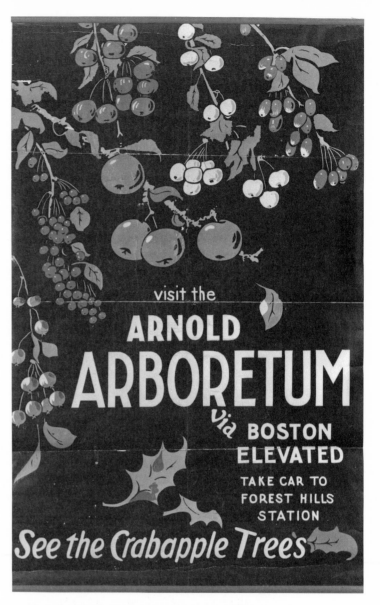

91. Once the elevated electric streetcar line was extended to Forest Hills in 1909, this poster (now in the Archives of the Arnold Arboretum) enticed people to take that route to view the multihued foliage and fruits in autumn.

Information, too, the Arboretum administration was heartened to see it was fulfilling an educational purpose. A recent change in transportation facilities for Boston's citizens undoubtedly contributed to an upswing of visitors at this time. In 1909, an extension of the elevated electric streetcar to Forest Hills was opened. Now people could ride from downtown Boston to Forest Hills in a few minutes at a cost of five cents. Not long afterward, the Arboretum's *Bulletin* carried the announcement that those who wished could hire carriages from two local stables to take them through the grounds. As another twentieth-century transportation mode came increasingly into use in the teens, the *Bulletin* was also obliged to admonish that automobiles were not permitted and were best left at the gates.

Less than ten years later, however, the Arboretum was forced to adapt to the inevitable. In 1918 a limited number of motorcars were allowed in by special permit, granted either by the park commissioner or the Arboretum director. Apparently the permits allowed one year's motoring through the grounds on weekdays between ten and two o'clock. By 1920 Sargent was so overwhelmed with applications that he was obliged to grant them only to persons who contributed annually to the Arboretum. The road surfaces were still hard-pressed gravel, and there is

92. Lilac Sunday, 1908. It became a tradition in Boston for thousands to visit the Arboretum on the Sunday in May when the lilacs were at their peak of bloom.

little photographic or other data as to the number of motorists that passed through. Although not many drivers could have passed through on any given day, the arrival of the automobile was a turning point, foreshadowing the much greater impact on the landscape that would come once roads were paved with asphalt.

Bostonians contemplating a visit to the Arboretum kept abreast of what to see every spring, as the sequence of bloom was also reported in the papers. In order, the flowering cherries, crabapples, lilacs, azaleas, and finally mountain laurels were the groups usually noted as worth the trip to Boston's great dendrological museum during each one's particular week of glory from April to June. The lilacs, especially, occasioned numerous articles each May when this most popular collection reached its height of blossom and fragrance. Lilac Sunday was not initiated by the institution, but arose spontaneously as visitor interest gained media attention. It quickly became a tradition for thousands to stroll the Arboretum grounds on the Sunday that the lilacs were at their best. Sometimes a reporter attended J. G. Jack's field classes, rendering more detailed accounts of Arboretum plant lore in succeeding columns. Wilson's accomplishments in China and the resultant expansion of the collections with rare plants from far-off lands made good copy and captured public attention.

Besides its publications and its instructive plant exhibition, the Arboretum staff's expertise was often shared directly with both amateurs and professionals. In addition to his popular field classes, Jack also at times conducted sessions on the Arboretum grounds for students from the newly formed departments of landscape architecture at MIT and Harvard. He led special classes of varying duration for Massachusetts Horticultural Society members, superintendents of the Metropolitan Park System, an informal group of Boston architects, and the Massachusetts Forestry Association as well. When Harvard started to offer an undergraduate course in forestry in 1903, Jack taught dendrology, arboriculture, forest pathology, and entomology. Once the Harvard Forest in Petersham was acquired in 1907, he provided instruction for graduate students in forestry. Special students also continued to

study at the Arboretum under the staff's guidance. One was Chen Huanyong (1890–1971; former spelling: Woon-Young Chun), from Shanghai, who stayed from 1915 to 1919 to gain knowledge of trees suitable for reforestation in his homeland. In the mid-1920s J. G. Jack guided the studies of Hu Xiansu (1894–1968; former spelling: H. H. Hu), another Harvard graduate student who returned to influence the development of botany in China.

Jackson Dawson was occasionally prevailed upon to lecture on propagation techniques and was always willing to show the Arboretum greenhouse and nurseries to journalists and visiting professionals. William Judd became part

93. John George Jack (left) points out the branching characteristics of trees to forestry students. Among those who studied the Arboretum collections with Jack were several botany students from China, as well as aspiring landscape designers.

of the public relations and education force even before he succeeded as head of the propagation unit. Alfred Rehder regularly assisted with the identification of unknown plants for amateur and professional growers and for his botanical colleagues. His continuing contributions to L. H. Bailey's publications, such as *The Standard Cyclopedia of Horticulture* (1914–17) and *Cultivated Evergreens* (1923), spread the Arboretum's expert knowledge to a wider audience.

Ernest Henry Wilson became an ardent publicist for the institution whenever he remained at his adopted home base. An important part of his work was to check on the progress of the plants he came to refer to as "his children." When he was not in the field, he enjoyed telling people about the beautiful and intriguing plants he had found. The plant explorer began writing articles about his adventures in China while still working for Veitch. Once he returned from his Arboretum expeditions, one of the first reports on these was published in *National Geographic* for November 1911 under the title "Kingdom of Flowers—China." Several of his excellent photographs of the land and people were included. Wilson authored many books on his adventures as well. After 1911, he had lantern slides made of some of his favorite images of China, Japan, and their floras that he used to illustrate the lectures he was often called on to give. Illustrated talks on the Arboretum itself became part of his repertoire. In the late teens and twenties garden clubs became a more frequent audience, and the assistant director encouraged the garden club movement as an important force for promoting horticulture and public gardens.

Wilson employed his genius for communicating about the value of plants to the life of humankind in writing *America's Greatest Garden,* a book about the Arboretum published in 1925. Even though a new edition of *A Guide to the Arnold Arboretum* by Sargent had been published in 1921, it was not much changed from the 1911 version except for mention of a few of the most successful introductions from the Far East. Apparently Sargent felt there was room for Wilson's interpretation of the living collections as well. "This little book," a 123-page volume, was written in response to public demand for a popular work

of moderate cost concerning the Arnold Arboretum. It included fifty photographs, views and individual plant portraits, mostly by the author. Wilson noted apologetically that it was not a definitive guide, but rather "a note of invitation to a banquet of flowers and fruit provided by an assemblage of the world's best hardy trees and shrubs." He was determined to appeal to "ordinary folk," whom he had come to know from years of lecturing to garden clubs, nurserymen's associations, and civic and other groups, and his book resembles a series of informal talks or tours through the grounds. Perhaps he sensed that the place was by then too complex to be explained by an all-encompassing guide such as Sargent's.

In the text, Wilson highlighted the Arboretum's assets in each season, entitling the relevant chapters "Spring Pageantry," "Summer Luxuriance," "Autumn Glory," and "Winter Beauty." These were followed by twelve chapters on the most important plant groups or collections: "Cherry Blossom Festival," "Crabapple Opulence," "Lilac Time," "Azalea Carnival," and so on. "Climbing Plants," "Border Plantings" (that is, the native shrubberies along drives), and "The Shrub Garden" were given separate chapters as well.

Despite Wilson's heady prose and unabashed sentimentality, *America's Greatest Garden* gives a glimpse of the Arboretum just two years before Sargent's administration came to a close. Wilson's style couched accurate botanical information in language colorful enough to interest almost any reader, whether a keen plantsperson or an admirer of "the landscape beautiful." He interpreted the Arboretum dually as a locus for enjoyment and for comparative study of the world's hardy flora. In every place and in every season Wilson pointed out the wealth of stimulation for all the senses. Also stressed was how the planted collections were artfully intercalated with the natural features of the site.

Throughout *America's Greatest Garden* Wilson freely gave information to aid and encourage the comparative observation for which the Arboretum was so suitably planned. Whether the goal was to become familiar with botanical classification and species identification or to evaluate their relative merits for use in gardens or parks,

1. Japanese wisteria as it grew on the trellis that edged the shrub collection from the 1950's to the early 1980's. The violet-blue flowers open sequentially in long pendant clusters that may attain more than twenty inches in length.

2. *Malus halliana* variety *spontanea,* a crabapple admired for its white flowers and vase-shaped form, was introduced from Japan into cultivation in America by the Arnold Arboretum in 1919. The species name honors George Rogers Hall, a Rhode Islander, who was the first to bring back a wide assortment of Asian plants to New England in the mid-nineteenth century, before the establishment of the Arboretum.

3. Despite Charles Sargent's early misgivings about placing showy garden shrubs prominently on the slope above Bussey Hill Road, the great mass planting of forsythia remains a welcome signal of spring's arrival. This location is adjacent to that of their relatives in the olive family (Oleaceae), the lilacs and ash trees, within the original Bentham and Hooker sequence.

4. In spring the buds of white oak *(Quercus alba)* expand to reveal felt-covered new leaves and green catkins. A subtle but major component of the Arboretum's vernal charm is contributed by the first delicate tints in the canopy of native forest trees such as oaks, maples, and birches.

5. *Rosa arnoldiana* cultivar 'Arnold' is one of many shrubs that brighten the palette in the Bradley Collection of Rosaceous Plants in June. This rose was selected by Arboretum propagator Jackson Dawson from hybrids he made between *Rosa rugosa,* the beach rose from Japan, and 'General Jacquiminot,' a popular hybrid perpetual with brilliant crimson flowers.

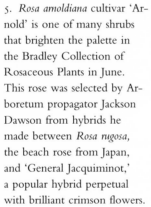

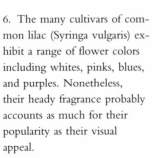

6. The many cultivars of common lilac (Syringa vulgaris) exhibit a range of flower colors including whites, pinks, blues, and purples. Nonetheless, their heady fragrance probably accounts as much for their popularity as their visual appeal.

8. The intriguing flowers of the harlequin glorybower *(Clerodendrum trichotomum)* appear from late August to September in the Arboretum. The hardiest species of a largely tropical genus, this *Clerodendrum* has stamens and style that protrude conspicuously beyond its white petals. Further color is provided by the red calyx, which later thickens and deepens in color as the small fruit it subtends turns metallic blue.

7. The rhododendron collection was extended along both sides of Bussey Brook in the early twentieth century. In this location, at the base of Hemlock Hill, these June-flowering, evergreen shrubs form an elegant foil for the native hemlock stand.

9. Below, left: Ernest Wilson wrote in *America's Greatest Garden* that during spring's pageantry, "bud-scales which have so splendidly protected the vital growing points of leafy shoots or shielded embryo flowers [all winter] are thrown aside, hastily . . . by some plants, tardily in others where they grow considerably and often become highly colored and conspicuous. . . . (Wilson, 1925, p. 9). The pinkish bud scales of striped maple *(Acer pensylvanicum)* are shown here as they open to allow expansion of leaves and drooping clusters of green flowers. 10. Below, left, middle: The fruits of the Yunnan crabapple *(Malus yunnanensis)* begin their transformation from green to purplish red in September, when this was photographed. As the season progresses, this crab stands out among its cohorts on Peters Hill, as its foliage turns from glossy green to orange and scarlet. 11. Below, right, middle: Small, blue, olive-like fruits are borne in profusion on the specimen of Chinese fringe tree *(Chionanthus retusus)* located along what is now known as Chinese Path on Bussey Hill. *Chionanthus* (the name is derived from the Greek words for snow flower, descriptive of its fleecy white blooms) is one genus that exemplifies the relatedness of the floras of eastern Asia and eastern North America. Despite its common name, *Chionanthus retusus* is native throughout China, Korea, Japan, and Taiwan, while the only other two species in the genus are native to eastern North America. 12. Below, right: The aptly named beautyberry *(Callicarpa japonica)* yields tiny, but densely clustered fruits of a pinkish purple hue, unlike any other in the Arboretum's fall display.

13. The high point of spring for many Arobretum visitors occurs in mid-May, when several hundred kinds of lilacs begin to bloom along Bussey Hill Road. When the recently developed, late-blooming hybrids and all the wild species are taken into account, the lilac season extends well into June, however.

14. The bold silhouettes of the cork trees (species of *Phellodendron*) along Meadow Road are enlivened early each spring when the yellow-flowered cornelian cherries planted behind them come into bloom. Not related to cherries at all, these two species, *Cornus mas* and *C. officinalis,* are Euro-Asian relatives of our flowering dogwood.

15. Right: Tupelo *(Nyssa sylvatica)* puts forth new leaves. When the Arboretum was initially aranged the witch hazel, aralia, and dogwood families were planted in the area around Rehder Pond. The tupelo was then considered a member of the dogwood group and several were planted just above the pond. Fortunately, this location coincided with the preferred natural habitat of this eastern North American species, on sites adjacent to wetlands and water courses.

16. Below, left: The mottled, exfoliating bark of Japanese Stewartia *(Stewartia pseudocamellia)* adds winter interest to this small, summer-flowering tree's many favorable attributes. The several species of *Stewartia* can be seen on the Bussey Hill plateau and near the Centre Street gate, two areas where newly introduced Asian plants were concentrated after the turn of the century. 17. Below, middle: The dawn redwood *(Metasequoia glyptostroboides),* a deciduous conifer discovered by scientists in China as recently as the late 1940s, was known previously only from fossil remains. Since its introduction into cultivation by the Arnold Arboretum, it has become admired for its rapid growth, pyramidal shape, delicate foliage, and the convoluted appearance of the trunk as it matures. 18. Below, right: Although conspicuous color on twigs is a character of some woody plants that may be noted all year, it is especially appreciated in winter when shrubs such as the red-twig dogwood *(Cornus alba* variety *sibirica)* stand out in the landscape.

19. The sassafrass and spicebush group (Lauraceae) on Bussey Hill Road opposite the lilac collection remains in the same position originally provided it when the Arboretum was laid out in the Bentham and Hooker sequence. The native American tree, *Sassafras albidum,* and the spice bush *(Lindera benzoin)* were joined by their relatives *Lindera obtusiloba* and *L. angustifolia* once the Arboretum began its active program of Asiatic plant exploration. The latter species is known for its brilliant red fall coloration, while its relatives turn shades of yellow and gold.

20. The golden-rain tree *(Koelreuteria paniculata),* a native of China long-cultivated in the West, adds color to the summer landscape along Meadow Road. Although individual flowers are small, they are borne in large, loose panicles which cover the trees for nearly a month from July into August. Equally ornamental are the bright green, bladder-like seed pods that follow. They persist through the fall, eventually turning a rich brown. Another much-admired summer blooming tree, sourwood *(Oxydendrum arboreum),* can be seen behind the golden-rain.

Wilson pointed out how the Arboretum collections could be used. He also managed to weave explanations of basic biological processes into the narrative in an attempt to awaken the most casual observer to a larger understanding of life.

His summary of institutional accomplishments in the final chapter reveals his view of the Arboretum's roles:

Assembles and nurtures all that is beautiful, interesting and hardy among woody plants from all parts of the world; distributes its surplus material among kindred institutions, nurserymen and garden lovers throughout the five continents. Seeks knowledge for the sake of knowledge. Garners information of every sort that concerns woody plants for the use of the specialist, the student and the tyro. . . . This is what the Arboretum does. (Wilson, 1925, p. 11)

Wilson, the world traveler, was not reticent about the moral and cultural role of a garden such as the Arboretum, either: "Toward bringing man nearer unto man this garden is a potent force. It exists for service, which service knows no boundary of race or creed. The Arnold Arboretum is worthy of the Nation and of the Nation's pride. Accept the invitation here set forth. Visit and judge of its value to the culture and amity of mankind" (Wilson, 1925, p. iv).

Since Wilson was well known as an author of books on his China travels and on trees, his ebullient account publicized the Arboretum as no previous work had. Even before *America's Greatest Garden* came off the press, Wilson had been marketing it, since the more widely distributed was the book, the more widely known was the Arboretum. He enlisted a few nurserymen who took special interest in the Arboretum to preorder quantities of the volume to sell to their customers. The book was noticed with several reviews. Sargent convinced the botanist Alice Eastwood of the California Academy of Sciences to bring it to the attention of Californians, writing, "Who can we get to write a notice of Wilson's book in the Pacific Coast papers? It is desirable to have it as widely known as possible in the interest of the Arboretum" (AAA, C. S. Sargent correspondence, 3 July 1925).

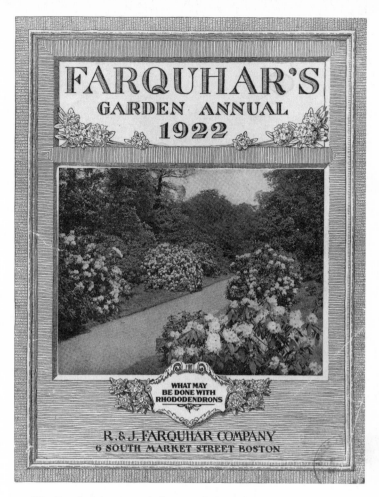

94. *The Farquhar Company, which was partly located in nearby Roslindale, was among the nursery businesses that gave financial support to Arboretum expeditions. It featured many of Wilson's plant introductions in its catalogs.*

PROFESSIONAL AUDIENCE

The nursery trade benefited greatly from the plant material and experience of the Arboretum. The institution long had close ties with commercial growers of woody plants in Europe and America. When the Arboretum began there was only a handful of American firms dealing in ornamental woody plants—the greater demand from nineteenth-century gardeners was for fruit trees and vines. Most estate owners, park planters, landscape gardeners, and public garden managers had to send to England or the Continent to buy trees and shrubs—even native American species were rarely grown and sold in the States.

This was part of the reason Sargent so emphatically stressed the importance of planting native American trees. The dominance of British and French growers over what was available to American planters limited the potential variety from which to choose. Even most American nursery businesses routinely imported understocks and other material for sale to their customers. Because the climate and growing conditions between eastern North America and Europe are not that well matched, much of the stock raised in Europe failed when planted in the eastern United States, and Americans were thus further limited in their selections.

These difficulties notwithstanding, growers in Europe were doing a thorough job of promoting the use of a wide variety of plants. English firms like Veitch and Sons had supported exploration and worked closely with botanical gardens to authenticate material offered to the public. The work of introducing and breeding superior ornamentals was prosecuted by firms such as that of Maurice de Vilmorin at Les Barres and the Lemoine Nursery at Nancy (both in France) and by H. A. Hesse at Weener on the Ems River in Germany, all suppliers of plants to the Arboretum. As the results of the Arboretum introduction program began to be seen, the knowledge of how woody plants could be more successfully grown in America came with them. A classic example is that of the Douglas fir, a conifer native to the Pacific Northwest. The tree rarely lived when New England conifer fanciers imported it from England, since English nurseries got their seed from the mild, moist forests of coastal Washington and Oregon. Once seed was obtained from Douglas firs of outlying populations growing in the colder Rocky Mountains, the resultant plants proved hardy in New England. The similar case of the cedar of Lebanon has been mentioned. In the course of woody-plant trials, the Arboretum staff noticed that the rootstocks used on the other side of the Atlantic for grafting certain genera such as rhododendrons or lilacs often caused poor performance in America.

As the Arboretum and like institutions began to make these observations known and publicize their discoveries of new plants, nurseries in the United States were starting to come into their own. After the turn of the century, as Americans came to rely less on producing food at home and suburbanization created a greater demand for landscaping materials, the trade in woody plants turned from fruit bearers toward greater diversity in woody ornamentals. The first federal plant quarantine legislation, passed by Congress in 1912, and made prohibitive in 1917, stimulated American firms to produce more of their own ornamentals. The effect of World War I on the European economy also led to the advance of American nursery businesses.

Arboretum publications, collections, and expertise were invaluable resources for the trade. In the Boston area, long-existing organizations such as the Massachusetts Horticultural Society were joined after 1900 by newer professionally oriented groups such as the Horticultural Club of Boston and the Gardeners and Florists Club, evidencing the heightened need for accurate information on the science of plant culture. The Arboretum could provide such information. When reasonable requests came, the institution readily shared its propagating material with nurserymen, often working out an exchange. The role played by arboreta and botanical gardens in educating the plant-buying public gave further impetus to the trade. Once his traveling ceased in the 1920s, E. H. Wilson actively carried on a campaign to gain the support of nurserymen for the Arboretum.

Landscape architects and garden designers likewise made use of Arboretum resources. Students of this subject, as well as professionals, learned the materials suitable for landscape improvements in New England and observed an excellent example of naturalistic design. Even after schools and universities began to offer courses of study and degrees in the subject, the Arboretum occasionally accepted special students intent on a career in landscape design. In the 1890s Sargent had overseen the early training of Beatrix Jones (later Farrand), who subsequently built a successful practice as a landscape gardener, the term she preferred for the work. At her suggestion, in the early teens, Sargent agreed to guide the instruction of two more would-be landscape gardeners. One of them, Gladys Brooks, described the contagious enthusiasm of her Arboretum mentors:

Mr. Jack, like Professor Sargent and probably every person working in that large administration building, lived for and within the world of woody plants that sent their roots deep beneath the soil and their branches toward heaven. The hearts of people such as these came alive when they were in company with a tree of fine symmetry, their blood coursing in apparent harmony with its sap; an example of the psychiatrist's term "identification," only that in the case of plant lovers their devotion was deep and profound and valid and tender and lifelong in its duration. (G. Brooks, 1962, pp. 74–75)

There was at least one other organization zealously introducing plants from overseas with which the Arboretum worked in concert at times. Although the United States Department of Agriculture's Office of Foreign Seed and Plant Introduction, headed by David Fairchild, had crops as its top priority (in contrast to the Arboretum's interest in trees and shrubs of scientific and potential aesthetic value), the two establishments exchanged plant material and information. Sargent and Fairchild coordinated the efforts of the Office's collector, Frank N. Meyer, with those of Wilson when both were in China. Fairchild, whose interest in Japanese flowering cherries was instrumental in the famous planting of these symbols of friendship around the Tidal Basin in Washington, D.C., encouraged Wilson's efforts to test them at the Arboretum.

THE ARBORETUM AND ITS PEERS

The Arnold Arboretum influenced and felt the influence of other American public gardens for botanical education in many ways. Although for several years it was unique among such institutions in being restricted to woody plants, the Arboretum shared some features and purposes with existing botanical gardens, notably the Missouri Botanical Garden and the New York Botanical Garden. Just as he had a hand in the development of the Arnold Arboretum, Asa Gray influenced the scientific program of the Missouri Botanical Garden in Saint Louis through his friendship with George Engelmann.

When successful businessman Henry Shaw (1800–1889) determined to create a public garden similar to the one at Kew, he consulted the Royal Botanic Gardens' director, who suggested he enlist another resident of Saint Louis, George Engelmann. Starting in 1859, Shaw planted and developed the garden with the intention of leaving it to trustees to be perpetually maintained for public benefit. Although Shaw personally designed the garden plantings, it was Engelmann and Gray who convinced him to provide for a scientific program, including establishment of an herbarium and a library. Engelmann obtained books and herbarium specimens while traveling in Europe, and the Missouri Botanical Garden became the repository for his own extensive herbarium. On Shaw's death the garden became an institution for horticultural education through its displays and research by alliance with Washington University. At the start the research focus was on systematic botany, Engelmann's field and that of the first director, William Trelease.

Unlike the Arnold Arboretum, there was no overall theme or scientific principle underlying the arrangement of the garden. This was often the case with gardens formed from private estates. Included on the seventy-five acres were conservatories, formal flower gardens, shrubberies, an arboretum, and a museum to house the library and herbarium, as well as exhibits of natural history and economic botany, Shaw's residence, and even his mausoleum. The arboretum was one small part of diverse plantings; it had no mandate for completeness such as enjoined the Harvard tree collection. In a description of a visit to "Shaw's Gardens" in 1874, the rather exacting editor of the *Horticulturist* stated: "It is not strictly an arboretum, for it is not complete. . . . It is more favorably considered as a pretty pleasure ground" (Williams, 1874, p. 227). Nonetheless, the Arnold Arboretum and Missouri Botanical Garden's joint sponsorship of the plant collector E. J. Palmer in the teens is typical of the scientific cooperation possible between gardens despite their different styles of living collections.

In 1889, the very year that the Missouri Botanical Garden came into formal existence under trustees of Shaw's will, some New York botanists started to put their impression of Kew Gardens into form. Nathaniel Lord Britton (1859–1934), professor of botany in Columbia College, and his wife, Elizabeth Knight Britton (1858–

1934), were so inspired by the royal gardens on a visit made the year before that they convinced members of the Torrey Botanical Club to form a committee to study the possibility of a similar institution for New York. The result was the establishment in 1895 of the New York Botanical Garden, to be located on a portion of Bronx Park. Once a substantial private subscription was raised, it was organized under the city government in association with Columbia University; these relationships were quite unlike those of the Arnold Arboretum. Planning was in the charge of a "commission of architects, engineers, gardeners and botanists" (Britton, 1898, p. 292), of which Britton

95. *Although the Rochester, New York, park system's most diverse woody-plant collections were located in Highland Park, extensive plantings were made in all its branches. Here the late-flowering prairie rose (*Rosa setigera*) rambles along a stairway in Seneca Park.*

was the driving force. The commission consulted Sargent, who had experience revamping the Harvard Botanic Garden in addition to creating the Arboretum, on the possibilities of the site, but apparently few of his suggestions were heeded. Native woodlands were preserved, and the first outdoor plantings were a fruticetum (shrub collection), pinetum, and a deciduous arboretum. Extensive collections of vines, orchids, and tropical plants were developed under glass. A museum, opened to the public soon after the turn of the century, housed exhibits on fossil plants, economic botany, and systematic botany. Under Britton's directorship, the New York garden developed as a center for research on the floras and economic botany of North America and the American tropics.

Neither of these institutions was an arboretum, however. They were the Boston arboretum's compeers in their scientific programs and general educational goals, but it took the urging of Sargent for other woody-plant collections open to the public to be created. One of the earliest was at Highland Park in Rochester, New York, the initial design of which was supplied by Frederick Law Olmsted. Olmsted, too, advocated the creation of arboreta whenever the opportunity arose. The partners of Mount Hope Nurseries, George Ellwanger and Patrick Barry, instigated the development of Rochester's park system by donating the first twenty acres for Highland Park; its plantings were developed by the horticulturist John Dunbar, who frequently corresponded and exchanged plants with the Arnold Arboretum. Dunbar formed extensive collections of lilacs, hawthorns, crabapples, and rhododendrons. Rochester's arboretum expanded with the addition of nearly five hundred acres given by George Eastman and Henry S. Durand, and here many of Wilson's introductions from Asia were planted. Following the Arnold's example, the Rochester park system maintained an herbarium to document and identify its plants.

When Sargent wrote the entry describing and defining "arboretum" for L. H. Bailey's *Standard Cyclopedia of Horticulture* in 1912, Highland Park was one of only two establishments (other than private collections and the Arnold) that he cited as examples. The other was the Central Experimental Farm at Ottawa, established in 1887, where the Canadian government's gathering of woody plants for hardiness testing also served as a park. By the time of

Sargent's writing, what could have been America's greatest arboretum had already failed. The Biltmore Arboretum, painstakingly planned by Olmsted and his sons, with Sargent's input, for the grounds of George Vanderbilt's estate near Asheville, North Carolina, was to be part of a great reforestation experiment and school of forestry. Here, on what ultimately became thousands of acres, the kind of mass planting and study of forest management for which the Arnold found itself unsuited were carried out on a plan conceived by Olmsted. It was his last work before illness forced his retirement in 1895 and finally his death in 1903. Along with his interest in social reform, Olmsted was greatly concerned that American forests should be managed by the most knowledgeable means.

Unfortunately, as Sargent feared during its planning, the interest and the fortune of Biltmore's patron waned after the turn of the century. Although the school of forestry continued until 1909, scientific oversight of the arboretum petered out before then. The Biltmore Arboretum fiasco reinforced Sargent's conviction that provision for permanence independent of an individual's lifespan or personal fortune was essential to the establishment of an arboretum. It became his standard advice when consulted about founding a new arboretum that some means to assure its institutional permanence was essential.

For the *Standard Cyclopedia,* Bailey urged Sargent to expand on his notion of a national arboretum or system of arboreta. The director commented:

The work the Arnold Arboretum attempts and the demands which are made on it are national in scope and extent, but for a national American arboretum a more temperate . . . climate than that of Massachusetts is desirable. . . . It is impossible, however, to cultivate in one collection the trees which grow naturally or can be made to grow in all the different regions of the United States, and the American national arboretum of the future must first of all be an institution . . . of long life and continuous control; and this central institution properly equipped with laboratories and material for research must be in a position to establish branches in Florida, Arizona, California and in some central regions of the continent, for in such branches managed by the central institute, it would be possible to collect and to study nearly all of the trees of the world suitable for different parts of the country, and so make possible . . . an arboretum really national in character.
(C. S. Sargent, 1914, p. 351)

This was a theme the director of America's first arboretum reiterated on occasion. The interest shown in the Boston tree garden, the requests for information that came to it from all over the country, indicated a need for more centers for the study of trees. The Arnold could not be the only one of its kind; its challenge was taken up by others. Although a coordinated countrywide system such as Sargent described was never implemented, a few new tree gardens, including a national arboretum, were started in the 1920s, and many more thereafter. Some were modeled on the Arnold, while others used its example as a point from which to develop or diverge.

In 1921 Joy Morton, the son of J. Sterling Morton, an early U.S. secretary of agriculture and originator of Arbor Day, started an institution he had dreamed of since he was a young man. He established an ample trust to create the "Morton Arboretum, for practical, scientific research work in horticulture and agriculture particularly in the growth and culture of trees, shrubs and vines" (Kammerer, 1963, p. 101). Like the Arnold, which apparently was Morton's chief inspiration, the new arboretum was to include every species, variety, and hybrid that would live outdoors in the climate of Lisle, Illinois, where the founder initially set aside two hundred acres of his country estate for the purpose. The acreage was subsequently much amplified and generic groups set out in harmony with existing woodland and varied topography, but no attempt at strict taxonomic sequence was made. Sargent took great interest in Morton's project; he visited the place several times and turned over duplicate books from the Arnold Arboretum library. An exchange of propagating material soon brought many Far Eastern hardy plants to the Illinois garden. The Morton remains rare among sister institutions as a garden pursuing scientific as well as edu-

96. Crabapple and tree peony collections at the Arthur Hoyt Scott Horticultural Foundation (now Scott Arboretum). In contrast to the encyclopedic approach of the Arnold, the Scott Foundation was instituted on the campus of Swarthmore College in 1929 to display a selection of the best in each woody-plant genus for Delaware Valley gardeners.

cational goals that has no university, school, or government affiliation.

The United States National Arboretum, for which appropriations were approved in 1927, had been under discussion for many years. Since before the turn of the century, personnel at the United States Department of Agriculture had desired an arboretum. There was a conservatory of tropical plants on display near the Capitol, the National Botanic Garden, managed by the Committee of the Congressional Library, but this did not serve the plant explorers of the USDA or the botanists of the National Museum. In 1914, David Fairchild and others at Agriculture formed a committee to push the matter through but they met with little success until such organizations as the Garden Club of America and the American Association of Nurserymen persistently took up the cause. After several years of debate by Congress, during which Sargent was consulted and testimony from many botanists and horticulturists was heard, a bill was passed authorizing purchase of the site and management of an arboretum "under competent scientific direction, in order to stimulate research and discovery."

The new arboretum's emphasis was to be on introducing garden forms and wild species for their potential as breeding stock and on setting standards for naming and identifying woody plants. Since the District of Columbia park system maintained large acreage of native trees, the National Arboretum concentrated on exotics. Establishment of plantings and facilities on the partially wooded tract along the Anacostia River in northeast Washington, D.C., took place gradually under the USDA's Division of Plant Exploration and Introduction. The design evolved as a series of generic and geographical groupings on what was ultimately some four hundred acres. It was not opened to the public until after 1949, however.

Another example of an arboretum conceived while Sargent was still at the helm was the Arthur Hoyt Scott Horticultural Foundation on the campus of Swarthmore College in suburban Philadelphia. Its establishment in 1929 was the result of eight years of deliberation by Scott and his wife, Edith Wilder Scott, both graduates of the college; they worked with Samuel C. Palmer, professor of botany, and John C. Wister, a landscape designer. Scott greatly appreciated the value of gardening as a counter-point to his industrial career. He desired to make a garden to demonstrate to his fellow Delaware Valley homeowners what they could grow with modest means and no special care, and he wished to promote the healthful avocation of horticulture. Both Scott and Wister had studied at Harvard, and they knew what had been accomplished at Jamaica Plain. Scott died before his dream could be implemented, but Swarthmore accepted an endowment from his family two years later, agreeing to create a practical garden that would encourage horticulture in its widest sense for students and the public.

In deliberate contrast to the comprehensive assemblages at the Arnold Arboretum and the botanical gardens in New York and Missouri, the foundation limited its collections to a selection of those genera, species, and varieties best suited for local gardeners. An exhibit of carefully chosen examples was deemed of greater educational potential than a bewilderingly vast collection. Nonetheless, the gardener's first director, Wister, planned to arrange the plants in a botanical sequence in a great clockwise circle among the college buildings. While the buildings, and the preservation of many fine old trees that the Quaker founders and subsequent graduating classes were fond of planting, precluded a completely naturalistic design, the style was decidedly informal and parklike. Plans were also made to restore strictly native species to an adjoining wooded creek valley as a demonstration of local flora. The selection of what was best in each cultivated group became the research function of this arboretum, and the notion later influenced Arnold Arboretum managers.

These twentieth-century tree gardens, as well as many others that sprung up across the country, each took up some or all of the components of the Arnold: the quest for knowledge, the desire to educate and communicate based on that knowledge, and the importance of beauty to humanity's well-being. While some emulated its comprehensiveness in collections and programs, others chose to limit themselves according to their environment, resources, or the interests of benefactors or trustees. Although there was to be no formal coordination of effort for many years, the arboreta tended to work in concert in the exchange of plant material, if not in scientific opinion or educational policy. Their evolution would come to influence that of the Arnold Arboretum.

CHAPTER SIX

TRANSITION, INTROSPECTION;
AMES MOVES FORWARD

B y the time Sargent and the Arboretum together reached the fifty-year mark with Harvard, the director must have realized what a rare opportunity it was for him to see the tree garden mature to such an extent. Not one for ostentation or institutional self-congratulation, Sargent marked the half century modestly with a report reiterating his hope that the institution grow and change with the times. When the inevitable change came with Sargent's death in 1927, the next administration sagaciously built new programs on the foundation he laid. Oakes Ames, Sargent's successor, managed to suffuse the Arboretum with a more biological approach in research and education, thus bringing it closer to harmony with the other botanical institutes of the university. Expansion into the fields of genetics and plant pathology was a step toward compatible utilization of the living-plant collections for research, while the taxonomic program diverged from what had been its traditional subjects.

The unexpected death of E. H. Wilson, only three years after Sargent, compounded the blow to complementary development of the grounds and living collections, however. With no one individual charged with care of the grounds, a period of stasis ensued that was exacerbated by the Great Depression and some of the worst weather since the Arboretum's establishment. Since Wilson's enthusiasm for plants outweighed his sense of the landscape's original composition, the innovations he made tended to diverge from Sargent's vision, although he probably did not know it. Thus, in terms of the style of the garden, links to the past were broken even before the death of Wilson.

AT FIFTY YEARS

The Arboretum's fiftieth year was complete in 1922. The director's account of the institution's accomplishments in the *Journal,* "The First Fifty Years of the Arnold Arboretum," emphasized its contribution to education, research, and the introduction of plants. Perennially conscious that its origins should be attributed to people who were as interested in natural history as in gardening, Sargent briefly explained the roles of James Arnold, George B. Emerson, and other colleagues. He extensively quoted the indenture of 1872 as a way of clarifying the institution's goals and purposes. Sargent's

97. Charles Sprague Sargent oversaw the Arboretum until a few years past its half-century point, as if his own longevity were inspired by the trees he planted. The incomparable institution he built had strengths as a sublime garden and as a center of knowledge.

98. When Sargent described the Arboretum's future needs in his fifty-year summary, he recommended a rock garden and a rose garden as two additions to the collections. He desired a rose collection of both garden forms and their wild progenitors that would be utilized in hybridization research. One rose species with a wide natural distribution is the Scotch rose, Rosa spinosissima, *of which the large white-flowered variety* altaica *is pictured here.*

A·39

statement in this article that the Arboretum was started on a "worn-out farm" has often been quoted, although he had made similar remarks earlier. It certainly made a good story as Sargent told it, but the transformation, the Arboretum's growth into a grand and gorgeous garden, did not need exaggeration.

It was also in this piece that Sargent ascribed to Frederick Law Olmsted the role of suggesting a union of the Arboretum with the city parks system. Most of the 1882 agreement between the park commission and the president and fellows of Harvard was quoted to elucidate that aspect of the Arboretum's development. Sargent reiterated what he had long considered one of the greatest advantages of this arrangement, that "by this agreement the location of the Arboretum was practically fixed for at least one thousand years." The founding director considered this permanence a unique asset that should be fully utilized to study tree growth over the long term.

In a concise description of the land and the collections, Sargent stated that the planted trees were arranged "in botanical sequence in family groups"; there was little further elaboration and no mention of George Bentham and Joseph Hooker, whose system had been used. Included in the report was a list of the 87 families and 334 genera of plants living in the Arboretum in the Engler and Prantl sequence. Since the classification system of Adolf Engler and Karl von Prantl was the generally accepted one in 1922, Sargent did not want to seem outdated by mentioning the Bentham and Hooker mode in his report. It would have been impractical to move the tree groups to keep up with changes in the science of classification. Nonetheless, the failure to mention Bentham and Hooker, combined with frequent use of the Englerian system in Arboretum publications, contributed to an eventual misunderstanding of the overall collections' arrangement during subsequent administrations.

The introduction of woody plants never before cultivated into the Arboretum and other gardens was the institution's most important accomplishment according to Sargent. Couched in terms of the original mandate to obtain, "as far as practicable, all the trees, shrubs and other plants which could be grown in West Roxbury," the introduction of new plants had made excellent strides—

some two thousand taxa were included in a nineteen-page list of all taxa introduced during the first fifty years.

Nevertheless, there was much left to be done. Sargent concluded his report with a summary of "Future Needs," including the four research programs that were described in chapter 4. He cited two specific gardens or collections as number five on his roster of future needs, a "Rose Garden and a Rock Garden." His use of capital letters was intentional, for he considered these excellent opportunities for large donations that could bear the endower's name. These two ideas had also been on the drawing board for some time.

Like Francis Parkman and Jackson Dawson, Sargent was fascinated with roses, whether the garden forms or their wild progenitors, but there had never been the room or the funds to maintain a complete collection at the Arboretum. The director had in mind "a scientific Rose garden in which all the roses of the old and new world could be planted and studied, and the breeding of new races of roses undertaken." While other great collections were in private hands, an Arnold Arboretum rose garden would have "greater permanency than can now be hoped for anywhere else in the world." The hybridization program was as important as the garden itself. Sargent apparently envisioned that the new collection would be placed on adjacent property rather than within the existing boundaries, for he included the cost of land when he outlined his rose garden proposal in detail to prospective donors.

A rock garden likewise could be a discrete, separately endowed unit. Even though the Arboretum had a display of naturally diminutive and alpine shrubs, it was quite modest and rather stiffly and artificially arranged, especially when compared with the rock gardens of Kew, Edinburgh, and many other British gardens known to Sargent. These two possible subsets to the living collections, he believed, would have wide visitor appeal, thus broadening the base of support and furthering the educational potential of the institution.

Sargent continued to oversee the Arboretum for another five years, doing what he could to assure its future success. More land was obtained to buffer its perimeter; the tree garden was brought to the attention of a wide audience through publication of *America's Greatest Garden;* Rock's expedition to north-central Asia was launched; and the director kept insisting that the scientific program, including the herbarium and library, should take up a worldwide focus in the face of the forest destruction taking place on almost every continent.

END OF AN ERA

The year 1927 began for the Arboretum with the publication of Rehder's *Manual,* a proud achievement. The edition was dedicated "To Charles Sprague Sargent, LL.D., who during more than fifty years as Director of the Arnold Arboretum has worked with untiring zeal and energy for the promotion of American dendrology and arboriculture as a token of admiration and gratitude." In late winter, nearly everyone in the administration building was helping to reorganize the "Conifer Room" to clear space for new herbarium cabinets. The dried-plant collection had outgrown its quarters in the rear wing of the building, and there was nothing to do but expand into one of the first-floor halls. While this work was under way, Sargent was taken ill with influenza and compelled to stay at Holm Lea.

The director's human "zeal and energy" could not go on forever. On 22 March, Charles Sprague Sargent died in his home after being ill for two weeks. He was nearly eighty-six years old, and, until this last illness, he had seldom missed a day at the Arboretum in fifty-four years except when traveling. Since the man and the institution had been synonymous for so long, the year 1927 truly marks a turning point for the Arboretum. Sargent had pertinaciously and single-mindedly shaped its course for a half century—5 percent of its projected term of existence if one took the thousand-year lease to heart, as Sargent always did. His life's purpose had been the Arboretum. Although he held his family and friends dear, they recognized the preeminence of the great tree experiment for him. He had outlived his wife, one son, and a son-in-law; as he wrote to a colleague, the Arboretum kept him going after so many were lost.

He built an incomparable institution in fifty-four years. The 120 acres of farmland had become 255 acres of carefully tended living collections, with a world-class library and herbarium housed in a dignified building on the site. Pasture, orchard, and second-growth woodlots containing 123 species of woody plants were transformed into an intricate exhibition of more than 6,000 kinds of trees and shrubs, woven together by native species and pockets of natural woods. The pastoral landscape of the Bussey farmstead, its varied parcels delimited by walls and fencerows, was now a series of sophisticated, graceful scenes evocative of forest or savannah, composed almost entirely of plants, the man-made elements entirely subservient to the vegetation.

Under Sargent's watchful eye, 30,000 lots of seed and 200,000 plants and cuttings had been brought to the Arboretum and nurtured in its propagation facilities over the course of fifty-four years. Not content to keep its plant treasures to itself, 63,000 lots of seed and more than 500,000 plants and cuttings had been distributed to institutions the world over from the Arboretum—an average of 1,200 lots of seed and 9,300 plants per year. Through its program of exploration, more than 2,000 species and varieties of woody plants had been introduced into cultivation from the far corners of the temperate world. The original $100,000 endowment had grown to $1,125,000 with Sargent's talents for enlisting financial support.

Not only were the tree-clad Jamaica Plain acres the ultimate school for the study of woody plants, they were one of the finest examples of what man can do to bring the elegant subtleties of nature into the urban environment. Sargent's plantings had matured to form the consummate naturalistic landscape design. It was a place sought by students and researchers from around the world and one held dear in the traditions of Boston's people.

Along with the grounds and plantings, Sargent developed complementary programs of education and research. For the public, the *Bulletin of Popular Information,* the *Guide to the Arnold Arboretum,* and regular staff-led tours interpreted the institution's diverse plant collections. Thousands of inquiries for plant identification and other information were answered by the staff each year. Frequent

99. View from Azalea Path on Bussey Hill looking west over the oaks and junipers to Bussey Brook valley, Hemlock Hill, and beyond. The Arboretum landscape, a careful blend of spontaneous vegetation and planted collections, was largely Sargent's creation.

contact with the press kept the Arboretum and its needs in the public mind. As part of a university, the Arboretum held its own as a research institution specializing in the taxonomy of the world's trees and related topics. This work was communicated through the *Journal of the Arnold Arboretum,* by 1927 in its eighth volume.

The tree garden's accomplishments up to this time reflected Charles Sargent's personal determination to build an institution unequaled in its precise field. Nevertheless, its excellence depended not on Sargent alone but on the staff he had gathered to it. He knew this as well as any of them. He had been fortunate to find capable assistants and win their commitment to his purpose. Sargent's unrelenting pursuit of his own research projects and administration of Arboretum affairs inspired their productivity. While he was often characterized as "autocratic" and said to be reserved, he made certain that his staff knew he appreciated their talents. Jackson Dawson worked with Sargent for forty-three years, Charles E. Faxon for thirty-five. When Sargent died, Alfred Rehder had been with the institution for twenty-seven years, John George Jack some forty years, Ethelyn Tucker twenty-three, E. H. Wilson twenty-one years, and William Judd ten. Each was as committed as his or her leader to the Arboretum. Although, as Alfred Rehder remarked to a colleague, they knew Sargent's life must end sometime, none was really prepared for the transition.

Upon Sargent's death, the papers carried obituaries and editorials noting his passing. In actuality, the Arboretum had not been his sole occupation these long years. He was a director of the Boston and Albany Railroad and the Ware River Railroad, vice president of the Massachusetts Hospital Life Insurance Company, and trustee of the Walnut Hills Cemetery. For years he served as a trustee of the Boston Museum of Fine Arts and the Brookline Library. His horticultural duties not involving the Arboretum included participation on the boards of the Massachusetts Horticultural Society and the Massachusetts Society for Promoting Agriculture. He was a commissioner of Brookline Parks. He served on many commissions and authored numerous reports urging the preservation of forests and the creation of national parks. The beauty of his estate, Holm Lea, was well known and admired as a superb

landscape in its own right. Yet the many summations in the press agreed that Sargent's greatest accomplishment was the building of the Arnold Arboretum and that the tree garden itself would be his most enduring monument.

At week's end, funeral services were held in Saint Paul's Episcopal Church near his Brookline home. There, family, colleagues, friends, and employees together participated in the rite. One paper reported:

Spread over the casket was a covering of Clivias, a flower of the amaryllis family which Professor Sargent imported from South Africa several years ago, and which won a gold medal when first exhibited. These flowers, cut from the greenhouses at "Holm Lea," were to have been exhibited at next week's exhibition of the Massachusetts Horticultural Society. (AAA Scrapbook 10. *Boston Herald,* 26 March 1927)

On opening day of the Massachusetts Horticultural Society's flower show the following week, at 3:30 in the afternoon, a bugle call resounded and the crowd of flower seekers stopped for a moment of silence to reflect on the memory of Charles Sargent.

When his will was made public later that spring, it was found to contain a legacy contrived for the distant future of the institution he had managed so long. Compared to the funds he had already contributed to the Arboretum's work year after year during his lifetime, Sargent's last bequest was a modest amount. Nevertheless, it was a generous sum, with its terms geared toward the permanence he was accustomed to think about. "Even in his will, Sargent took the long view regarding the destiny of the Arboretum" (Robinson, 1929, p. 365). Sargent left $30,000 to the institution, two-thirds of which was to be invested, its income immediately available solely for the support and development of the library. The other $10,000 was to be invested, with its yield reinvested annually for one hundred years. After a century, half of the accumulated funds would become available for the care of the Arboretum, while the remainder would have to accumulate for an additional hundred years. Typically, Sargent's gift to the institution was on the timescale of the life of trees rather than of humans, an attempt to influence its destiny farther forward in time than any of his children's generation would know. His legacy was in essence

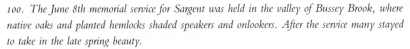

100. The June 8th memorial service for Sargent was held in the valley of Bussey Brook, where native oaks and planted hemlocks shaded speakers and onlookers. After the service many stayed to take in the late spring beauty.

101. Episcopal bishop William Lawrence of Boston (left) gave the benediction, and Harvard's president, A. Lawrence Lowell (right), spoke at the Sargent memorial service. Lowell said that Sargent had "built his thoughts into" the lovely landscape that surrounded the listeners. It was Lowell who choose Oakes Ames to succeed Sargent as administrator of the Arboretum.

a challenge: if his successors could keep the Arboretum viable for one hundred years, Sargent would contribute measurably to assure its existence for the following two centuries.

The university paid tribute to Sargent at a memorial service held outdoors at the Arboretum on the afternoon of 8 June. The service was planned and arranged by a committee headed by John S. Ames, one of the Arboretum's staunchest supporters. Invitations were extended by the "President and Fellows of Harvard College and the Friends of the late Charles Sprague Sargent." In the valley of Bussey Brook, a temporary platform for speakers and honored guests was erected below the juniper collection; it was shaded by an ancient white oak and a native beech grove that Sargent had nurtured from the first years of his stewardship of the land. The spot was said to have been one of his favorites on the grounds. To accommodate the listeners, chairs were placed on the opposite side of the brook. Behind them grew forty-foot hemlocks planted by Sargent in the 1880s, and beyond those stood the cherished native stand on Hemlock Hill. Shining green masses of mountain laurel, heavy with buds ready to burst, glowed beneath the hemlock shades. Nearby, the rhododendrons and azaleas displayed their many hues.

More than three hundred persons sat on camp chairs or on the grass on the mild, late-spring day to hear tributes to Sargent. Roger Wolcott, an overseer of Harvard and for many years chairman of the Arboretum visiting committee, presided. Representing the institution's many constituencies, testimonials were voiced by A. Lawrence Lowell, president of Harvard; William C. Endicott, trustee of the Massachusetts Horticultural Society and chairman of a memorial-fund drive just initiated; E. H. Wilson of the Arboretum staff; Mrs. John A. Stewart, Jr., president of the Garden Club of America; and J. Horace McFarland, former president of the American Civic Association and editor of the Rose annual. While each bespoke personal admiration for the man and the value of his accomplishments at the Arboretum and in the horticultural world, President Lowell best characterized Sargent's genius in creating the living collections:

It is rare that anyone but an artist is able to leave a material memorial of himself that endures, and it is still rarer that anyone is able to leave a memorial that is the product of his whole life instead of a series of smaller and shorter ones. Professor Sargent built his own memorial but he built it without the slightest idea that it was to be a memorial to himself. He built his thoughts into what you see before you just as much as the artist builds his thoughts into the picture that he paints with his brush.

There is not a tree that did not have its place in his imagination before it stood in the ground as you see it today. The whole of this great place here is simply his mind, his thoughts, his imagination crystallized into actual shape. He thought it all out. He thought how it would look and he built it to look as he intended it should.

This must remain a memorial to him. The trees may die. They may grow old. In one hundred years many will perhaps be replaced. But the general design and the general kind of plant will always be the same and they will always represent his thought.

His great idea was the idea of a living museum. It is a great contribution, and we are wise in coming here to remind ourselves how great and valuable it is. (AAA Scrapbook 10. *Boston Evening Transcript,* 9 June 1927)

Accompanied by the sounds of birds, the brook, and wind in the boughs, the service was "simple, solemn, dignified and yet signally beautiful" (Anonymous, 1927b, p. 235).

After the benediction, many of those present lingered among the groves and wandered the paths of Sargent's masterpiece.

AMES, THE DIPLOMAT, TAKES CONTROL

Saddened by the loss of their leader, the Arboretum staff carried on with the routine and the immediate tasks at hand that spring. Assistant Director Wilson acted as interim administrator while university officials considered the challenge of a replacement for Sargent.

Little more than a month after Sargent's death, President Lowell and his advisers made their decision on a new administration for the Arboretum. With the goal of bringing the Arboretum and the other independently endowed botanical institutions of the university into closer cohesion, Oakes Ames, professor of botany, was made chairman of a new Council of Botanical Collections. This consisted of the Arnold Arboretum, the Harvard Botanical Museum, and the Atkins Institution at Soledad, Cuba, as well as the Gray Herbarium, the Harvard Forest, and the Farlow Herbarium. Ames's role with regard to the Arboretum was termed supervisor, and he held similar posts in relation to the museum and the Cuban garden as well.

Professor Ames was an excellent choice for the role of bringing harmony among the several botanical establishments; during his twenty-nine-year Harvard career, he had held positions at a number of them. Born in 1874, Oakes Ames graduated from Harvard College in 1898. The following year, he received his master's in botany, studying with George L. Goodale, who took charge of the botanic garden after Sargent left it in 1879. Even before entering college, Ames started a collection of living orchids in the glasshouses on his father's estate in North Easton, Massachusetts. When Ames completed his master's degree, Goodale named him assistant director of the botanic garden and instructor in botany. Ames took over as director of the botanic garden when Goodale resigned this post in 1909 to devote full time to the Botanical Museum, a collection of specimens illustrating the world-wide economic uses of plants that was housed in the same building as the Museum of Comparative Zoology. Goodale was instrumental in the addition of the fabulous Ware Collection of Glass Flowers to the museum's holdings. For many years, Ames and Goodale advised the sugar grower Edwin F. Atkins on the development of his Cuban estate near Soledad as an experimental garden; it was turned over to Harvard in 1926.

In 1900 Ames gave a large portion of his living orchids to the New York Botanical Garden, choosing to devote his time to scientific study rather than cultivation of orchids. He and his wife, Blanche Ames Ames, built a stately home, Borderland, on an ample property in North Easton. The house was provided with a large wing where Ames assembled his own botanical library and herbarium of orchids. Oakes Ames's intensive study of orchids led him to publish the results of his research in a journal he and his assistants produced. He also contributed to Liberty Hyde Bailey's *Standard Cyclopedia of Horticulture*. Under Goodale's influence, Ames became interested in economic botany. Whenever he traveled to pursue orchids, he obtained samples to document plant materials as the source of products useful to mankind.

In 1915, Ames was appointed assistant professor and began teaching the taxonomy of economic plants to graduate students at the Bussey Institution, moving his botanical collections to Jamaica Plain. He also lectured on medicinal plants to medical school students. In 1922, Ames resigned his position with the Harvard Botanic Garden, intending to devote more time to orchid research. However, the death of his mentor, Goodale, the following year brought a call for Ames to oversee one more of his alma mater's collections, the Botanical Museum. Reluctant as he was to take another Harvard post, his loyalty to Goodale's cause was greater. Ames moved his own assemblage of economic plants, his library on economic botany, and his orchid herbarium to the museum building on Oxford Street. There he revamped the exhibit areas with the aid of a donor and rearranged the glass flowers into an evolutionary rather than alphabetical sequence.

In 1926 university authorities promoted Ames to full professor and made him chairman of the new Division of Biology. President Lowell apparently thought him some-one whose diplomacy could unify competing factions within the department. Indeed, Ames was not partial to any one of the botanical units, while he appreciated the value and importance of all. It was his nature to be fair and open-minded. Since Ames's primary interest was his orchid research, often accomplished on his own while he held his various posts, he maintained a healthy objectivity toward Harvard's numerous botanical units. Although he was prodigious in his botanical output, his scientific pursuits never became all consuming.

Ames was a scholar in the broadest sense, making it a point to read daily from the classics, great poets, Shakespeare, or the Bible. An avid sportsman, he went riding or walking regularly, and he joined his wife and their employees in improving the landscape of Borderland. His live orchid collection, his richly planted estate, and his oversight of Atkins's plantation and the botanic garden in Cambridge gave him a wealth of experience in the cultivation of plants and the management of grounds. With so much to occupy him, Ames kept somewhat aloof from the academic politics of the biology department. Although he modestly termed himself a "botanical playboy," he was painstakingly thorough in his orchid work and had great respect for the endeavors of his colleagues.

Ames's style of administration was termed "remote control" by the Arboretum historian S. B. Sutton, and he did keep his distance at first. This is not surprising, since he was overseeing several institutions simultaneously. He discharged most affairs from his office in Cambridge or his laboratory in North Easton, making weekly visits to the Arboretum. He relied on Wilson to manage the daily affairs on the grounds, while keeping a careful eye on finances and scientific developments. Ames soon won acceptance among the staff for allowing past traditions to carry on while making innovations to expand Arboretum programs in the directions Sargent had outlined in his fifty-year report.

Although Sargent's bequest was an agreeable addition to the institution, its future income did not solve the financial situation facing Ames and Harvard authorities in 1927. If the Arboretum was to maintain its present stature, as well as to grow and to develop new programs relevant to twentieth-century biological inquiry, it needed a

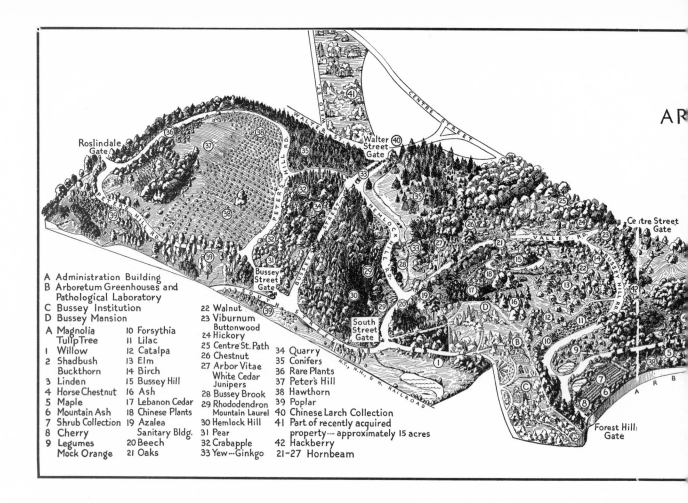

A Administration Building
B Arboretum Greenhouses and
  Pathological Laboratory
C Bussey Institution
D Bussey Mansion

A Magnolia	10 Forsythia	22 Walnut	
TulipTree	11 Lilac	23 Viburnum	
1 Willow	12 Catalpa	Buttonwood	
2 Shadbush	13 Elm	24 Hickory	
Buckthorn	14 Birch	25 Centre St. Path	34 Quarry
3 Linden	15 Bussey Hill	26 Chestnut	35 Conifers
4 Horse Chestnut	16 Ash	27 Arbor Vitae	36 Rare Plants
5 Maple	17 Lebanon Cedar	White Cedar	37 Peter's Hill
6 Mountain Ash	18 Chinese Plants	Junipers	38 Hawthorn
7 Shrub Collection	19 Azalea	28 Bussey Brook	39 Poplar
8 Cherry	Sanitary Bldg.	29 Rhododendron	40 Chinese Larch Collection
9 Legumes	20 Beech	Mountain Laurel	41 Part of recently acquired
Mock Orange	21 Oaks	30 Hemlock Hill	property···· approximately 15 acres
		31 Pear	42 Hackberry
		32 Crabapple	21-27 Hornbeam
		33 Yew····Ginkgo	

greater endowment immediately. With Sargent gone, a new scheme for raising money was needed. A few of the late director's financial allies and members of the visiting committee joined Ames and university officials in a campaign to enlist contributions to a Charles Sprague Sargent Memorial Fund, with a goal of $1 million. One of the first, cornerstone contributions was $50,000 voted to the memory of C. S. Sargent by the trustees of the Massachusetts Society for Promoting Agriculture, an organization that had been supporting botany and horticulture at Harvard since the botanic garden began in 1806. Sargent had been president of the MSPA for the past thirty-two years.

Committees of the Charles Sprague Sargent Memorial Fund were established in Boston, New York, Philadelphia, and Chicago. A private firm specializing in fundraising, with which Sargent had held discussions shortly before his death, was engaged to prepare written material, set up meetings, supply information to the press, and more. The fund's executive offices were in New York, and J. P. Morgan acted as treasurer. The Boston chairman

was William C. Endicott, a long-serving member of the Arboretum visiting committee.

For most of 1927, the Arboretum's energies were turned to the fund drive. The committee produced a booklet, *The Arnold Arboretum and Its Future,* which addressed many of the new goals outlined by Sargent during his last ten years. Illustrated with black-and-white photographs and the first aerial photograph of the grounds, its text vividly pleaded for support to continue what Sargent had started:

It remains for those who knew and honored the leader who has died, for those who would encourage research and the collection of exact knowledge in the field in which he labored, and for those who would foster the important arts of arboriculture and gardening in our country to see to it that the Arboretum shall go on uninterrupted to a full realization of the vision Professor Sargent saw steadily and whole. (Anonymous, 1927a, p. 4)

The pamphlet explained the institution's strengths as

YE VIEW
THE
RBORETUM
THE
STITUTION
Massachusetts
934

ROUTE 1

TRAFFIC CIRCLE

Jamaica Plain Gate

W N S E

102. This illuminating bird's-eye view, originally prepared to accompany Sargent Memorial Fund literature, was revised in 1934 to include the Bussey Institution grounds.

these years when the American economy was booming, before the 1929 stock market crash. Indicative of the tree garden's role as a national resource was the support it garnered from outside New England.

NEW RESEARCH, PATHOLOGY AND CYTOLOGY LABORATORIES

Even before the memorial fund was complete, Ames considered ways to expand the Arboretum's research program that would concurrently unify some of the botanical units under his wing. The adjacent Bussey Institution was a logical starting point. Since its 1907 transformation into the Bussey Institution for Research in Applied Biology, it had flourished as a center for studies in genetics, entomology, plant anatomy, and economic botany. Genetics itself was at the forefront of biology, and the Bussey's scientists were pioneers. Between 1907 and 1929, some 130 graduate degrees were awarded to students under the Bussey's close-knit faculty of five or six; at least three of these graduates would influence the future affairs of the Arboretum. Ames described the academic life at the Bussey as ideal—a fraternity of biologists with ample time for research, plus the stimulation of graduate-level students. The new director hoped to bring Bussey and Arboretum researchers into one circle, enhancing the intellectual give-and-take for all.

Of the subjects suggested by Sargent, Ames chose plant breeding and plant pathology as appropriate fields by which to broaden the scope of research. In statements to the press relative to the fund campaign, he repeatedly stressed the importance of doing work in genetics that would improve the inherent quality of trees for timber production as well as for ornamental use. While Ames was considering potential appointees, he initiated construction of a facility for them in February 1928. Some months earlier an agreement had been signed to sell the Orchard and Prince Street property; the Arboretum had until July 1928 to move its nurseries, glasshouse, and rock garden. The plan was to relocate the propagation unit, together with new research facilities, on the Bussey grounds. Toward this end the well-known firm of Lord and Burnham

threefold, lying in the fields of "science, economics and culture." The value of its collections and its expertise for advancing knowledge, the potential for contributing materially to human welfare through improved timber, fruit, and ornamental crops, and the beauty of its landscape were all cogently argued as reasons to contribute to the Arboretum.

Sargent's friends and the institution's followers rose to the challenge. By early November, half the $1 million had been pledged. At the beginning of December, the New York committee's campaign began in high style with a dinner at Sherry's, where Oakes Ames, David Fairchild, William Endicott, and the Reverend William Lawrence spoke. Within days, $110,000 had been contributed, and one donor promised to give an additional large sum if the rest of the country matched the amount already raised in Boston. By March 1928, Ames could inform reporters that three-quarters of the goal had been raised, and it was not long after that the entire amount was secured. It is most fortunate that the endowment drive took place during was contracted to build a glasshouse sited near the original Bussey greenhouse, where propagation for the Arboretum had begun. The new structure included a workroom for William Judd and his assistants, an ample indoor growing area, and a cold pit. Two acres of nursery ground nearby were prepared for Arboretum accessions and experimental subjects. A laboratory, fully equipped for research in plant pathology and genetics, was built into the facility. A Bussey Institution professor, Edward M. East, an expert in genetics, advised Ames on furnishing the labs.

While Judd established propagation operations at the new site in the summer of 1928, two new appointees joined him on the Bussey "plainfield." Joseph H. Faull assumed the duties of professor of forest pathology on 1 July. A native of northern Michigan, Faull graduated from the University of Toronto in 1898 and received his doctorate from Harvard in 1904, where he pursued research under William G. Farlow and Roland Thaxter, American pioneers in the study of fungi. Like Farlow, Faull became chiefly interested in those fungi that cause disease conditions on living higher plants. From the year he was awarded his Ph.D. until his appointment at the Arboretum, Faull held posts at the University of Toronto, ultimately becoming full professor and head of his department. Under the auspices of the Canadian government, he established its Department of Forest Pathology in 1918 and a field laboratory in the Lake Timagami Forest Reservation in Ontario.

Faull was held in high regard as a teacher, and many of his students gained responsible positions in botany and forestry throughout Canada. His reputation for exacting research in a field still considered new led to many requests for his expertise from lumber companies, government agencies, and other enterprises concerned with the growth of healthy trees.

When he accepted Ames's proposal, Faull looked forward to the decreased load of academic administration that work at the Arnold and Bussey entailed. Nevertheless, his aspirations remained broadly based, including not only study of "immediate pathological problems presented by trees in the Arboretum" but also discovery of the etiology, and the means to control, many diseases that affect "our untamed and abused American forests" (Faull, 1962,

103. Joseph H. Faull came from the University of Toronto to the Arboretum in 1928 to head the new phytopathology laboratory initiated by Oakes Ames. While Faull's specialty was certain fungi that cause disease in some North American forest trees, many additional problems in plant pathology were studied under his direction.

pp. 225, 226). Some of Faull's graduate students came with him to Harvard, and he soon attracted others. He also gave instruction in plant pathology to undergraduates in Cambridge. Not since J. G. Jack taught forestry students had an Arboretum staff member given regular courses of instruction at the college. In his first year, Faull, his assistant, and his students tackled no less than six projects; some were continuations of previous work, others were new.

Ames's second research appointment, Karl Sax, officially joined the Arboretum staff on 1 September as associate professor of plant cytology. Sax, too, had previous training at Harvard, and the link was even closer since he had earned both master's (1917) and doctor of science (1922) degrees from the Bussey Institution. Born in 1892 in Spokane, Sax graduated from Washington State College in 1916 with a bachelor's in agriculture. After serving in the army in the Panama Canal Zone during the Great War, Sax took a post at the agricultural experiment station at the University of Maine at Orono. Here he acquired an interest in experimental breeding and cultural practices on apples that continued throughout his life. He did the work to complete his doctorate in Maine while being supervised by the Bussey's geneticist, Edward East.

Sax was among the first to study extensively the number and the behavior of chromosomes, the rod-shaped elements that organize in the nucleus of every living cell during the process of division and contain the mechanism of inheritance, or genes. Decades later biologists would work out the chemical structure of the chromosomal material and dub it DNA, but in the teens and twenties biologists were eagerly studying the chromosomes in hope of gaining greater understanding of inheritance and evolution. That each species has a constant, characteristic number of chromosomes and that this number could be doubled or trebled during the cross-fertilization process was one phenomenon upon which Sax's work yielded information. The number of chromosomes could shed light on the relatedness of species and on the chances for successful hybridization; studying chromosome numbers of the results of deliberate crosses also yielded valuable knowledge.

The Arboretum collections, with some sixty-five hundred species and varieties, offered a wealth of material for surveys of chromosome numbers. This was the work Sax took up as soon as he settled into his laboratory overlook-

ing the grounds. The following spring he started the first of his many breeding experiments utilizing Arboretum species.

After a year of operation it became apparent that the new greenhouse was too small for Judd's Arboretum propagules, Sax's hybrid seedlings, and Faull's disease cultures. Ames authorized expenditure for a second greenhouse and directed Wilson to work out the detailed requirements of Faull and Sax with the Lord and Burnham agent. By year's end this new glasshouse was ready to accommodate the plants under study by Faull and Sax, who readily adjusted to improved work areas.

By initiating these two programs and strengthening the taxonomic staff in the herbarium, Ames modernized the Arboretum's research program. It was a start to the process of coordinating the work of the Arboretum faculty with that of other Harvard botanists. For the remainder of Ames's eight-year administration the accomplishments of the phytopathology laboratory, the cytology laboratory, and the herbarium were extraordinary.

104. The greenhouses built after 1928 on the Bussey Institution grounds housed the Arboretum propagation department and experimental work in plant diseases and hybridization. In the surrounding nurseries, the results of Karl Sax's breeding of woody ornamentals were grown for evaluation.

Faull's program for the pathology department encompassed teaching, research, and what he termed extension services. The topics tackled by Professor Faull, his assistants, and his students had bearing on practical problems in the Arboretum and the North American forests; they also contributed to the larger understanding of the biology of the organisms involved. Faull and his colleagues found that a devastating disease of beech trees in the Maritime Provinces of Canada was caused by a fungus infection that followed the attack of an introduced insect. Although the beech-bark disease advanced south into eastern Massachusetts, the Arboretum laboratory worked out effective control measures in cooperation with the Boston Parks Department. Fungi that alternately parasitize species of *Juniperus* and such pomaceous genera as apples and hawthorns, known commonly as cedar-apple rusts, also came in for scrutiny by Faull's graduate students, and again it was possible to recommend effective control measures once the study was complete.

The Arboretum phytopathology laboratory worked on one of the best-known tree afflictions of our time, Dutch elm disease. While Christine J. Buisman, a mycologist from the Netherlands, was working with Faull in 1930, she identified an organism that was killing trees in Europe on samples of American elm from Ohio. With this information, authorities were alerted to the potential threat, and for a time it was thought the outbreak had been checked. When the disease turned up a few years later in New Jersey and southern New York, the Arboretum "took an active part in the campaign for control and elimination of Dutch elm disease, emphasizing the view that complete eradication of affected trees was the surest means of saving America's elms" (O. Ames, 1936, p. 276). Two other organisms investigated at the Arboretum laboratory were the rusts of conifers that spend part of each life cycle on certain kinds of ferns and subterranean fungi that associate beneficially with forest trees, the latter collectively known as mycorrhizae.

Faull and his colleagues made considerable effort to answer the many queries that came from public agencies and private individuals on the identification and treatment of plant diseases. His program was one that ideally meshed the resources and mission of the Arboretum with those of the university. Faull was one of the first on the Arboretum staff to teach graduate students regularly and continually, and his students often excelled, attaining fellowships for travel and study and high-ranking positions when they sought employment. The living collections of the Arboretum, as well as its expertise in propagation and cultivation, were readily available to the pathologists.

When, in 1932, a new building for biology was completed in Cambridge, laboratory space was allocated for Faull's graduate students there. This allowed them to be fully immersed in university life while drawing inspiration from the Arboretum, as Ames put it. The institution's contacts with the outside world materially aided phytopathologic investigations. The identification requests and queries that came in each year brought useful specimens to its reference herbarium and sometimes pointed out problems worthy of detailed study. At least two major grants were received from individuals who had long supported the Arboretum, one to aid Dutch elm disease study and the other for the cedar-apple rust project.

Working in his lab on the Bussey Institution grounds, Karl Sax and his students and assistants pursued research relating to the mechanism of heredity. The fundamental discoveries of the Arboretum cytogenetic laboratory had to do with the behavior of chromosomes during meiosis—in other words, observation of the changes genetic material goes through when passing from one generation to the next. Each year the chromosome number for several species in a few genera of Arboretum plants was investigated, yielding information used to evaluate traditional taxonomic grouping. A program of hybridization also aided in studying the relationship between taxa while potentially developing new garden forms. Under Ames's administration Sax made hundreds of crosses in such groups as crabapples and other pomoideae, lilacs, philadelphus, species roses, honeysuckles, elms, and magnolias. Although a nursery area on the Bussey grounds was created in 1935 for Sax's hybrids, it would be a few more years before many would reach flowering age and their evaluation for various attributes would yield results.

HERBARIUM AND LIBRARY; CONTINUED GROWTH

The herbarium staff carried on much of the activity under way during Sargent's administration, with more staff members studying taxonomic problems and a steady growth of the collections through exploration and exchange. On the significance of this department, Ames stated:

The Herbarium, in its highly specialized field, is one of the great botanical treasures of the world. It is not a collection of dead things brought together simply through a love of acquisition; it is an indispensable biological tool. In the final analysis no identification is reliable that does not rest on a type or on critically named material, and in close and careful work the Herbarium is the most helpful guide we have in arriving at the identity of plants used in research either for the interpretation of structure or for the investigation of function. (O. Ames, 1933, p. 251)

Not long after Ames assumed leadership, the herbarium added a staff member, Clarence E. Kobuski, to assist Alfred Rehder in managing the collections. Kobuski came to Boston from Saint Louis, where he had just earned his Ph.D. at Washington University and the Missouri Botanical Garden under Jesse M. Greenman, renowned teacher of systematic botanists. Although the tall young botanist's outgoing and sunny personality contrasted with Rehder's quiet, retiring disposition, the two got along well. Both were meticulous and shared a great knowledge of languages. Kobuski eagerly tackled the backlog of undetermined and uninserted collections in the Arboretum herbarium, assisted in editing the *Journal*, and pursued taxonomic research.

One of the first tasks on which he worked with Rehder was the enumeration of plants collected by Joseph F. Rock on the Arboretum expedition to northwestern China and southeastern Tibet. Rock and all of his botanical and ornithological specimens reached Boston safely in the fall of 1927. Although some of the areas to which he had been assigned turned out to be disappointingly bare of trees and shrubs, Rock managed to collect twenty thousand herbarium specimens and several hundred packets of seeds and

to expose three hundred photographic plates for the Arboretum. The staff welcomed Rock back to Boston in September, but the explorer had already made plans to return to China under the auspices of the National Geographic Society. Rock spent the next twenty-five years in China collecting natural-history specimens and studying the culture and languages of certain tribes. His 1924–27 expedition was the last to be sent out from the Arnold Arboretum to the Far East for many years.

Rather than send staff on expeditions overseas, Ames and Rehder adopted a policy of financing, wholly or in part, the efforts of in situ institutions and collectors. In this way pressed plants continued to arrive in Jamaica Plain. In any one year the Arboretum might receive its share of specimens from collecting trips in China, southern Asia, Australia, Polynesia, Africa, or Central and South America.

After Rehder and Kobuski completed enumeration of Joseph Rock's collections, they turned to organizing and naming material coming from Chinese collectors in some ten provinces of China under the auspices of the University of Nanking, the Fan Memorial Institute, and Lingnan University. To aid the work on these collections, Rehder traveled to Europe twice. In the summer of 1930 he attended the International Congress of Horticulture and the International Botanical Congress as a seasoned expert on the problems of plant nomenclature. On this trip he began an extensive search of European herbaria for the type specimens of Chinese plants, an investigation continued on a return visit in 1932. Rehder recorded most of these nomenclaturally important specimens with photographs, which were added to the herbarium on his return.

Arboretum staff members still explored North America and adjoining regions in their research, obtaining specimens for the herbarium and to be used for exchange. Ernest J. Palmer took several trips in the eastern United States in an attempt to simplify Sargent's treatment of *Crataegus,* and he continued to collaborate with Missouri botanists to catalog the flora of that state. John G. Jack pressed specimens every winter in the vicinity of the Atkins Institution in Cuba.

After publication of *The Lilac,* Susan D. McKelvey began a project inspired by a botanizing trip she and Alice

Eastwood took in New Mexico and Arizona in the fall of 1928. She subsequently turned to investigation of two similar genera of desert plants, *Yucca* and *Agave,* which she pursued over the next fifteen years. McKelvey covered much of the American Southwest in search of yucca and agave and their moth pollinators, assembling a complete collection of these difficult-to-preserve plants.

Ecologist Hugh M. Raup joined the Arboretum staff in 1932 as research assistant. A graduate of Wittenberg College in Springfield, Ohio, he served as instructor there while earning a doctorate from the University of Pittsburgh. Once at the Arboretum he began a phyto-geographic study of the little-known Peace River region of Alberta and British Columbia, returning from each field trip with thousands of specimens. Raup also floristically surveyed the vicinity of the Harvard Forest in Petersham, extending the Arboretum's interdepartmental cooperation to that institute.

In the summer of 1931 the herbarium gained Ivan M. Johnston as research associate to round out the taxonomic staff. A Californian, Johnston graduated from the university at Berkeley in 1919 and earned a master's degree there in 1922. During his student years Johnston did extensive field work in California and spent summers at the Alpine Laboratory of the Carnegie Institution in Colorado. He became interested in plants of the family Boraginaceae, which inhabit alpine and desert areas of the Western Hemisphere. After corresponding with botanists at the Gray Herbarium, Johnston came to Cambridge in 1922 for graduate work. He received a doctorate from Harvard in 1925 for his thesis on the borage genus *Cryptantha* in North America.

He remained with the Gray for the next six years, receiving grants to explore for plants in Chile and to study at European herbaria. In his role as coordinator, Ames transferred Johnston to the Arboretum staff to enable him to continue his excellent work on the borage family. Johnston stayed with the Arboretum for the remainder of his career. Aside from Ames, he was the first botanist at the institution whose specialty had no representation in the living collection. With a grant from the Guggenheim Foundation, Ivan Johnston planned an expedition to Chile in 1932; unfortunately, political conditions there precluded his return to the South American deserts. Instead, he spent another year in Europe examining collections of borages in botanical institutions and arranging for important collections to be sent to the Arboretum on exchange.

With its hand in so many successful expeditions and research projects, the Arboretum herbarium grew by about fifteen thousand specimens per year during Ames's stint as supervisor. Even though the herbarium staff prepared and distributed thousands of duplicate specimens to other institutions, expansion space for the material retained was a great need. The equivalent of about eighty-five full-sized new cases was installed in the administration building in 1931 and 1932, but this would provide for only about ten years of growth at the rate specimens were then being added. The latest installation of cases usurped the room where wood had been displayed, putting an end to public exhibits in the building. The scientifically valuable wood specimens were moved to the new biology building recently constructed in Cambridge. Each year hundreds of Arboretum herbarium specimens were loaned for study by botanists at kindred institutions, and more and more often scientists visited the Arboretum to examine its valuable herbarium and library.

The library also was beginning to show signs of overcrowding. Despite cutbacks in the purchase of old and rare books during the worst years of the Great Depression, the library had to grow to keep up with modern science. Ethelyn Tucker was forced to shelve many books on the fourth floor of the herbarium wing. With both collections competing for space, Ames thought the solution would be to build a fireproof addition to accommodate the library under improved physical conditions, thus freeing the entire Hunnewell building for herbarium operations. In annual reports and publicity statements, he repeated the need for funds to better house the library. When James B. Conant succeeded Lowell as Harvard's president in 1933, Ames informed him that better housing for the library was the Arboretum's most pressing need.

WILSON KEEPS THE COLLECTIONS

At Ames's suggestion, E. H. Wilson assumed the title of Keeper, a British term analogous to curator. He had conducted affairs in Jamaica Plain for more than a month in the interim after Sargent's death, and the new title was a reflection of the emphasis to be given to care of the living collection as distinct from other Arboretum matters. The new director relied on Wilson's ability to inspire the grounds staff with his wit and energy. One of his first actions was to dismiss the grounds superintendent, Van der Voet, quite justifiably by all accounts. Louis Victor Schmitt, in charge of labeling the plants and mapping the grounds since 1906, was made superintendent. Together, he and Wilson overcame past inertia to update methods of care for trees and shrubs.

The spring of 1927 brought an interruption to the customary floral progression that could have been misconstrued as a gesture of mourning for Sargent, but was, according to written records, simply the result of horticultural practice. It was announced in volume 1, number 1, of the *Bulletin of Popular Information:* "There will be no Lilac display this year. Owing to impoverished conditions it has been necessary to prune the Lilac bushes severely and liberally fertilize the soil. . . . For the public's sake it is a pity that the lilacs had to be given a year's grace but there is a limit to the endurance of even the good natured Lilac."

Rejuvenation of old lilacs and shrubs of similar habit by pruning back to the ground is an accepted horticultural procedure. The plant is lost to the landscape for one year, its blossom display is sacrificed perhaps for two, but the ultimate result is a vigorous plant with new wood capable of producing abundant flowers. As Victor Schmitt recalled some years later, the work of pruning the lilacs commenced in the winter of 1926, but it was not until spring that the decision to cut them entirely down was made. Although the newspapers announced "Lilacs at Arnold Arboretum Will Not Be Out Until 1928," visitors were urged to go see the flowering cherries and crabapples instead.

The lilacs were given special attention all summer. To encourage their return to vigor, the ground around each plant was carefully cultivated and fertilizers were judiciously applied, something that had rarely been done at the Arboretum. When fall planting concluded, Wilson and McKelvey together scrutinized the lilacs for signs of recovery. Relieved, they observed that all the severely pruned plants had sent up stout shoots from the base and that many appeared to have formed flower buds.

Aside from the drastic treatment of the lilacs, the only other significant change in the grounds Wilson made in the spring of 1927 was to have a new collection of cherry species installed among the birches on Bussey Hill. The reason for siting them in this location, on the northwest slope to the east of the summit section of Bussey Hill Road, has been lost to the record. Providing them a microclimate favorable to protection of early spring blossom seems the best explanation. Perhaps the cherries in the birch collection were meant to link with the large group of double-flowered forms from Japan planted farther up the drive, on the plateau, a few years earlier.

Wilson and the new superintendent spent a good part of the summer surveying the grounds and concluded that certain areas were overcrowded—namely, the Chinese shrub collection and Hickory Path, now more often called Centre Street Path. Wilson directed an ambitious amount of transplanting, or "spreading out and regrouping" as he termed it. This was not the addition of young plants from the nurseries into the collections, but rather moving and resiting mature plants.

It was Wilson's first attempt at managing living-collections operations, inasmuch as fall planting had never been part of the Arboretum routine. The azaleas and the Chinese shrubs on Bussey Hill came in for special attention. The former were apparently rearranged so that each kind made up one mass rather than being distributed in smaller clumps throughout the area. At its lower end, the course of the path was moved east, and another path splitting off to the west was installed. Some of the plants in the Chinese shrub beds were moved out and planted among the pears and cherries nearby. In anticipation of giving up the

105. *The Arboretum staff and Alfred Rehder's family paused to celebrate his seventieth birthday in September 1933. First row, left to right: Oakes Ames (supervisor), Alfred Rehder (curator of herbarium, editor of journal), Mrs. Rehder, Ethelyn Tucker (librarian), Miss Rehder, Katherine Kelley (assistant librarian), Eleanor Stuhlman (secretary), Miss Judge, Heman Howard (plant records), Caroline Allen (herbarium assistant), Ethel Anderson (business secretary), and Louis Victor Schmitt (grounds superintendent). Second row: Ernest Palmer (botanist), Harald Rehder, John Jack (botanist), John Morse (student assistant), Heriklea Yeranian (herbarium technician), Clarence Kobuski (assistant curator), Edmund Wilberding (plant labeling), and Alfred Fordham (propagation assistant).*

Orchard and Prince Street locus, two sizable Higan cherries were relocated from there to the Forest Hills gate area. In the vicinity of the administration building, three large magnolias were moved and fifty Chinese spruces and silver firs were brought from the far side of Bussey Hill to the top of the slope above the lawn. A new planting of Asiatic crabapples introduced into cultivation by the Arnold Arboretum was installed in front of these Chinese conifers. Along Meadow Road, beds of native azaleas were added on the west side, in the foreground of the lindens.

106. Although the lilac collection was so severely pruned in 1927 that few plants flowered that spring, within a few years the collection had returned to its former vigor. The late lilac (Syringa villosa), a native of northern China, blooms at least a week after the cultivated varieties of the common lilac (S. vulgaris). It is one of several species that together extend the lilac-blooming season into June at the Arboretum.

The following spring, as warmth brought on the opening of buds once more, Wilson had several reasons to be pleased. The winter of 1927–28 was extremely mild, and ample snows had fallen when most needed. This weather proved favorable to the extensive experiment of fall planting. All the large trees, and even the difficult royal azalea (*Rhododendron schlippenbachii*), came through with no harm. In his usual anthropomorphic tone, Wilson stated in the *Bulletin of Popular Information* that the royal azaleas "seem to have enjoyed the experience."

When the lilacs finally opened their blossoms late in May, many of the rejuvenated ones flowered freely. Although the plants had not yet regained full stature and had fewer flower trusses per plant than formerly, each cluster bore an unusually large number of flowers. Another blooming "event" of that spring was the production of flowers on the dove tree (*Davidia involucrata*) on Bussey Hill. This was the first time that an Arboretum dove tree had bloomed since the initial planting some two decades before. It was to obtain this species that Wilson had traveled to China on his first expedition for Veitch. Unfortunately, the Bussey Hill dove tree flowered two years too late for Sargent to see. He must have had faith that it would bloom, however, because he knew of a specimen planted on an estate in Newport, Rhode Island, that set forth its languid white flower bracts in 1926.

SPREADING AND REGROUPING; BULBS FOR COLOR

Satisfied with the success of the previous fall's transplantings, Wilson and Schmitt continued their work in September and October 1928, this time overhauling the deutzias, spiraeas, and barberries. William Judd noted in his diary that numbers of *Deutzia* species were removed from the beds along the Centre Street Path and planted under the nearby hickories and that many weigelas and spiraeas from the same beds were moved out nearer Valley Road. What Wilson meant by the "Berberis group . . . being entirely rearranged" is more difficult to ascertain. From about 1910 onward, during Sargent's lifetime, the

Arboretum's barberries were grown in three areas: the shrub collection, the Chinese plant bed on the plateau of Bussey Hill, and in the mix of shrubs along Centre Street Path. After 1929 many barberries were to be found along the west side of Bussey Hill Road, undoubtedly the result of 1928 "regrouping."

As evidenced ten years later (from which time the only detailed maps of the plantings exist), barberries of many sorts were lined up in single file, paralleling the summit section of Bussey Hill Road, between it and the side path. Many of the plants had recently been accessioned, most likely moved straight from the nursery at Orchard and Prince Streets when this site was vacated, while others were older plants moved from other locations to relieve overcrowding. This new planting of Wilson's must have looked like a hedge defining the drive's edge from its junction with Valley Road all the way to the point where Azalea Path began, although it is certain the plants were never clipped formally. Unlike the original roadside shrubberies, this one was composed of individuals of many different, mostly exotic species instead of masses of a single native species. It is possible that the moving of plants away from the Centre Street wall was done in anticipation of the proposal to widen the street that was under discussion in the late 1920s.

Regrouping also took place among the conifers. Wilson and Schmitt directed their labor crew to move "several hundred" specimens from the supplementary pinetum on Peters Hill to "the Pinetum proper." Wilson remarked that these moves, too, were made to afford more space for each plant to develop its full potential. Most of these were placed on the low land on both sides of Bussey Brook close to the Bussey and Walter Street gates, land that had formerly been left unplanted to provide a panoramic view of the older conifers on the upper slopes. The following season several *Taxus* from the new bed along Walter Street and from the original *Taxus* collection near the junipers and arborvitaes were moved to give room. Most were placed under the Douglas firs along Bussey Street, with a few being positioned close to Bussey Brook.

Another type of planting, hitherto untried, was added to the Arboretum landscape as the leaves turned and began to drop in 1928. Four thousand spring-flowering bulbs purchased by the Arboretum were planted under the magnolias in front of the administration building. These were mainly smaller sorts: grape hyacinths, snowdrops, squills, and some narcissi. When they bloomed early the following spring, Wilson was pleased with the "touch of color" they provided—so much so that in autumn of 1929 he accepted a gift of twelve thousand daffodil bulbs pro-

vided by a Long Island nurseryman. These were planted on the lawn area immediately beyond the administration building.

While Sargent had encouraged the Arboretum's native herbaceous undergrowth, it is quite possible he would have disapproved of these installations made for their colorful effect. In fifty-four years he had never made such plantings in the Arboretum, yet he had naturalized spring-

107. Spreading and regrouping. Under E. H. Wilson's oversight, the grounds staff moved several large conifers from Peters Hill into the original pinetum above Bussey Brook in 1929. Among them was this limber pine (Pinus flexilis), a species native to western North America along the Rocky Mountains.

flowering bulbs at Holm Lea. Perhaps it was the vulnerability to foot traffic and unwelcome picking in a public place that kept him from adding these colorful little flowers. Maybe he feared it would divert attention from the Arboretum's mainstay, flowering trees and shrubs. The tree garden still abounded in wildflowers, as Ernest J. Palmer attested in a 1930 report, "The Spontaneous Flora of the Arnold Arboretum."

The additional funds available for grounds operations since the completion of the Charles Sprague Sargent Memorial Fund enabled Schmitt not only to increase the labor force but also to purchase fertilizers and soil amend-

108. Mountain laurel (Kalmia latifolia). Under the boundary plantation of red and white pines on the South Street bank, E. H. Wilson put in nearly a thousand plants each of mountain laurel and torch azalea in 1929 and 1930, creating an elegant combination of the eastern North American and eastern Asian shrubs favored by Charles Sargent.

ers as had never before been possible. The recovery of the lilacs after their 1927 pruning was proof of the value of fertilization and soil improvement. Starting with the crabapples, Japanese cherries, hawthorns, and species lilacs, different groups received fertilizer each spring for several years.

The rhododendrons were subject to improved cultivation methods over several seasons. The collection was the result of years of hardiness testing of species and hybrids from as many likely sources as Sargent could find. The soil around each plant was carefully cultivated and renewed by the addition of peat moss and leaf mold purchased for the purpose. Native species of oak were planted among the rhododendrons along the brook to provide high shade. An additional planting of 150 new hybrid rhododendrons was made on the ridge of Hemlock Hill in 1929 as well. Sargent and Wilson had long urged that American nurserymen take up the breeding of better rhododendrons, suited to eastern America's growing conditions. Few of the ones bred in Europe, where most of this type of ornamental-plant improvement was taking place, were suited to the colder, drier climate of the Northeast. Despite the potential beauty of the flowers, some of the species used by English, Belgian, and Dutch hybridizers were not sufficiently cold tolerant to produce hardy offspring.

The plants installed on the western end of Hemlock Hill were unnamed hybrids developed by a retired industrialist, Charles O. Dexter (1862–1943), of Sandwich, Massachusetts. Dexter's breeding project involved the tender Fortune rhododendron from eastern China, a large-flowered, delicately colored species. He was determined to create a hardy race that had the flowering attributes of *Rhododendron fortunei.* Surrounded by a protective hemlock canopy, the hill was a good place to test some of Dexter's new plants in the Arboretum.

Across Valley Road from the main rhododendron collection, beyond the beeches on the slope to the east of the South Street gate, was one of Sargent's earliest Arboretum plantings—a wide border of red pine interspersed with a few indigenous white pines and American elms dating at least to Bussey's time. Over the two fall planting seasons of 1929 and 1930 Wilson established 750 mountain laurel plants (*Kalmia latifolia*) and 1,000 torch azaleas (*Rhododendron obtusum* var. *kaempferi*) in broad masses under these bordering pines. This was possibly the most inspired, successful, and enduring of Wilson's changes to the grounds. Charles Sargent was particularly fond of the mountain laurel, a native American shrub that was a most reliable and charming broad-leaved evergreen. He never failed to let *Bulletin* readers know when the mountain laurels were ready to bloom and considered them the last great floral show of the Arboretum's spring. Wilson's new *Kalmia* planting had the effect of echoing the Hemlock Hill *Rhododendrons* and *Kalmias,* framing the South Street visitor entrance with their deep, cool cover. The torch azaleas were a favorite of Wilson's, and he often sang the praises of this glowing, orange-pink flowered shrub. In so doing he always pointed out that this wonderful plant had been introduced to American gardens from Japan by Sargent himself during his 1892 trip. The color of blossom remained more intense when this azalea was planted under high shade, and for years some had been growing just above the mountain laurels along Hemlock Hill Road. The bloom period of the two barely overlapped, but the cool tones of mountain laurel, even out of bloom, served as an excellent foil for the flamboyant heat of torch azalea. The mass planting of them both on the South Street bank was more of a good thing. Consciously or not, it was Wilson's own memorial to Sargent.

WILSON AND PUBLIC RELATIONS

Fortunately, Wilson and Ames complemented each other's abilities in the oversight of Arboretum affairs—the orchidologist set scientific and financial policies, while Wilson managed horticultural and public relations. Wilson wholeheartedly continued to advocate the Arboretum to its many constituencies and took time to show its great assemblage of woody plants to visiting professionals from around the world. He authored eighteen numbers of the *Bulletin of Popular Information* each year, carrying on Sargent's formula of up-to-date reports on sequence of bloom, effects of weather, and detailed discussions of generic groups at the time when each was most worthy of a visit for firsthand observation. One innovation Wilson made was the inclusion of black-and-white photographs; previously, the *Bulletin* had not been illustrated. The publication was indeed a bulletin, coming out on an irregular schedule as seasonal advance brought noteworthy plants into bloom or fruit. By 1930 there were more than nineteen hundred subscribers.

Wilson also contributed voluminously to garden periodicals of wide distribution, such as the Massachusetts Horticultural Society's *Horticulture; Garden,* a New York magazine; *House and Garden,* under the editorship of Richardson Wright; and *Ladies' Home Journal.* The majority of his articles were on woody plants: the best for spring bloom, street trees, home gardens, and more. He drew not only from his travels but from nearly two decades of observation in the Arnold Arboretum to inform and encourage garden enthusiasts to plant a wider variety of trees and shrubs. His articles were often illustrated with photographs of Arboretum specimens and contained frequent mention of its work.

Between 1926 and 1930 Wilson also managed to publish four books: *Plant Hunting* (1927), in two volumes; *More Aristocrats of the Garden* (1928), a companion to a work published first in 1917; *China, Mother of Gardens* (1929); and *Aristocrats of the Trees* (1930). Moreover, while the fund campaign was in full swing Wilson traveled to New York, Philadelphia, and other cities to speak on the Arboretum, and he continued to be sought as a lecturer by garden clubs and plant societies. In 1929 Wilson entered the "radio age," giving a talk on his travels and the Arboretum broadcast by a New York station. Largely through his efforts, the Arnold Arboretum was probably as well known among British garden lovers as American. Wilson contributed to the horticultural press across the Atlantic, and his plant introductions from the Far East created great enthusiasm there.

Determined to follow up on the success of his introductions and evaluate their performance over the long term, he surveyed the numerous recipients of plants from his Arboretum expeditions. Each grower of Wilson introductions was sent a list of the plants he or she had received, with columns in which to note largest dimensions obtained, date planted, and dates when flowering and fruiting occurred. There were also many estate owners and private collectors on the East Coast whom he visited to observe the plants the Arboretum shared with them. To a grower in Cornwall, England, Wilson wrote: "No parent ever read a school report of his favorite child with greater pleasure than I did the questionnaire you kindly filled in and returned. As you must have long since sensed, my plants are as my children to me and the knowledge of their well-being is all important" (AAA, E. H. Wilson to J. C. Williams, 7 March 1929). Wilson found the information that trickled in interesting and planned to use it to give more accurate evaluations of his introductions in his writing.

Wilson worked especially hard to educate and enlist the interest of "that hardboiled set," the nurserymen, in the Arboretum and its plants. Through personal contact and correspondence he exhorted the commercial growers to study the Arboretum collections for themselves and make use of the *Bulletin*s and other publications to become familiar with "a better class" of plant material than they had yet offered to their buying public. Many took his advice, and several were loyal supporters and advocates of the Arboretum in turn. Wilson was quick to point out to the nurserymen that the Arboretum collections served not only as demonstrations of what could be grown but also as a bureau of standards by which the correct identity and names of plants could be determined. It was for this role that nurserymen had championed the establishment of the National Arboretum as well. While he was gratified to see "his plants" become available in the trade, Wilson never failed to urge the utilization of other worthies, whether

109. Group planting of river birch (Betula nigra) *on the north slope of Bussey Hill. Although a bridle path was opened along the Arboretum's western boundary in 1928, visitors interested in viewing the tree and shrub collections still came most often on foot. Nonetheless, a limited number of automobiles were admitted by permission of the director and the park commissioner.*

they be native Americans or the introductions of other explorers and institutions.

Change, inevitable with a new administration, came to visitor use of the Arboretum grounds in the spring of 1928. No mention was made of it in Arboretum reports, but the *Boston Herald* carried an article on 8 April with the headline "New Arboretum Bridle Path Affords Fine Canter," accompanied by a map showing the route through the grounds. The new course connected an equestrian path already established along the Arborway with one following the West Roxbury Parkway (now the VFW Parkway), thus allowing traffic-free horseback riding from Boston out to Milton, Dedham, and the Blue Hills. Apparently, riding enthusiasts had long petitioned for such a path, but "previous administrations of the Arboretum were opposed to the scheme." (One can just hear Sargent saying, "Over my dead body!") Ames, quite a keen rider himself, and Wilson apparently were sympathetic enough to permit the project to go forward. The equestrian club responsible for the idea made a contribution to meet the cost of constructing the dirt track that skirted the northwestern boundary.

While horseback riding had always been permitted on the Arboretum drives, their hard-packed stone surfaces did not afford pleasurable riding, and besides, the new route was a much more direct connection to suburban regions than the meandering Arboretum drive. Wilson expressed the hope that riders would refrain from damaging the plants and trees, since injury to Arboretum plants, brought from all over the world, "means much more than it would to native trees and plants on an ordinary woodland road." Such a remark reveals the potential for misuse inherent in permitting unrelated recreational activities within the tree garden. While not nearly as drastic as the threat that the introduction of horse racing brought to Central Park in the 1890s, the Bridle Path nevertheless represented a divergence from the kind of use for which the Arboretum had been so carefully planned, looking at plants and leisurely taking in the landscapes they composed.

Boston parkmakers had tried to provide for such sporting activities elsewhere in the system; there were, for example, equestrian ways planned for Franklin Park. But lifestyles and priorities had changed since the 1890s, and facilities for active recreation were commonly added to urban parks after the turn of the century. At least the Bridle Path was kept away from the main drive for most of its length. Not one to lose an opportunity for favorable press, Ames saw to it that the article announcing the opening of the Bridle Path included his pitch for the need to complete the one-million-dollar Sargent Memorial Fund. The placement of the equestrian route along the Adams Nervine border led to the eventual abandonment of the grassy way known as Linden Path, however. Pedestrians came to prefer the Bridle Path, too, over the less obvious grass path, especially once the high-level maintenance of edged shrub beds gave way to simple mowing around shrub masses in succeeding years.

That spring, the City of Boston officially paid tribute to the memory of Sargent on the anniversary of his birth, 27 April. On that day Park Commissioner William P. Long presided at a commemorative tree-planting ceremony in the heart of the Jamaica Plain neighborhood. The Arboretum staff assisted with the installation of a tulip tree in Sargent's honor on the grounds of the Eliot School, a community resource since 1660 when the fabled Reverend John Eliot established a school at this location for the edification of the Roxbury district's Native American population. The mayor and his wife attended the ceremony, and William Judd spoke on behalf of the Arboretum.

Other aspects of the institution's relationship with the city were given new emphasis during Ames's administration. In response to increased requests to view the collections from motorcars, Wilson and Ames worked out a better system with the Boston Parks Department for granting permits for their use on Arboretum drives. Wilson complimented the parks department's maintenance efforts in his 1929 annual report, no doubt inspired by several improvements made by the department during this period. Three drinking fountains were installed in 1928, and in 1929 the most frequently traveled footpaths, those flanking Forest Hills Road down to the junction of Meadow Road, were paved with asphalt.

110. *In June 1930 Ernest Wilson received an honorary doctor of science degree from Trinity College, Hartford, Connecticut. Afterward, the botanist-explorer posed for photographs in front of regal lilies, his most famous introduction from China, growing in his backyard in Jamaica Plain.*

LINK WITH SARGENT BROKEN

Although he will always be remembered more universally for his explorations, Wilson took to his responsibilities as Keeper of the Arboretum. In 1929 he wrote one of his many overseas colleagues, Takenoshi N. Nakai, in Japan, "My own duties seem to increase every day but as I enjoy them all they weigh lightly on my shoulders" (AAA, Wilson correspondence, 17 May 1929). Perhaps this man who had spent three days being carried on the shoulders of others down a steep and narrow mountain track with an untreated leg fracture gained a better perspective on what weighs heavily in life than most.

Wilson received many honors and awards for his botanical explorations, plant introductions, photographs, and authorship. In June 1930 Trinity College of Hartford, Connecticut, conferred on him an honorary doctor of science degree. Apparently the college was convinced to

make the move by one of its trustees, Wilson's longtime literary associate Richardson Wright. Happily, his colleague Arthur Osborn, curator of the arboretum at Kew, was there to share the day. Wilson had many reasons for satisfaction and every expectation of carrying on the work of the Arnold Arboretum and disseminating its findings.

In the preface to his *More Aristocrats of the Garden* Wilson wrote:

Trail-blazing is never an easy task but it is an essential one and I propose, until fate decrees otherwise, to pioneer onward as I have done since 1899. Introducing new plants is not in itself final or sufficient. Time must prove them of quality, superior and desirable, and means must be found of placing them in the gardens where they belong. The more widely they are known and cultivated, the more pleasure they give to others, the greater the satisfaction to those who cull them from distant lands. (Wilson, 1928, p. vii)

This is a likely summary of Wilson's agenda at fifty-odd years of age.

"Fate decreed otherwise," and quite suddenly, on 15 October 1930. That afternoon Wilson and his wife died in Worcester, Massachusetts, when their car skidded off the road and dropped down a steep embankment. They were returning from a trip to Geneva, New York, to visit their daughter, Muriel, who had married a pomologist with the agricultural experiment station a little more than a year earlier. Apparently, rainy conditions and an awkward curve were responsible for the accident. The tragic and sensational news, carried by the Associated Press, made headlines up and down the East Coast.

Wilson's friend Edward I. Farrington, secretary of the Massachusetts Horticultural Society and editor of *Horticulture,* oversaw funeral arrangements. A double service for the Wilsons was held at Boston's Trinity Church in Copley Square the following Saturday. The church was decorated with flowers sent by bereaved colleagues from all over the country. The Arboretum staff surrounded and covered the caskets with sprays of colorful berries and foliage of some of Wilson's most treasured plants: crab-apples, honeysuckles, cotoneaster, beauty berry, heather, yew, the dove tree. Honorary pallbearers and distinguished guests included Oakes Ames, A. Lawrence Lowell, Richardson Wright, the Massachusetts commissioner of agriculture, the president of Massachusetts Agricultural College, heads of the horticultural societies of New York and Philadelphia, and many more of Wilson's friends in American and Canadian horticulture. Alfred Rehder, Victor Schmitt, J. G. Jack, and William Judd participated in the ceremony, as did Harlan P. Kelsey and W. Ormiston Roy of the Montreal parks system, the latter a fellow Kewite.

As the stunned and saddened Arboretum staff returned to work, letters of condolence poured in from around the world echoing their disbelief and shock. Arthur Osborn wrote Judd from Kew: "It does not seem possible I shall not see Wilson over here next year. We had built so much on his visit." And some weeks later: "What a marvellous man Wilson was. I can still picture Mr. Wilson at the Arboretum, and feel that any moment I shall wake up and find the tragic accident was a bad dream" (AAA, Judd papers, A. Osborn to W. Judd, 17 October 1930, 4 December 1930).

Wilson had managed the Arboretum collections alone, without Sargent, for a scant four years, although before 1927 there were another few during which he carried a lot of the weight. Whether the changes Wilson made were part of a plan or trend that would have measurably diverged from Sargent's conception will never be known. Even Sargent himself had made alterations from the "final plans" Olmsted and his associates gave him in 1885 and 1895. Most of Sargent's own alterations had to do with the placement of shrubs and coping with the problem of greater numbers of plants once the early-twentieth-century explorations were under way. The decision to leave the shrub garden "as is," the addition of special shrub beds, the creation of "supplementary collections" of certain groups such as *Ribes, Cornus, Philadelphus, Viburnum, Sorbus, Prunus, Malus, Crataegus,* and so on, were all alterations of the original scheme. They were necessitated by the mandate to make the collections as complete as possible. But Sargent made these changes with full knowledge of the original scheme. In his mind, the overarching system of tree groups from magnolia to larix in the Ben-

tham and Hooker sequence remained the backdrop against which other groupings were accommodated.

The naturalistic charm and the botanical organization could be preserved if there were a limit to these changes. While Wilson knew the system, he had never known Olmsted, never been in on the original dreams and plans. His forte was plants and plant collecting, and landscape composition took second place. In a sense, he was one step removed from the original plan. The changes he made to the landscape tended to involve filling in open spaces established and maintained as such for their breadth of view. Japanese double-flowered cherries were lined up in rows filling in the lawn on the Bussey Hill plateau, spruces and firs were moved into the lower slopes of Bussey Brook's valley, barberries were lined single file along the summit section of Bussey Hill Road. As will be seen, none of these plantings remain today. Yet these changes were undertaken because of the need for room rather than for their landscape effect. Nevertheless, most observers felt Wilson and Ames were admirably carrying on Sargent's traditions in their oversight of the grounds.

In the 1890s, Sargent had said that the real solution to the placement of shrubs would be ten more acres of level land. By the time additional land came with the acquisitions of the South Street, Weld-Walter Street, and Centre Street parcels from 1919 to 1927, Sargent lacked the stamina and financial resources to realize their full development. His energy was turned to raising funds for developing the herbarium and for establishing a program to keep up with modern science. Had he lived, Wilson might have furthered the utilization of outlying parcels for Arboretum collections, thus preventing the crowding that became more troublesome later. Having been with Sargent during the negotiations for the land, he had a better sense of its importance to Sargent's conception of space allocation than did those who followed. In any case Wilson, like his successors, would have had to face the monetary restraints that the Great Depression brought on.

Among Wilson's more successful and enduring installations were the extension of azalea plantings under the oaks on Bussey Hill's western and southern slopes and the torch azalea–mountain laurel mass planting on the South Street bank. The latter, a serendipitous combination of exotic and native, soon gained the appearance of having naturalized on the site. Unfortunately, it was the last work he oversaw before heading off to visit Geneva, New York, that October.

INTERIM FOR THE GROUNDS

Out of concern for the institution and its staff, who had only just gotten over the loss of Sargent, Ames put off finding a replacement for Wilson. In early November he asked J. G. Jack, William Judd, Victor Schmitt, and Susan D. McKelvey to work together as a committee to oversee the grounds. For the next five years there would be no one person to take up the lead that Wilson had barely begun. The committee members, and the person ultimately chosen to fill in, all had other interests and obligations that precluded the full-time commitment, and the perception, needed to carry on management of the living collections.

Jack, employed at the Arboretum since 1886, had the greatest length of service in connection with the living collection. He had overseen the plant records for years and filled a crucial role as interpreter of the Arboretum to the public with his spring and fall field classes. Since 1926, however, when Harvard took up oversight of the Atkins tropical garden in Cuba, Jack had been sent there twice a year to collect herbarium specimens to document the flora of the garden and the nearby Cuban forests. These field trips continued until 1935, when Jack officially retired. Schmitt and Judd each had around twenty years' experience at the Arboretum. Schmitt, although fully occupied with the day-to-day supervision of the labor force, knew firsthand the pressing needs of collections and grounds maintenance.

As propagator Judd oversaw the operations by which new plant material came in to the institution and was nurtured until ready to be set out on the grounds. There were the greenhouses, cold frames, and nursery areas to see to, including assisting with Sax's and Faull's experimental work. And there were nurseries on the South Street and Weld-Walter tracts where plants were grown to larger sizes. Additionally, the yearly distribution of seed,

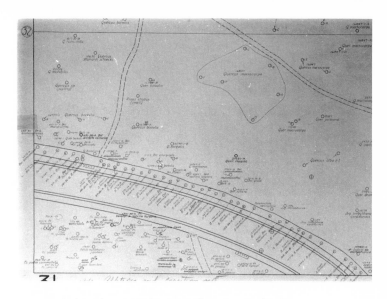

111. *Detail of records map number 32, quadrant c, 1942, showing part of the dense, linear planting of barberries (several species and varieties of* Berberis) *sited by Wilson in the late 1920s. This is one of the oldest existing plant records maps; apparently, nothing drawn before 1940 was retained.*

112. *Surprisingly, hay was still harvested in the Arboretum as late as 1931, when Ernest J. Palmer took this photograph. These haystacks were located on the recently acquired Adams Nervine parcel near 1090 Centre Street.*

cuttings, and other propagation material from Arboretum accessions to institutions the world over was an important part of Judd's work. While McKelvey had unshakable enthusiasm for and understanding of all the Arboretum's work, she had by this time embarked on a revisionary monograph of agave and yucca, and spent several months each year traveling in the desert. Records of how the committee operated are scant. Apparently they all had worked so long under Sargent's perfectionist standards that they were hesitant to make alterations. They tended to add plants to the generic groups as left by Wilson and Sargent.

Almost at once, however, the committee faced an unwelcome disturbance to the grounds. In 1931, after years of debate by city authorities and the public, Centre Street was widened. This two-lane road that skirted the Arboretum's northwestern boundary had become a busy thoroughfare to points south of Boston, known as Route 1, since the automobile had become the most important transportation form. At first the proposal was to widen the section from the Arborway to the VFW Parkway to one hundred feet and make a four-lane road with a tree-clad center median to conform with those two parkways. In the Arboretum's interest, Harvard officials objected, since a number of mature specimen trees would be lost—notably, a fine example of the Japanese raisin tree (*Hovenia dulcis*), one of the only trees in the buckthorn family hardy in the Arboretum, and a group of wing-nuts (*Pterocarya*), an Asian genus in the walnut family.

The project subsequently was reduced to an eighty-foot width, perhaps as much because of the economic situation in the 1930s as to save trees. During the widening of Centre Street, the wall separating it from the Arboretum was completely rebuilt, causing considerable disruption to the plantings in the area. The grounds committee was forced to delay replanting, however, because of budgetary constraints.

After some months Ames decided another staff member was needed in Jamaica Plain. Rather than fill the position of Keeper, Ames created a new position, arborist, for which the primary responsibilities were care of the living collections and furthering of the Arboretum's relations with the public. A geneticist from the Missouri Botanical

Garden, Edgar Anderson (1897–1969), accepted the post and came to the Arboretum in the summer of 1931.

It was not his first time in Jamaica Plain. After receiving a bachelor's degree in horticulture from Michigan Agriculture College, Anderson spent the years 1919 to 1922 at the Bussey Institution, earning a master's and a doctor of science with Edward M. East. Like many Bussey students, Anderson lived in the dormitory on the Bussey grounds overlooking the Arboretum. He attended the economic botany course given by Oakes Ames and became acquainted with Karl Sax. He made full use of the Bussey's greenhouse and growing areas for experiments in hybridization. At times he called on John George Jack and studied specimens in the Arboretum herbarium. After receiving his doctorate, Anderson joined the staff of the Missouri Botanical Garden, where he held several posts. Although Anderson had gained a wide base of experience when he returned to Jamaica Plain to take up the duties of arborist, he retained a strong interest in theoretical genetics.

Once established at the Arboretum, Anderson pursued research in hybridization and genetics, publishing technical papers alone or jointly with Karl Sax. Apparently, the "committee on outdoor operations" under the chairmanship of J. G. Jack continued to oversee the care and development of the living collections with some input from Anderson. Moreover, the committee was rarely inclined to revise any of the plantings. Between 1930 and 1935 two factors worked to bring development of the grounds to a standstill: the economy and the weather.

By the beginning of fiscal 1932–33, Ames answered President Lowell's request for alternative budgets showing reductions of 10, 20, and 30 percent in Arboretum expenses. The Great Depression had by this time reached its worst point, and the Arboretum, despite its recent financial gain, faced loss of income just like all other university departments. Early that year Ames had been forced to cut back the labor crew's working time by two hours a day as an economy measure. With funds and workforce depleted, progress on the grounds was limited.

These few years of lessened activity had little effect on the plant collections, however, when compared with the destruction wrought by the severe cold of the winter of

1933–34, the coldest winter ever recorded for Boston. On 29 December the temperature dropped to seventeen degrees below zero Fahrenheit. In late January temperatures went to two below with strong winds blowing, and from 7 to 10 February the temperature never rose above minus two and went as low as eighteen below. The monthly mean temperature for every winter month except January was below the averages that had been recorded over the preceding forty-seven years. There was no protective covering of snow on the ground during any of the extremely cold spells. It was a test of hardiness such as had never been given Wilson's introductions during his and Sargent's lifetimes. Those plants that survived with flower buds intact were indeed the ones to rely on for future breeding and widespread planting.

In 1934, spring bloom at the Arboretum was the exception rather than the rule for many groups. Among flowering cherries, most lost all their flower buds except the Sargent cherry, *Prunus sargentii*; only three species of forsythia bloomed, *Forsythia ovata, F. europaea,* and *F. japonica.* With regard to these three, Anderson reported: "In an ordinary year they are relatively inconspicuous members of the collection. This spring, as full of bloom as ever, they stand out in dramatic contrast to the bare branches of the commoner sorts" (E. Anderson, 1934, p. 9). As the season of growth progressed, the extent of the coldest-ever winter's damage revealed itself. All spring and summer Schmitt, Jack, Faull, and Judd scrutinized the entire collection for damage and wrote up a report that filled four numbers of the *Bulletin.* Listing the affected plants according to three categories of extent of injury, the staff found some 27 kinds that were killed outright. Plants of some 248 taxa were killed to the ground but were able to send up new growth from the roots. Another 225 were injured to a lesser extent, many having dead branches or killed flower buds. In all, some 500 kinds, the majority of them shrubs and vines, were killed or disfigured. Thus, approximately 10 percent of the taxa grown in the Arboretum were afflicted by the winter's cold that year.

Schmitt supervised the rehabilitation of hundreds of shrubs and trees. After cautiously waiting for signs of life, pruning of deadwood commenced. The object was to improve the appearance and to minimize the risk of further injury by fungal and insect pests that might beset decaying stems and branches. The pathology staff was on the alert for such diseases in the aftermath. What manure and fertilizer were available with the Arboretum's decreased budget were used before mid-June to stimulate new growth on the plants with the best chance of recovery. While many injured plants successfully recovered, in some the damage initiated a decline that impaired their performance for years. The condition of some was worsened by the winter of 1934–35, which was almost as severely cold as the previous one. Fortunately, a heavy snowfall in late January, which covered the ground for most of the rest of the winter, lessened the extent of the damage. The collection most permanently affected by this double dose of frigid weather was that of the trees in the genus *Prunus.* The double-flowered Japanese cherries were hardest hit, but other species succumbed as well. Rhododendrons and azaleas received a large share of losses the second time around, while the lilacs and crabapples were relatively unscathed. It is no wonder the last two groups became the favorites of the succeeding horticultural staff.

ECOLOGICAL SELF-EXAMINATION

To encourage interdisciplinary study of the Arboretum and to widen its educational role, Ames initiated ecological surveys of the Jamaica Plain grounds. Collector Ernest J. Palmer had long been interested in the natural history of the Arboretum property and occasionally collected specimens of wildflowers he found there. Under Ames's direction this study was formalized into a comprehensive inspection, published as "The Spontaneous Flora of the Arnold Arboretum" in volume 11 of the *Journal of the Arnold Arboretum* in April 1930. Palmer found 608 species and varieties in 87 families to occur within the grounds. He documented the wild and weedy, noncultivated plants with herbarium specimens and discussed their distribution through the Arboretum in relation to soils, topography, geology, and amount of past human disturbance. Palmer noted that introduced plants constituted about 30 percent of the flora and predicted that this proportion would increase in coming years, since he had already observed

113. Before the increased mechanization of grounds maintenance procedures and the pressures of urbanization became too great, the Arboretum contained a rich mixture of native and naturalized herbaceous plants. Ernest Palmer found nearly eight hundred kinds of plants in his survey of the spontaneous flora between 1924 and 1935. Shown here is Leitneria Path, which skirted the base of one of the North Woods knolls behind a boggy depression where corkwood (Leitneria floridana), a native of the southern United States, was planted. In this vicinity Palmer found ostrich fern, jack-in-the-pulpit, skunk cabbage, clearweed, star-of-Bethlehem, and wild garlic in the wet ground. Whorled loosestrife, spreading dogbane, Indian tobacco, and woodland sunflower were among the plants of the woods' margins.

native plants give way to more aggressive foreign weeds and precarious natural habitats be destroyed by increased visitor activity and other forces of urbanization.

While Palmer's survey was mainly intended as a guide for students and visitors interested in the Arboretum's native plants, it serves as a valuable record of many aspects of the landscape and shows that the institution still valued its natural areas once Charles Sprague Sargent was gone:

In the Arnold Arboretum this blending of the natural and the artificial has been part of the plan from the beginning, and while the principal object has been the bringing together of a collection of such trees and shrubs, from all parts of the world, as can be grown in the open in the climate of New England, the most careful consideration has also been given to landscape effect. Remnants of the native woods and open spaces of meadow and grassy slopes have been left at intervals as examples of the wild flora, and these bits of the primitive have been so happily blended with the planted groups as to add greatly to the general result. It is thus possible here within the boundaries of a great city not only to enjoy the extensive collections of exotic trees and shrubs, but also for those who are interested in the native trees and wild flowers, either from the standpoint of the scientific student or merely as lovers of them for their beauty and charm, to see a considerable representation of the spontaneous flora. (Palmer, 1930, p. 65)

In this treatise Palmer used names to designate the several natural areas within the Arboretum such as North Woods and Central Woods. Although a useful innovation, this practice was not generally adopted for use in guides or on maps until the 1980s.

The disastrous winters of 1934 and 1935 caused another round of Arboretum "self-examination," this time in terms of microclimate. While the institution had maintained a weather station for many years, in the fall of 1934 temperature-measuring stations were set up in twelve localities throughout the grounds. During the subsequent winter, differences of as much as twelve degrees were observed in different niches of the property at the same time. Hugh Raup, the forest ecologist, further analyzed the report made by the staff on 1933–34 winter injury to see if plants in certain areas of the grounds were more badly damaged than those in others. Raup took the species and varieties on the lists in the report that were repre-

114. Perforator (drill) found at Spring Brook village site (left) and projectile point found on Peters Hill. More than half of the Native American artifacts E. J. Palmer collected in the Arboretum were found in the vicinity of Bussey Brook, especially on the plateau planted with junipers overlooking the junction of Bussey and Spring Brooks, a site he named the Spring Brook village. From 1945 to 1980 these stone implements and fragments were exhibited in the vestibule of the administration building. More than forty years after Palmer made his observations, archaeologists studying prehistoric human activity in the Charles River basin found many of them still valid.

sented in more than one locality on the grounds and compared the extent of the injury on the same kind of plant from one locale to another. The results, published as "Comparative Studies on Winter Injury in the Arnold Arboretum" in the *Bulletin of Popular Information* in 1935, confirmed long-held views about the suitability of certain areas for less hardy plants and correlated with the differing temperatures measured. The section of the Arboretum in which plants seemed to suffer most from cold damage was the low-lying shrub collection; plants were less seriously injured on the Bussey Hill plateau and along Centre Street Path. Raup did not find enough common examples to make a meaningful comparison between Bussey Hill and Centre Street, however.

Raup also studied wind patterns in the Arboretum in an attempt to pinpoint which areas would be more suitable for growing untested or less hardy plants. He reasoned that the dry northwest winds were one of the most important causes of winter injury to plants in the Arboretum. When a deep snowfall was followed by below-freezing temperatures and steady northwesterly winds, Raup seized the opportunity to read the snow surface and record the extent of drifting in the Arboretum. The resulting map revealed the areas hardest hit by this frequent

weather pattern: the North Meadow; the northwest side of Bussey Hill, where the birch and elm collections stood; the same side of Peters Hill, where he also noted that many hawthorns had asymmetrical crowns shaped by the wind; and all of the Weld-Walter tract. The Centre Street Path area, where many introductions from eastern Asia had been growing, was again found to be one of the most protected sites.

Besides studying the geologic and meteorologic forces at work on the Arboretum, the history of human use of the land was also investigated during Ames's administration. After Palmer found one stone Indian artifact in the 1920s, he began a persistent search for such objects during his frequent walks in the Arboretum. William Judd and other staff members often accompanied Palmer on these evening and weekend explorations of the Arboretum's surface. Over several years he accumulated some sixty stone implements or fragments—projectile points, knives, scrapers, and other tools of the Indians' making. While not a trained archaeologist, Palmer was sufficiently knowledgeable to keep accurate records and to write up his observations for the *Bulletin*. He discovered that there were certain areas where these finds seemed to be concentrated, and from this he gained insight not only into the lives of the prehistoric peoples but into the physical conditions of the land itself:

Aside from the sentimental and romantic interest of these bits of prehistoric art, their chief value lies in the deductions that can be made from them in regard to the life and customs of the people who made them and in locating the sites of habitations occupied probably long before the coming of the first white colonists. And this in turn helps us to reconstruct some picture of what the local conditions must have been in those times and of the significance of the changes that have ensued. (Palmer, 1934, p. 61)

Hugh Raup next took up the study of "human modification of primeval conditions." His findings were published in a long and intriguing article, "Notes on the Early Uses of Land Now in the Arnold Arboretum," in the *Bulletin of Popular Information* in 1935. He sought information in the registered deeds of conveyance and in the probated records of wills, inventories, and divisions of estates for the lands involved. He also consulted the early planting records of Sargent's time and observed the physical evidence in tree rings and in species distribution in the natural-woods areas. The history of forest utilization and farm practices in New England, and their effects on present-day forest composition and edaphic conditions, became a specialty of Raup's. His Arboretum study was a microcosm of the findings of greater New England land-use history. The repeated cutting and regrowth of forests and the cultivation of orchards and extensive animal pasturage (with relatively small areas devoted to intensive plowland for crops) had continued from the colonial era down to the time Harvard took over from Benjamin Bussey's heirs.

Meanwhile, E. J. Palmer continued his inventory of the spontaneous flora; by 1935 he had found 173 new species and varieties, including 7 families not recorded before. A supplement to the spontaneous-flora paper was published in the *Journal*. He seemed pleased that more than half of the additions to the list were natives that had previously been overlooked, even though many were exceedingly rare in the Arboretum, represented by single individuals or small colonies struggling to survive.

Ames hoped that these studies of the tree garden's ecosystem would be useful for future planning of the collections. Since many plants were only slowly recovering from two years of winter injury, the losses called for steps toward replacement. Rehder and Jack kept up the correspondence with other institutions and longtime colleagues to obtain seed of plants likely to grow at the Arnold Arboretum. After the two devastating winters, Anderson took a collecting trip to test his theory that the place with the most rigorous climatic conditions in Europe was the Balkan Mountains. He collected propagating material of common lilac (*Syringa vulgaris*), English ivy (*Hedera helix*), flowering ash (*Fraxinus ornus*), and boxwood (*Buxus sempervirens*) from wild populations that could yield hardier strains than earlier introductions. Sax's hybridization program promised new plants for the Arboretum collection once evaluation and further breeding took place. In 1935, Judd was sent to Europe, where he selected some six hundred kinds of plants from nurseries to be additions to or renewals of the Arboretum's holdings.

PUBLIC OUTREACH CONTINUES

After Wilson's death, it immediately fell to John George Jack to finish out the *Bulletin* for the year. The final two numbers of 1930 reflected Jack's perspective on education with two articles that presented more practical horticultural tips than had usually been included. By explaining the Arboretum's grounds maintenance procedures in terms of biological functions and ecological factors, Jack gave recommendations and rationale for fall tree planting and winter protection that *Bulletin* readers could apply in their own gardens or grounds.

Oakes Ames also became more directly involved with Arboretum affairs once Wilson was gone. He personally took up the documentation of the Arnold Arboretum plants and landscape with photographs, something that had previously been one of Wilson's pursuits. With the 1931 volume of the *Bulletin of Popular Information* both Ames's photographs and Blanche Ames's fine drawings of plants were frequently used to illustrate the text. Until Anderson arrived, Jack carried on the main load of writing for the publication, although Oakes Ames contributed articles, as did Karl Sax and one of his assistants.

Despite Anderson's title of arborist, Ames expected him to act as liaison between the Arboretum and the public. Both men agreed that his top priority was to further the Arboretum's relations with horticultural clubs and societies. For the next three years Anderson contributed the majority of articles and notices for the *Bulletin* and gave tours and lectures for the public. Under his direction the publication continued to broaden its subject matter. It was no longer simply a guide to the plants and when to see them, but included articles of historical and biological, as well as horticultural, interest.

The biological scrutiny of the grounds also provided information for interpreting the Arboretum and its research program to the public. A further step toward interpretation was taken with the publication of a new guidebook, *Through the Arnold Arboretum,* in 1934. Production of the book was overseen by the publicity firm of Melvin and Ronald. This is one of very few times in the institution's history that a guide was produced by someone other than the staff. One of the partners, Mary Armstrong

*115. The wall garden at the Morris Arboretum of the University of Pennsylvania, Chestnut Hill. In 1932
Lydia Thompson Morris willed her estate of 160 acres and an endowment to create the Morris Arboretum.
Both she and her brother, John T. Morris, who died in 1915, had cooperated with Charles Sargent in funding the Wilson expeditions and in evaluating the plants obtained. The Morrises' diverse collection of trees
and shrubs, modest formal beds, herbaceous borders, and native woodlands along Wissahickon Creek were to
be the basis of an institution devoted to the study of botany, horticulture, and landscape design.*

Melvin, worked closely with the Arboretum staff to assure accuracy and strove to make the guide more "user-friendly" than previous ones. The main description was organized by season, as in Wilson's *America's Greatest Garden*. Scientific plant names and the research undertaken at the Arboretum were carefully explained. Sections on bird and animal life and on geology were included at the suggestion of Edgar Anderson, as well as a list of fifty of the institution's important plant introductions. A brief account of the historical background and statement of future needs concluded the book. While the beauty and significance of the collections were featured, no attempt was made to explain the overall arrangement or the naturalistic style of design, almost as though these were taken for granted.

An intriguing inclusion was a foldout map tipped inside the back cover. It was an updated depiction of a bird's-eye view that had been produced in 1927 as part of the publicity for the Sargent Memorial Fund drive. For the 1934 version, the Bussey Institution grounds and the extended boundary near the Adams Nervine Asylum were added. The existence of aerial photographs undoubtedly aided the rendering of this enhanced depiction of the Arboretum landscape. *Through the Arnold Arboretum* was published in two editions—a deluxe copy with high-quality paper and a cover featuring a floral design by the artist Charles Capon (which showed the influence of the arts-and-crafts movement) and a less costly edition printed on glossy paper. Visitors bought the guide through mail order or at the administration building, where postcards were also sold.

The ranks of the Arnold Arboretum's fellows grew in the early 1930s with several arboreta established under the auspices of colleges and universities or city parks systems. Among them, two are notable for the direct influence of the Jamaica Plain collections on their development. The Morris Arboretum, bequeathed to the University of Pennsylvania by Lydia Thompson Morris, was dedicated in June 1933. Morris and her brother, John T. Morris, owned adjoining estates north of Philadelphia that they developed with a great variety of woody plants. The grounds also featured herbaceous plantings and greenhouses, where a collection of ferns was particularly extensive. The Morrises were among the sponsors Sargent enlisted for Wilson's expeditions, and they raised many fine specimens of Asiatic plants as a result. To the university Lydia Morris left not only the well-established plantings of both estates but an endowment, a library, an herbarium, and the directive "to grow and propagate plants; to provide a place for the scientific study of trees and woody ornamentals; and to exhibit such plants in attractive surroundings for the benefit of the general public" (Wyman, 1959, p. 47). In contrast to the Arnold Arboretum, the Morris was ready-made at its inception, although as private grounds it had no taxonomic principle in its arrangement.

In 1933, Ames answered a letter from former staff member Ralph W. Curtis, now professor of ornamental horticulture at Cornell University, requesting information on the costs of maintenance of the living collections and grounds, including propagation and record-keeping operations. Curtis desired the information to assist in establishing an arboretum at Cornell, a notion that had been under discussion for nearly two decades since its suggestion by L. H. Bailey. To aid his instruction of students in woody ornamentals, Curtis visited the Arnold collections often to take photographs. To provide the Ithaca campus with a greater variety of correctly identified, living specimens of native and exotic trees and shrubs would be the new arboretum's first priority.

Curtis, together with others on a committee that included professors of forestry, landscape architecture, and botany, soon won the university's approval of their statement of objectives and policies, but it was many years before substantial funds were available to plant and maintain the collections. In the 1930s and 1940s, the Civilian Conservation Corps constructed a system of trails and roads under the direction of the interdepartmental committee. In 1944, when the arboretum expanded its scope to include all of the campus gardens, Bailey coined the name Cornell Plantations. Curtis, who continued as a member of the committee after his retirement as professor in 1945, asserted that his early years at the Arnold Arboretum inspired his contribution to the founding of collections at Ithaca.

AMES YIELDS THE REINS

As Sargent's successor began to feel some stress from his multiple posts, personnel changes took place in 1935. Edgar Anderson left the Arboretum and returned to the Missouri Botanical Garden, where he stayed for the rest of his career, serving as director there from 1954 to 1957. John George Jack, who was seventy-four years of age, retired and became assistant professor of dendrology, emeritus. During his fifty-year association with the Arboretum, Jack's greatest contributions were careful keeping of the plant records, obtaining many fine, hardy plants for America's gardens during a trip to Korea and northern Asia, overseeing the training of special students, and giving his perennial walking tours of the grounds. In consideration of these last, the Arboretum staff wrote a resolution in appreciation of Professor Jack: "As teacher, consultant and lecturer, he succeeded in interesting a wide circle of men and women in the work of the arboretum and gained for it many friends who, in turn, were of great help to the institution financially and in other ways" ([Rehder, Faull, and Raup], 1936, p. 201). Although after retirement Jack busied himself with the care of his orchards and home grounds in Walpole, Massachusetts, he volunteered his help at the Arboretum. In 1948 he wrote "The Arnold Arboretum, Some Personal Notes" for publication, a detailed and fascinating account of his views, especially on Sargent's administration.

Ames's administrative role still covered not just the Arboretum but several of the other botanical collections. Weary of this balancing act, he requested to be relieved from his multifarious administrative duties in 1935, and was appointed research professor of botany to carry on his orchidological studies. Ames left the Arboretum officially in October when his replacement, Elmer Drew Merrill, came from the New York Botanical Garden. Ames continued his association with Harvard and studied orchids for many more years. He took on one last administrative post in 1937, that of director of the Botanical Museum. He retired as professor in 1941, and stepped down to become associate director of the museum in 1945.

Ames briefly summarized his own direction of the Ar-

boretum in his ninth and final report to the president for the year ending 30 June 1935. In the midst of the Great Depression, by making serious and continuous efforts to keep the public informed about the Arboretum and engaging the interest of those with the means to give major support, the institution received gifts of more than half a million dollars between 1930 and 1935. Together with the Sargent Memorial Fund campaign, these contributions more than doubled the Arboretum's endowment.

Ames had taken several steps toward greater unity and efficiency among the botanical units under his charge, which he felt strengthened them all. Members of the Arboretum staff now regularly taught in the college; Faull and Sax advised graduate students as well. In 1934, Ivan Johnston was appointed lecturer so that he could offer instruction in plant systematics to Harvard students. The

116. *Oakes Ames continued to study orchids for many years after giving up the post of supervisor of botanical collections. He directed the Harvard Botanical Museum from 1937 to 1945 and served as associate director from 1945 until his death in 1950. Research on the Orchidaceae, one of the largest and most interesting plant families, has been carried on by succeeding botanists charged with the care and development of the Oakes Ames Orchid Herbarium, at Harvard.*

close interaction of Arboretum and Bussey Institution staff was enabled by his early decision that the programs should share their physical facilities. Ames had headed a committee to investigate ways to avoid duplication in the herbaria and libraries of the several botanical institutes. All in all, he had tried to make the Arboretum staff less isolated from their Harvard colleagues and was pleased with the start made.

Looking at Arboretum science from its beginning under Sargent and Gray, its traditional focus on taxonomy was gradually turning toward an evolutionary approach, adding investigations in genetics and other fields to the basis for classification. As Faull worked out the life cycles of pathogenic fungi, he also found out more about their relationships and was able to propose better classification in his genera of specialty. The herbarium now collected, and its staff studied, more than woody north-temperate plants; herbaceous taxa and tropical flora had become part of the repertoire. From the point of view of the monographer, the lines between temperate and tropical, or between ligneous and herbaceous, do not necessarily delimit a group of related entities. Although no one seemed conscious of it, arboretum taxonomists were starting to be less directly involved with the living collections than had been the case during Sargent's administration. The trend was to use dried specimens that document wild populations rather than cultivated individuals in order to obtain a complete understanding of a species, genus, or larger group. In the study of species and classification, the living collections would henceforth have greater use as an educational resource than as a subject of original taxonomic research. Yet Sax, the geneticist, utilized Arboretum plants to a great extent in experimental cytology and breeding.

The Arboretum staff now had a greater role within the university in terms of scientific research and instruction. The Jamaica Plain grounds had been subjected to the same kind of scrutiny given Sax's chromosomes, Faull's pathogens, or Raup's forest systems. Much information had been gathered to be used for future plans and planting. Unexpectedly, Wilson's program to further test and evaluate newly introduced plants for cultivation in the United States and to educate growers on their use had been halted. The introduction of plants was still considered an important role for the Arboretum, although little effort had been made in recent years. Jack, Judd, Schmitt, Anderson, and Rehder had together attempted to manage the living collections, but it had been a very trying period for this element of the institution. While these staffers took a great interest in the living collections, no one of them took the lead in their development. This was one of the challenges the next administrator would meet.

117. Elmer D. Merrill (in the rear) tours the Atkins Institution in Soledad, Cuba, with David Sturrock, supervisor of the Cuban garden, and Thomas Barbour, director of Harvard's Museum of Comparative Zoology. When Merrill came to the university in 1935 as administrator of botanical collections he already had thirty years' experience in the management of botanical research institutions. He also had a vast knowledge of the tropical forest biology of eastern Asia.

CHAPTER SEVEN

MERRILL AND SAX; HORTICULTURE EMERGES

The administrations of Elmer Merrill and Karl Sax can be considered together for several reasons. For the Arboretum's research program their tenures were marked by the perpetual obstacle of overcrowding in the herbarium and library. The nearly twenty-year course of finding a solution to this problem brought the tensions of uncertainty and impending loss of unity to the institution. Despite the impediments, Merrill built up the Arboretum's scientific contribution in taxonomy with new evolutionary and biogeographic emphasis. The strong tradition of excellence in publication was continued through the *Journal* and many bibliographic and monographic works. When Sax stepped in from the ranks of staff to take over the administration, the start of reorganization among Harvard's botanical units had created a crisis for the Arboretum. In the face of attrition rather than growth, the best the geneticist could do was to continue Merrill's programs.

Fortunately for the living collections and the landscape, the person Merrill added to oversee them provided continuity in this department that lasted through Sax's term and beyond. With the appointment of Donald Wyman to the position of horticulturist, and in his interpretation of the role, the care and development of the woody-plant collection and its utilization for pleasure and education were more definitely separated from botanical research. "Horticulture" came into its own at the Arboretum as a program that integrated collections policy with a frank commitment to educate the gardening public. Taking a route that sister arboreta were likewise following, Wyman attempted to educate Arboretum users by emphasizing selective subjects rather than the all-inclusive whole.

Horticultural research took the form of an investigative approach to plant propagation, selection of the results of hybridization, and rigorous evaluation of ever increasing numbers of cultivated varieties for suitability in home and public landscapes. Although the addition of new exhibits, the worst natural disaster ever to strike, labor shortages during World War II, lack of expansion room, and well-intentioned but inapt advice all posed threats to the character of the landscape, its integrity as a series of collections displayed in naturalistic style was sustained. Merrill's ability and willingness to delegate responsibility for the development of activities in horticulture and public relations came from years of experience managing organizations with diverse functions.

ELMER DREW MERRILL, BOTANIST, ADMINISTRATOR

The new administrator arrived to take over from Ames in late October 1935, having just returned from the International Botanical Congress at Amsterdam, where he undoubtedly conferred with Rehder, who was also in attendance. Merrill was well known to the Arboretum staff as an expert in the forest flora of Southeast Asia and as a capable administrator of scientific institutions. He had visited Sargent back in 1915 and had exchanged specimens with the Arboretum for many years. His reputation was such that Oakes Ames had actually suggested Merrill as a possible replacement for Sargent before his own appointment to oversee Harvard's botanical collections was made.

Merrill was born in East Auburn, Maine, a village of farmers and shoe-factory workers, in 1876. It was the work and the pleasures of rural life that shaped his character, as he recalled years later: "Swimming, boating, fishing, hunting, tramping in the woods—many things were more appealing to us than work, but when there was work to be done it always came first" (Merrill, 1953, p. 338). Yet the youth found time to collect birds' nests, eggs, rocks, minerals, Indian relics, wood samples, shelf fungi, and plants that he pressed. Unlike their other siblings, Elmer and his twin, Dana, decided to go on to high school; from there, they both entered Maine State College at Orono, which became the University of Maine in 1898, the year they graduated. Elmer Merrill studied as much biology as he could and independently collected an herbarium of some two thousand specimens, which he later presented to the New England Botanical Club. Like George B. Emerson seventy-five years before and probably most other self-respecting New England botanists of his day, Merrill tramped and botanized on New Hampshire's Mount Washington. He likewise explored Mount Katahdin in northern Maine, a more difficult journey in those days. The Spanish-American War determined Dana Merrill's career choice. He enlisted in the army and stayed in the service, attaining the rank of brigadier general in 1935.

In contrast to his twin, Elmer Merrill remained at Orono after graduation. With a modest assistantship in natural science, he took additional courses and single-mindedly studied systematic botany, even though no formal training was offered in the discipline. Later, in 1904, the University of Maine awarded him a master's degree for this work. In 1899 Merrill went to Washington as an assistant agrostologist in the Department of Agriculture. Although this job was rewarding and expanded his understanding of plant taxonomy, especially its published literature, Merrill was still undecided about a career. With time on his hands at night, he had completed one and a half years of medical school when an offer of employment in the Philippines turned him permanently in the direction of plant science.

Among many programs started in the Philippines after the Spanish-American War, the United States government established an Insular Bureau of Agriculture in 1901, and the next year Merrill was persuaded by his superior at the USDA to accept the post of botanist in Manila. Within a couple of months of his arrival, Merrill observed that the equally new Bureau of Forestry was even more in need of a botanist and convinced authorities to give him a joint appointment. Merrill stayed in the archipelago for the next twenty-two years, with periodic leaves to study in herbaria in Europe and the United States. While continually working toward the goal of a published flora of the Philippines, he traveled to every large island and numerous smaller ones and visited many adjacent regions. From almost nothing on his arrival, Merrill built up an herbarium of 275,000 specimens and a corresponding botanical library. In connection with the identification of the flora, Merrill corresponded with European and American botanists familiar with certain Asian tropical groups. Among the latter was Oakes Ames, who worked on all the orchid specimens Merrill could supply.

The botanist was also prevailed upon to teach at the University of the Philippines. After serving as division chief and as acting director of the Bureau of Science, he was appointed its director in 1919. In this capacity Merrill successfully directed work in a wide variety of areas outside botany: medicine, public health, chemistry, weights and measures, materials testing, geology, mining, fisheries, zoology, and anthropology. Although Merrill's administrative ability won him the respect of many in the government service, he continued to pursue his own research on the Philippine flora and its phytogeographic relation to the adjacent territories of China, Indochina, and the Malay Archipelago.

In 1924 Merrill accepted an offer to become dean of the College of Agriculture at the University of California. Although he regretted leaving Manila, he thought the post offered a greater guarantee of permanence and retirement benefits, a chance to limit his administrative responsibilities in favor of research, and, most important, the opportunity to live together with his wife and children, who had been residing in the States since 1914. Merrill tackled the duties of dean with the same "quickness of perception, unfailing patience and courtesy, great store of common sense, promptness in taking action, approachability, and consideration for others" that had served him in the Philippine capital (Burkill, 1956, pp. 687–88). As in Manila, he continued his work with Asian plants by extending his day, taking the early hours or weekends to work in quiet in the herbarium at Berkeley. For six years Merrill led 350 faculty members at several sites through a reorganization that stressed an upgrading of academic training of the staff

118. When Donald Wyman took this picture on "Crabapple Sunday" in May 1936, his first spring as horticulturist, he must have been pleased to find a public so interested in the floral displays. Crabapples became a genus of special interest to him, although "Crabapple Sunday" never gained the status that Lilac Sunday has.

and a greater emphasis on fundamental research. His goal became a "university education on the basis of Agriculture, [as] opposed to vocational instruction in agriculture" (Merrill, 1953, p. 363).

Although in 1924 he thought his move to Berkeley would be permanent, Merrill changed position and locale once more in 1929, when he agreed to become director-in-chief of the New York Botanical Garden. Administering a strictly botanical institution, rather than one that was for all of science or for agricultural science, appealed to his abiding desire to study plants. Merrill rose to the challenges presented at New York, especially that of procuring funding during the difficult depression years, and accomplished much there, ingeniously employing workers from emergency work-relief programs. Nevertheless, after six years in New York, the opportunity to return to New England was one he could not pass up.

Ames's successor brought to Harvard a mature capacity to mediate and integrate seemingly conflicting interests, a direct and practical approach to problem solving, and a vast knowledge of eastern Asian botany. It was this last that made him lean most toward the Arboretum of the eight botanical "orphans," as he called them, that were placed under his guidance. No less important was the Arboretum's potential to raise funds for botany through its excellent rapport with donors. Dubbed a "tiny dynamo" (he was five feet four in height) by one Arboretum historian, Merrill's style was straightforward, and he expected others to work as hard as he did. (Nowadays he would probably be called a workaholic.) His attempts at fairness, and his phenomenal knowledge of plants, were inspirational, nonetheless.

Although the administrative shift in 1935 set off rumors of great change in botany at Harvard, Merrill was quick to discount them. The eight separately endowed units, in order of founding, were the Botanic Garden, the Gray Herbarium, the Bussey Institution, the Arnold Arboretum, the Botanical Museum, the Harvard Forest, the Atkins Institution, and the Farlow Reference Library and Herbarium. Half of them had extensive facilities outside Cambridge: the Arboretum and the Bussey Institution in Jamaica Plain, the Harvard Forest, located seventy miles west in Petersham, and the Atkins garden in Soledad,

Cuba. While Merrill agreed with his predecessor, Ames, that physical separation and restricted endowments rendered absolute consolidation impossible, he continued to coordinate the work of the several units and to integrate further their work with the operations of the Division of Biology. For the Arnold, the Bussey, and the botanic garden, Merrill handled all details of direction; each of the other units had its own executive officer with whom he worked.

As Merrill took stock of the situation, he had to make some tough decisions. The Harvard Botanic Garden was worst off. Its endowment income was not even sufficient to keep its obsolete glasshouses heated, prompting Merrill to distribute the indoor plants and tear the structures down. The outdoor collections were put in reasonable condition by eliminating areas needing the highest maintenance, but future development did not look promising.

Even before Merrill took charge, steps had been taken to consolidate the Bussey's teaching and degree-granting functions with those of the biology department. Most of the Bussey's staff moved to Cambridge when the new biology building was built in the early 1930s. Henceforth, Bussey scientists would use the buildings and grounds at Forest Hills only for research projects. Another use had been made of the Bussey's physical plant since the turn of the century, when Harvard began cooperating with the Massachusetts health department, allowing utilization of buildings and providing animal-keeping facilities. The commonwealth eventually located a center for the study and production of antitoxins there, and this use was continued. Nevertheless, the Arboretum's employment of the area for propagation, breeding experiments, and special displays expanded. The Harvard Forest and the Cuban garden needed better-defined goals and funding sources, but this could develop in time.

The Gray, the Arnold, and the Farlow were flourishing in their research and teaching programs, yet, at the first two, the collections were expanding beyond the capacity of their facilities. Merrill reported in 1936:

It is a fact that Harvard University maintains three of the largest botanical libraries in America and three great herbaria. There is thus a very considerable and unavoidable duplication in purchases of both reference material, books, subscriptions to periodicals, and a considerable amount of triplication in the library situation. Could some equitable plan of concentration be developed whereby the three libraries and the three herbaria could be assembled in one building, very material savings could be effected in maintenance charges, as well as in the operations of the herbaria and the libraries; with such a change a single library would serve all three herbarium units. . . . The herbarium facilities of both the Gray Herbarium and the Arnold Arboretum have about reached the point of possible expansion and, in the not distant future, we must face the problem of additional construction for the normal expansion of these two units. If additions be made to the two present buildings, about seven miles apart, the present unsatisfactory situation will be perpetuated. It is hoped that some practical plan of concentration may be developed whereby the herbarium and library facilities of the Gray, the Arnold, and the Farlow may be concentrated in one locus. (Merrill, 1937, p. 293)

While the consolidation idea was being considered, Merrill put some efforts in motion to eliminate the duplication right away. A new policy for the Gray and the Arnold was adopted whereby both institutions would no longer obtain subscriptions to the same set of specimens from a collector or an expedition. It was agreed that the Gray would mainly seek specimens from the New World and the Arnold from the Old World. The Arnold Arboretum also began transferring specimens of herbaceous plants to the Gray, since space in Jamaica Plain was so limited and woody plants had always been its primary focus. The Farlow Herbarium received cryptogams (fungi, lichens, mosses, algae) from both institutions, as had long been the practice. Specimens of the orchid family, especially those from the tropics, were transferred to the Botanical Museum, where Ames and his associates could make the best use of them.

Among the "orphans," the Arboretum and two of the other units were distinguished by having educational exhibits. The Harvard Forest was in the midst of production of twenty-two models, illustrating the history of forests in New England and silvicultural practices, intended for public display at Petersham. The Botanical Museum's glass flowers and economic-botany specimens were popular public exhibitions in Cambridge. Merrill took steps to strengthen the Arboretum's display and public education functions by delegating their supervision to an assistant.

HORTICULTURIST AND HORTICULTURAL EDUCATION

Merrill wasted no time putting the living collections of the Arboretum under firmer control. Within two months of his arrival at Harvard he arranged to hire Donald Wyman to fill the new position of horticulturist at the Arboretum. A native of California, Wyman came east to study at Pennsylvania State College, a land-grant institution with strong departments in agriculture and related plant sciences. He graduated in 1927 and served as an instructor there for two years before going to Cornell. At Ithaca, Wyman earned a master's in agriculture in 1931 and a doctorate in horticulture in 1935. For the duration of his studies at Cornell he also served as an instructor in horticulture; while he was there, Ralph W. Curtis and his colleagues were getting the arboretum started. His schooling at two universities with active extension services, and his own inclinations, led Wyman to value the communication of practical information to the public.

That Merrill had confidence in the young horticulturist is indicated by the first memo he sent Wyman: "I am disposed to leave the development of your activities at the Arnold Arboretum largely at your own discretion. There are, however, certain items that you will be responsible for" (AAA, Merrill papers, 16 January 1936). He then outlined a program that Wyman readily adopted. The grounds committee idea was not scrapped; rather, Wyman was made chairman, with Judd and Schmitt the only other members. Nonetheless, Wyman took charge of maintenance and development of the collections, subject to Merrill's approval. The *Bulletin of Popular Information* was to be Wyman's number-two priority. Merrill wished it to have "as wide [an] appeal as possible to the supporters of the Arboretum" and felt it should compare with the publications of like institutions such as the Morton Arboretum, Brooklyn Botanic Garden, and Missouri Botanical Garden. Next in importance was answering requests for plant material and for information on the cultivation of woody plants; Wyman would handle these or refer them to the appropriate staff. The herbarium personnel continued to answer the numerous queries on plant identification.

Publicity and fostering the educational use and appreciation of the "outside collections" was assigned to Wyman as well. He was to maintain contact with newspapers, radio, and magazines; give lectures to groups that called for them; act as official guide; and develop contacts with garden and horticultural organizations, among them the Massachusetts Horticultural Society, Massachusetts Nurserymen's Association, and local garden clubs.

In connection with publications, lectures, and publicity, the preparation of photographs, both prints and color slides, was another of Wyman's assignments. Fortunately he took a great interest in photography, and Merrill promised to make funds available to improve the supplies and equipment of the Arboretum's existing darkroom, especially for the newer color techniques.

Since Wyman already had some thirty-five published

articles to his credit before coming to the Arboretum, this aspect of his new job gave him little pause. Right away, he steered the articles he and other Arboretum staff wrote for the *Bulletin* toward practical subjects useful to the homeowner. While many of his pieces were based on experience at the Arboretum, Wyman brought with him the latest information from studies at experiment stations throughout the United States. He also continued the traditional *Bulletin* topics on the plant collections themselves, such as what to see at certain times of year and discussions comparing the merits of several kinds within important ornamental groups. To these he brought a fresh viewpoint, with a more decided emphasis on aiding the would-be planter in choosing his material with a mind toward affordable maintenance. Bibliographies and suggested further reading were frequently incorporated as well. The *Bulletin of Popular Information* also became a bit more "newsy," reporting on staff activities, awards, and projects that would interest a general audience. Beyond readers of

119. *The stately exterior of the administration building and herbarium wing in this spring 1936 photograph belies the struggles within to find space for the ever increasing herbarium and library collections. Magnolias bloom in front of the building, and a robust specimen of katsura (Cercidiphyllum japonicum) is just coming into leaf on the right in the foreground.*

the *Bulletin,* Wyman soon extended the Arboretum's coverage by writing horticultural articles for the Boston and New York papers.

During his first spring, Wyman offered Saturday morning field classes to observe plants in the Arboretum, more or less a continuation of what John George Jack had started under Sargent. An enthusiastic crowd of some fifty people returned each week to hear Wyman and to see the plants. He led the field classes for the next thirty-five years, changing to Friday mornings once two-day weekends were the norm.

Wyman's first few years with the living collections were marked by modest changes and the addition of two entirely new collections. In the autumn of 1936 he convinced the grounds committee that removal of one length of the trellis surrounding three sides of the shrub collection would be an improvement for several reasons. This, the south side of the concrete-post and wire construction supporting the vines, was in poor repair. It was covered with wisterias, which often lost their flower buds because the spot was the coldest in the Arboretum. They were moved up to the Bussey grounds, where a rustic arbor of heavy timbers was constructed for them. With the wall of wisteria eliminated, Wyman expected better air circulation for the shrub collection. He thought the ability to see this grouping from Forest Hills gate was a change for the better, whereas Sargent had sought to hide the formal arrangement of rows from view.

In the winter of 1936–37 part of the forsythia collection was subject to treatment similar to that given the lilacs in 1927. A large planting of showy border forsythia (*Forsythia intermedia* 'Spectabilis') on the bank near the lilacs had become so dense and overgrown that it bloomed poorly; 50 percent of the thicket was deadwood. This mass was cut to the ground to renew its vigor. The other specimen plants of numerous kinds did not receive rejuvenation, but were kept in bounds by a program of pruning out the oldest stems over several years. The mass of forsythia recovered to flower again to everyone's satisfaction within two years.

Wyman struck a new course for Arboretum exhibitions in late 1936 by planning and planting a hedge demonstration plot. Wyman had been collecting information on

120. *One of Wyman's first innovations was a collection of more than a hundred kinds of deciduous and evergreen woody plants maintained as hedges. In order not to disrupt the naturalistic style of existing plantings, the hedge demonstration was placed on the Bussey Institution grounds, where Arboretum nurseries and experimental plantings were also located.*

hedges and other shelter-providing plantings for years. Aside from the lawn, hedges were the most common element in home landscape plantings of the day (and probably still are). The variety of plant material suitable for such delineating, screening, and sheltering of rows was greater than most people realized, and Wyman had been investigating the possibilities. The Arboretum hedge plot was to be not only a demonstration but an experimental ground, since some of the kinds installed had rarely been treated as hedges. The site, laid out with 115 kinds each in twenty-five-foot rows, was an area defined by the semicircular drive fronting the original Bussey Institution building on South Street, conveniently near the Arboretum greenhouse and nursery facilities. Wyman stated the reason for using the space: "Because of the necessary formality of the planting, it was not laid out near any of the lovely informal plantings so enjoyed by the public" (Wyman, 1938a, p. 79). Here, since the hedges were

121. *The Larz Anderson collection of Japanese dwarf trees, or bonsai, was presented to the Arboretum in 1937 by Anderson's widow, Isabel Weld Anderson, as a memorial to Charles S. Sargent. A rectangular, warm-weather display house of cypress slats was erected for the miniature, container-grown trees near the Arboretum propagation greenhouses. Here, the plants could easily receive the extra care and watering they required.*

grown under nearly identical conditions, it would be possible to compare their performance and their visual qualities.

There were few such hedge experiments in the Western Hemisphere, the only extensive ones being at the Dominion Experimental Farm, Ottawa, Ontario, where some of the hedges were fifty years old, at the Morton Arboretum, Lisle, Illinois, and at Cornell University. For homeowners, landscape designers, and nurserymen, the Arboretum display would greatly aid the choice of a woody plant for a particular need. It would expand visitors' appreciation of the versatility and importance of woody plants in their everyday environment. Eight nursery firms contributed nearly all of the one thousand individual plants needed to create the hedges; five of the businesses were in Massachusetts, three were from out of state. The Arboretum had received plant material and financial support from members of the trade in the past, and Wyman, with

Merrill's backing, regularly sought collaborative relations with this constituency. The donors' material, well grown and on display in a public garden, could not help but create homeowner demand for better woody plants.

It took a few years for the hedges to become established, and it was not until 1938 that *Bulletin* readers were informed that the display was ready for view. At that time, Whittlesey House of New York announced publication of Wyman's first book, *Hedges, Screens, and Windbreaks*. The book mentioned the recently set-out Arnold Arboretum demonstration and discussed the selection, planting, and care of hedges. Illustrated with wonderful black-and-white photographs of hedges, mainly in home grounds throughout the United States, the volume was dense with useful lists and an alphabetical treatment of the best species. It established a format used in Wyman's subsequent works.

The next addition to the Arboretum collections was

also placed on the Bussey Institution grounds because its distinctiveness would blend in nowhere in the Arboretum proper. Besides, its specialized cultural requirements and its vulnerability to weather or to human misconduct required a well-guarded location. This was the Larz Anderson collection of Japanese dwarf trees presented in the fall of 1937 by his widow, Isabel Weld Anderson. Mrs. Anderson was descended from the same Weld family who had originally been granted much of the land on which the Arboretum and the Bussey Institution were founded. Charles S. Sargent had been friends with her father since Civil War days. The Welds lived in Brookline near the Sargents, and when Isabel married Larz Anderson, an American diplomat, the couple took over care and management of the gardens of Weld, the hilltop estate, now a Brookline park. The Andersons kept up a cordial friendship with their neighbor, Sargent, after the death of Isabel's father.

In 1912 Larz Anderson was sent to Tokyo by President Taft as ambassador to Japan. There he acquired, and received as gifts, several dwarf trees cultivated in ceramic containers according to the centuries-old art of bonsai. When Anderson returned to Weld from his tour of duty he brought more than fifty of these diminutive woody plants with him. It was before the days of plant quarantine and import regulations, so the miniature trees were shipped to the United States in their containers, with the soil and the moss that protected their roots intact. Some of the plants were a century old when Anderson obtained them. To the Japanese, however, what is important is not the actual age of a bonsai, but its appearance of age. One idea behind the art is to convert the image of venerable survivors of woodland and mountaintop to a size manageable in the home or in the small space available to most of that nation's gardeners. The Andersons cared for their bonsai with the help of a Japanese doctoral candidate studying in America, occasionally loaning them for the Massachusetts Horticultural Society's flower shows.

After Larz Anderson's death, his wife decided to give twenty-nine of the plants to the Arboretum as a memorial to Charles S. Sargent. With funds from Isabel Anderson, the Arboretum staff built a display house of cedar posts and cypress slats to shade and protect the container-grown trees while they were on display for the spring and summer months. It was placed near William Judd's growing areas so that the plants could receive the extra attention needed. The bonsai plants were wintered in the below-ground "pit house" to protect the roots from freezing temperatures and the dormant crowns from desiccating winds. During the growing season the plants required frequent watering and, over the years, careful pruning of tops and roots when renewal of soil became necessary. When the institution's horticulturist wrote the announcement of Anderson's gift for the *Bulletin,* he included an explanation of the intricacies of starting, training, and maintaining bonsai, a process as new to most American gardeners as it was to the Arboretum staff. Merrill pointed out in his annual report for fiscal 1937–38 that the Arnold Arboretum had the only extensive collection of these plants in an institution open to the public in America.

CURATING THE LIVING COLLECTIONS

Bent on improving other aspects of the living collections in the late 1930s, Merrill and Wyman initiated projects to upgrade acquisitions, records and mapping, and nomenclature. On the acquisitions side, Wyman began a systematic scrutiny of catalogs of American nursery firms for plants not represented at the Arboretum in an effort to make the collections as complete as possible. Judd traveled once more in Great Britain and Europe to locate additional desirable woody plants for the Arboretum and to arrange for their shipping. Sax was urged to continue his breeding program.

As the new material came in at an unprecedented rate the nursery space on the Bussey grounds was expanded, and a large area for lining out young plants for trial was developed on the Weld-Walter tract at the extreme western side of the property. By 1939 the number of species and varieties in the permanent collections had risen from a long-standing average of five thousand to some seven thousand, with another thousand kinds undergoing propagation.

Although Wyman knew his woody plants well, he had not been at the Arboretum for decades like Judd, Schmitt, Rehder, and Palmer. He was not trained as a taxonomist, and besides, the Arboretum's taxonomic arrangement was by now a little-used system. Therefore, his knowledge of which plant could be found where on the grounds depended on accurate records and mapping. The older staff had come to know where most things were and relied on various broad references in the records. These included "in group," meaning with others of its genus; "Peters Hill," which covered seventy-five acres; "overlook," which at some times referred to the very top of Bussey Hill and at others to the plateau on the south side where the Chinese collection, the double-flowered cherries, and the azaleas were planted; and "in order" or "shrub collection," for the area containing several rows of hundreds of plants. There was a mapping system, but apparently it had not been kept up-to-date and it still required a knowledge of the whole layout for someone to know which map to look on for the particular plant in question. The index card for each plant did not refer to numbered maps but only to the rather vague groupings just mentioned. To Wyman and Merrill it was obvious that a change to a more enduring system of mapping was needed—one that required less prior knowledge of the grounds, and of an old taxonomic system, to use.

The two men decided to start from scratch and create new maps using simple surveying techniques, rather than revise the old system. As vast an improvement as this new approach was, it is unfortunate that the old maps apparently were discarded. Although they were not helpful for everyday operations once the new set was in use, for historians they would have shed valuable light on the collections as they were during the administrations of Sargent, Wilson, and Ames. The person Merrill hired to conduct the survey for the maps was Leon Croizat, a taxonomist who had earned a law degree from the University of Turin before turning to botany. Croizat emigrated from Italy and got a job on the grounds of the New York Botanical Garden, doing research in the herbarium on his own time. He maintained a similar pattern at the Arboretum, mapping the grounds for his salary while studying the family Euphorbiaceae on the side. Heman

Howard, on the Arboretum staff since he had graduated from high school in the late 1920s, drew the maps from Croizat's surveys and rechecked them in the field.

For the new system the Arboretum was divided into a grid of some seventy-four rectangles, each measuring four hundred by six hundred feet. The "hardscape" features, such as roads, paved paths, benches, fences, and buildings, as well as all the accessioned plants and borders of natural-woods areas, were surveyed in the field, drawn on tracing paper at a scale of one inch to twenty feet, rechecked in the field, and then dated and reproduced for use. Each of the maps was designated by number, and each was also divided into quadrants designated a to d. The map number and the quadrant designation were recorded on every plant's card in the main index, so that locating any individual specimen became a matter of finding it in a two-hundred-by-three-hundred-foot piece of ground. Certain densely planted areas, such as the shrub collection, the lilacs, the upper portions of Bussey Hill, and the Hickory Path/Centre Street beds, were further subdivided and drawn at larger scale so the reduced maps could be easily read.

The system of record cards was revamped simultaneously, and Merrill urged Palmer and Rehder to update nomenclature and verify the identity of recently received plants. E. J. Palmer was assigned to procure herbarium specimens of every plant in the collections, and his voucher specimens now included the plant's map number and quadrant as well as its accession record number.

THE HURRICANE OF 1938

After three years of becoming familiar with the Arboretum grounds, Merrill and Wyman saw it undergo a drastic change in just one or two hours during the infamous hurricane of 1938. (Names were not assigned to tropical storms by the National Weather Service until 1950.) The director called it "the greatest single catastrophe that has happened to the plantings in the Arboretum since its establishment in 1872" (Merrill, 1938b, p. 70). Wyman "happened to be peacefully sleeping on a pullman car somewhere in the Middle West" when the hurricane hit

122. *On the afternoon and evening of 21 September 1938, one of the worst hurricanes ever to hit New England destroyed 1,500 trees in the Arboretum. Scenes like this one of flattened white pines behind the administration building were repeated in many locations, especially on slopes facing south and east. Many venerable native trees were toppled, and a large area of the ancient forest on Hemlock Hill was laid waste.*

but got back to find fifteen hundred Arboretum trees on the ground (AAA, Wyman correspondence, 3 November 1938). He wrote an account of the storm for *Bulletin* readers:

Rain had been falling rather consistently for four days when on September twenty-first, over large areas in New England, the downpour assumed the proportions of a deluge. Rivers in Massachusetts were at flood stage, and everywhere the ground was soggy from excessive rain. By late afternoon the rain slackened and the wind increased to a gale. At 4:50 p.m. when the lights went out in the Administration Building staff members expected a "blow", but certainly did not anticipate the hurricane which caused frightful damage throughout New England. (Wyman, 1938b, p. 71)

Merrill was on the scene, since he now lived adjacent to the grounds. Having exposed himself to considerable personal risk, he was able to give details of what must have been an unnerving couple of hours for someone charged with the care of the sixty-six-year-old tree collection:

The storm was intense at 5:00 p.m. and gradually increased in violence. The worst damage was done in the Arboretum between about 5:30 p.m. and 6:30 p.m. A tour of the grounds at 5:30 p.m. revealed relatively slight damage; for example there were only three or four trees in the extensive pine grove back of the Administration Building that were down or showed signs of weakening at that time; an hour later nearly all the trees in the entire planting were prostrate. The sound of rustling leaves, breaking branches, and creaking trunks was at times almost deafening. The worst of the storm was over by 8:00 p.m. (Merrill, 1938b, p. 70)

The hurricane was the worst in living memory; in fact, reports generally agreed that none so devastating had hit New England since 1815. Winds of up to 150 miles an hour were measured at the Blue Hills Observatory just south of Boston, while the highest wind velocity in the city was officially reported as having been 87 miles an hour. The storm hit the northeastern United States harder than most because it never made landfall until it reached Atlantic City, New Jersey, an unusual course for hurricanes.

Wyman thought a newspaper estimate of some 100 million trees downed in Massachusetts alone not unreasonable. The material damage everywhere was great. The Arboretum was a mess, and Merrill feared the worst as he made his way over leaf-strewn ground, past fallen trees and broken limbs. Nonetheless, once the damage was assessed on a species-by-species basis (and compared with the economic damage and personal loss sustained by all of New England), Wyman saw a "bright side." Of all the destroyed trees, only a dozen were not duplicated elsewhere in the Arboretum. "By far the majority of injured or destroyed trees were native in the Arboretum, trees which added materially to the natural beauty of the plantings, but which were not prominent in the collections," that is, the accessioned plants (Wyman, 1938b, p. 72). The reason so many of the natives were affected might have been that the oldest trees were more prone to injury or windthrow. The soggy soil condition after four days of rain was thought to have caused many trees to be uprooted by the raging winds.

Certain areas of the Arboretum were hit especially hard, mainly those with a southeast exposure: the original poplar collection and adjacent oaks on Peters Hill, the southeastern side of Hemlock Hill, the lower slope of Bussey Hill above South Street, and the hillside behind the administration building. Other collections with heavy losses were the conifers, along with the pears and crabapples near Forest Hills gate. These groups together accounted for half the losses, while an additional 550 trees scattered throughout the grounds were also destroyed.

The cleanup began at once, and little of the routine fall and winter work was accomplished. The visiting committee sent out an appeal for donations to a Hurricane Rehabilitation Fund in early November, and budget plans were adjusted to allow Wyman and Schmitt to retain a full complement of grounds staff all winter, a time many men were usually dropped from the payroll. A few of the smaller uprooted trees were replanted in hopes that they would reestablish themselves, but most were too large to handle. The grounds crew spent all winter sawing wood: first, the driveway and paths were cleared; then, they moved on to cut and remove fallen logs from less accessible or visible areas.

To aid the process, Merrill authorized the purchase of the Arboretum's first powered portable saw, a tractor and low wagon for moving wood, and other new equipment. Also added was a storage tank and pump for gasoline, so

123. To clean up after the hurricane of 1938, the Arboretum purchased its first powered portable saw. By spring, most of the fallen trees had been cut up and removed, but there was still considerable pruning to be done. Replanting the worst-hit areas began in the fall of 1939.

that it could be purchased in bulk, thus curtailing trips to obtain it. By April, Wyman reported that most of the fallen trees were gone, but the slow process of removing the smaller stumps and much pruning of broken branches on healthy trees remained to be done. The larger stumps, cut as close to the ground as possible, would have to remain to rot; even with the new tractor, removal was too time consuming. Considering the destruction that had occurred, the grounds were in quite presentable condition for the May display of bloom.

The acquisition of a tractor proved not just useful for hurricane cleanup—it started a trend toward mechanization that would eventually alter the appearance of the grounds. A cutter-bar attachment was purchased so that grass could be mowed by the Arboretum's grounds staff. During Sargent's day, local farmers paid the Arboretum for the right to come and cut hay once or twice a year. Later, when the demand lessened as the auto replaced the horse, contractors did the mowing in exchange for hay. More recently, the Arboretum paid someone to cut and take hay away. One tractor did not make a big difference at first, but as motorized equipment became more common the trend was toward more frequent mowing and a

less lush appearance to the grounds during the summer. The practice of keeping closely mown paths through taller grass and wildflowers was eventually abandoned, as was consciousness of the ecological role of meadow vegetation.

The grounds committee made plans to replant certain areas denuded by the hurricane's force. Most of the plants to be used were purchased with rehabilitation gifts, although some would come from the Arboretum nurseries. To assure success, the planting was done in stages over at least two fall seasons. Behind the administration building, three- and four-foot white pines replaced a grove of the same species that had been flattened. Wyman also added a mass planting of forsythia nearby to create a spring display of bloom visible from the Arborway. Closer to the tulip trees, the remaining exotic spruces placed on the bank by Wilson were rearranged, and other mixed conifers, including red pine and Carolina hemlock, were added to fill out the evergreen planting that screened the boundary.

On the South Street bank, where nine years before Wilson had planted masses of torch azalea and mountain laurel, most of Sargent's original boundary plantation of red pine was killed. Since the much admired torch azaleas' blossoms faded in bright sun, red pines were planted there once more in hopes of restoring the protective canopy.

The destruction on Hemlock Hill was more devastating, since this was a native stand that had developed over centuries. Young hemlocks ideally take root and develop under the shade of other trees or in northern exposures, where their roots are kept cool and moist until their own crowns provide protection. Since few "nurse trees" remained in the wake of the hurricane, successful replanting would be difficult. To make matters worse, the southern end of the hill slope had already been suffering for many years from soil erosion due to too much foot traffic. During Ames's administration, a chain-link fence had been installed behind the rhododendron collection, at the foot of the slope, to divert climbers from the most threatened part of the hill, but now the thin soil of the rocky prominence was exposed. The steep terrain precluded any possibility of supplemental watering. Over at least two years, the attempt was made to place young hemlocks with as much soil as possible in spots where they might flourish, but Wyman was not optimistic: "It will take the better part of a century before the magnificent grove on Hemlock Hill will again approach its perfection of September 1938" (Wyman, 1939, p. 2).

The last area planned for "replanting" was not a restoration of existing collections but, rather, a divergence from Sargent's original plantings. Where the destruction of most of the poplars and some auxiliary oaks left the southeastern slope of Peters Hill nearly bare, plans were made to place an extensive collection of new crabapple varieties. There already were crabapples nearby on the northeast, and so the new planting extended this arrangement. Many new varieties in this group had been developed since Sargent's 1905 crabapple installation, and Wyman took great interest in the group as the best all-around small flowering tree available in such a wide variety of size, shape, and color of flower and of fruit. Since the slope faced the tracks of the railroad, Wyman thought it a good area to demonstrate Arboretum displays to commuters. Nothing was said about replacing the poplars. Years later, Wyman stated that susceptibility to disease and breakage in storms and a lack of "ornamental" attributes were factors in giving up on a poplar collection. A few trees remained on Peters Hill and the adjacent South Street tract, and in time some new poplars were grown on the Weld-Walter parcel.

Before the hurricane, replacement of the large willows that lined the Arborway side of the North Meadow had already begun. Many of these relatively short-lived trees were old and decrepit, with lost limbs and unsound trunks, and some had been cut down in the winter of 1937. Wyman planned to take out a few each year and slowly replace them with tree species expected to perform better under the meadow's conditions. The hurricane so damaged those willows remaining that most were removed in 1939. A border of red maple and tupelo, both native to America, was planted in their stead. Both are plants of wet soil, and, according to the horticulturist, they were also chosen for their fall color effect. It is not certain whether these species were picked intentionally to continue the use of natives for boundary plantings. A new area for willows was opened up on the other side of the meadow behind the roadside planting of native shrubs, shadbushes, silver maples, and cork trees, but only smaller species were planted there.

JUDICIOUS THINNING

The hurricane showed Wyman that eliminating plants could be as much of an improvement as adding them. He hinted at this notion as early as October 1938: "Individuals intimately acquainted with the Arboretum plantings will note trees lacking here and there in the grounds. Those who occasionally visit the institution, after the winter cleanup is completed will note few changes. The conspicuously beautiful landscape features for the most part remain essentially unchanged" (Wyman, 1938b, p. 74). Thirty years later Wyman revealed his thinking by stating: "When the big hurricane of 1938 blew down 1500 trees in the Arboretum, it was obvious even to the most critical observers that judicious thinning out of old and decrepit specimens could add much to the beauty of the grounds" (Wyman, 1970, pp. 82–83).

Once the mapping and the card files were carefully

checked, critical examination of many of the planted trees and shrubs revealed more duplication of species and varieties than anyone previously suspected. The original group plantings of certain native American trees were one source of multiple representation; so were the New England shrubs originally placed along the drives. The receipt of the same entity under different names from several sources was another. Wyman turned his attention to eliminating "unnecessary duplicates as well as certain unsightly species of trees and shrubs that were represented elsewhere on the grounds by good specimens." Some of the "unsightly" or unnecessary trees were those in group plantings. Their initial purpose as forest tree demonstrations had not been completely realized once the Arboretum turned from forestry early in the twentieth century, yet Sargent had not given them up. To Wyman and others, however, the Arboretum by the 1940s was just too thickly planted; certain areas needed light and air, and space was deemed necessary to try more new kinds rather than keep mass plantings of one type.

Starting in 1940 certain areas or collections were evaluated for removal of duplicates. More than 50 plants were removed from among the conifers, many of them because of long-term defects from the winter of 1933–34 or hurricane damage. Many removals were made from the chamaecyparis and arborvitae collections, where Wilson had spread things out a decade before. About 150 plants were eliminated from the Centre Street beds and Bussey Hill plantings. This group-by-group evaluation for duplication was to continue over many years. Some of the plants were given to Harvard, New England colleges, and the Boston Parks Department.

Wyman also began checking the cultivars (horticultural varieties), as well as species, in such genera as *Weigela, Philadelphus, Deutzia, Chaenomeles, Rosa,* and *Spiraea* for accuracy of identity and evaluation for duplication. Another frigid winter in 1942–43 prompted the decision to give up the attempt to grow all plants in Rehder's *Manual* with a hardiness zone rating of seven and those zone-six plants that were known to have been repeatedly injured by winter cold. The horticulturist thought it better to concentrate acquisitions on plants known to be reliably hardy. Since sources for Arboretum material were now most often nurseries, a few botanic gardens, and expeditions for herbarium specimens, there was little control over the geographic provenance of material coming in. The idea of deliberately seeking out plants from the northernmost or coldest part of their natural range seems to have been dropped in favor of obtaining garden forms selected for hardiness and ornamental qualities at commercial nurseries, experimental stations, sister arboreta, and Sax's own breeding program.

By the early 1940s most of the Arboretum was on its way to recovery from the hurricane, and oversight of the collections was greatly enhanced by completion of the survey work for the new mapping. At this point the three-

124. At Merrill's urging, William Judd and his assistants began experimentation with new methods in asexual (nonseed) propagation, such as the use of hormones to induce rooting of cuttings. An additional goal was to make more of the Arboretum's rare and difficult-to-propagate introductions available to the public through a cooperating-nursery-owners program. Here Judd examines the progress of cuttings in the greenhouse.

man grounds committee was given up at Merrill's request. Wyman took on all responsibility for collections development, grounds maintenance, propagation, and distribution. The director also urged that programs of horticultural research be strengthened.

One field of experimentation that Merrill suggested was studies in asexual propagation. New rooting chemicals and other methods were being tried at the United States Department of Agriculture, the Boyce Thompson Institute, various state universities, and some enterprising nursery firms. The Arboretum administrator wished his institution to be equally up-to-date, if not a leader, in this new field. Space in the greenhouses was made available, and two assistants, Lewis Lipp and Alfred J. Fordham, were assigned to help. Judd started on some of the rarer Arboretum plants whose propagation had never been successfully worked out.

From this new effort developed what came to be known as the cooperating nursery program. This, too, was the "tiny dynamo's" idea, although Wyman evolved the details of a workable arrangement over several years. In order to get some of the Arboretum's rarest and finest introductions into circulation, well-established plants from a selected list were offered to participating nursery companies each year. As a way of limiting the participants to companies with the commitment to get their plants into circulation, the firms that received the plants had to publish a catalog. They agreed not to sell the specimens received from the Arboretum, but to use them as stock from which to propagate the rarities. The plants they raised could then be marketed to the buying public. Judd and his assistants shared their findings on the best propagation methods with the participating nurseries.

Despite his preoccupation with botanical research and the constant demands of administering the several entities under his wing, Merrill worked on the Arboretum's relations with the City of Boston during the early years of his tenure. When a revival of bicycling reached nuisance levels in the spring of 1937, causing numerous complaints by pedestrians, the captain of the local division of police, together with the Arboretum administrator, convinced the park commission to forbid bicycling in the Arboretum. A few years later two of the Arboretum staff were

125. *Alfred Rehder officially retired in 1940, the same year a revised, second edition of his* Manual *was published. Nonetheless, he continued to come to the building almost daily for the next nine years to work on a bibliography to complement the manual.*

"deputized" as special police to deal with misconduct or perpetrators of vandalism.

When a city agency diverted Bussey Brook into a storm sewer, halting its flow through the valley of Hemlock and Bussey Hills, there was dismay at the aesthetic loss and consternation over potentially permanent damage to the native hemlocks through the elimination of moisture supplied by the brook. Fortunately, after official protests from the Arboretum, the brook was allowed to flow once more. The parks department also extended and repaired the boundary fences and improved the drainage of the North Meadow in the late 1930s. To improve conditions for visitors, the Arboretum carried on "an extensive poison ivy eradication campaign" from 1938 onward, although no mention was made of the means employed. A change that could not be resisted occurred in 1940, when Boston authorities widened Bussey Street and rebuilt the stone wall on the Peters Hill side. While this had little effect on the Arboretum's accessioned plantings, it did make the perceived distance between the two sections of the grounds seem greater to pedestrians, thus adding somewhat to the isolation of Peters Hill.

RESEARCH: EBB AND FLOW

After 1940 the Arboretum's nonliving collections and botanical research program moved toward coordination with some of the other units. In the interest of keeping the best investigators and instructors in plant science at Harvard, the units pooled their resources at times. This meant that the selection of personnel and topics more frequently reflected the interests of biology as a whole, in contrast to the earlier taxonomic scope of the Arboretum. The demise of pathology investigation after Joseph Faull's retirement and the addition of wood anatomy were the result of such collaborative decisions.

For the Arboretum, taxonomy was still the major topic of research. The first year of the decade was an eventful one for Alfred Rehder. A revised, second edition of his *Manual* was published, and, having reached the then designated retirement age of seventy-five, he became associate professor, emeritus, after forty-two years at the Arboretum. As of his retirement he had published 980 papers, 82 of them in the *Journal of the Arnold Arboretum*. Rehder had been editing the *Journal* since its start in 1919, in recent years with Clarence Kobuski's assistance. Rehder agreed to continue this task until a replacement was on board. Indeed, he came daily from his nearby Jamaica Plain home to the Arboretum for the rest of his life, working on a new project, a bibliography to complement the *Manual*.

Merrill called Albert C. Smith, one of the curators at the New York Botanical Garden, to take up similar duties at the herbarium of the Arnold Arboretum in 1940. A graduate of Columbia University who also received his doctorate there, Smith was an expert on tropical American and Fijian floras. After a year he was given responsibility for the *Journal* as chairman of an editorial committee. Smith took up taxonomic study of certain families important in Papuasia and worked on collections from Fiji, an effort that was a logical extension of Merrill's own scrutiny of Philippine and related floras. It was mainly Smith, Kobuski, and a recently hired assistant, Caroline K. Allen, who directly managed herbarium operations.

Another retirement in 1940 was that of the librarian, Ethelyn M. Tucker, who had served forty years, first under Charles Faxon's direction and later on her own. Her assistant was promoted to oversee the ever expanding book collection but resigned within a few years. In 1944 Lazella Schwarten came from the New York Botanical Garden, where she had worked with Merrill during the Great Depression, to head the Arboretum library.

126. Shiu-ying Hu and Mary Tai sort specimens in the herbarium wing. Merrill continued the tradition of encouraging scholars from China to work with the Arboretum collections, especially those in the herbarium. Hu became a staff member after receiving her doctorate from Radcliffe.

Throughout Merrill's administration, no fewer than nine staff members were involved in research on various topics in systematic botany in the herbarium. He continued to work on plant collections from the Philippines, China, and Southeast Asia. With his assistant, Lily M. Perry, Merrill enumerated collections from the Archibold expeditions to Papuasia and New Guinea. A native of New Brunswick, Canada, Perry earned a master's from Radcliffe in 1925 and a doctorate from Washington University in Saint Louis in 1932. She worked at the Gray Herbarium until Merrill hired her in 1936. Merrill also started an extensive bibliographic survey of the work of C. S. Rafinesque, an erratic nineteenth-century botanist who had named many North American plants. Ivan Johnston studied collections from southern South America and Mexico in addition to pursuing his interest in the borage family. Kobuski's specialties were the camellia family and the genus *Jasminum*. Caroline Allen completed research on the Lauraceae of eastern Asia and then turned to New World members of the family. Leon Croizat's special group was the euphorbia family, on which he published extensively.

When E. J. Palmer was not checking the identity of Arboretum plants in consultation with Rehder, he worked on *Crataegus* and *Quercus* of North America. Hugh Raup, an expert on northern Canadian ecology, worked on specimens from the Mackenzie Mountains and headed up botanical exploration of the Alaska Military Highway once it opened. Susan D. McKelvey brought out the results of her studies of the yuccas of the southwestern United States in two parts (1938 and 1947). Merrill, Bailey, Sax, Johnston, and Raup were additionally involved in instruction and the supervision of graduate students at Harvard.

Joseph H. Faull, after twelve years of teaching and running the Arboretum laboratory for research on the causal organisms of plant disease, also officially retired (as professor of forest pathology) in 1940. Faull, like Rehder, continued to pursue his research for many years beyond retirement age, moving to the biology laboratories in Cambridge. A replacement was not immediately sought, since another vacancy on the staff of the Bussey Institution was not yet filled. World War II caused further postponement in budgetery decisions, and the position of forest pathologist was never filled again.

Although the field of plant pathology was given up, the Arboretum had taken on another topic that had ramifications for systematic botany, as well as for forestry. As of 1938 Merrill annually reported on accomplishments in wood anatomy, mainly as pursued by Irving W. Bailey. Bailey had earned a master's degree from the Harvard Forest in 1909, then joined the staff of the Bussey Institution. After 1931 he worked in Cambridge studying the structure of wood, occasionally offering courses and guidance to graduate students. There, the wood specimens of the Arboretum, integrated with those of the Bussey Institution and the biological laboratory, provided a basis for studies of comparative anatomy. In order to better understand evolutionary relationships and trends, A. C. Smith and I. W. Bailey collaborated on studies of a group of plant families thought to be primitive.

World War II caused certain disruptions to botanical research, but Arboretum staff were able to contribute to the war effort in several ways. Kobuski was drafted within days of reaching the upper age limit and served in the Army Medical Corps for the duration. Johnston was asked to work on a classified project for the Chemical Warfare Service and traveled to Panama. Others on the research staff cooperated with the armed forces in studying the use of plant material for camouflage and by supplying information about poisonous and emergency-food plants, especially for soldiers in the Pacific theater of operations. Merrill made plant identifications from material forwarded by servicemen stationed in the Old World tropics and sent the information back immediately, much to their benefit. He authored a survival manual for the Pacific forces and traveled to Washington regularly to lecture on poisonous and emergency-food plants to trainees at the Army Medical School. Despite war in Europe and in the Far East, living and dried plant material still arrived at the Arboretum from both areas.

PLAN FOR CONSOLIDATION RECONSIDERED

Even though Merrill and his staff had taken many measures to increase efficiency and better utilize the space left in the building in Jamaica Plain, the herbarium was bursting at the seams in the mid-1940s. Moving the wood collection to Cambridge allowed a little expansion room, but this was soon taken up by pressed plants. Merrill initiated the breaking up of the herbarium collection into a more detailed geographic arrangement than the earlier Old World/New World distinction. The new system divided the Western Hemisphere into four subsets and the Eastern into ten, thus allowing botanists and curators to find particular taxa or areas of interest more easily. When space ran out in all available herbarium cabinets, cardboard cartons, designed by Merrill to hold specimens and to open from the front, were used on top of the cases, on work tables, and in stacks from floor to ceiling. A second sequence of families had to be started in the cartons, since shifting the original arrangement to accommodate additions became impossible. From then on it was often necessary to look in at least two places to consult all the material of a genus or family.

At the outset of Merrill's administration, he had worked closely with the heads of other herbaria at Harvard and with the staff of the Division of Biology to come up with a tentative plan to combine all the biological libraries and preserved collections in one building to improve efficiency, communications, cooperation, and teaching strength. But the outbreak of war caused the contemplated building project to be set aside. When an end to World War II was in sight, the Harvard administration looked again at the problem of space for the several herbaria.

Since the prewar plan now seemed outmoded, in late 1944 Irving W. Bailey was assigned by Harvard's provost to make a general study of the total botanical situation; after less than a year, he submitted a report, which was confidential at the provost's request. Six months later it was put in a form that was circulated to the board of overseers, the faculty and staff of biology, and the separately endowed botanical units. After a month or two of meetings and discussions, the faculty and the teaching staff of the botanical institutions agreed to carry out the Bailey Plan, as it became known; the Harvard Corporation gave final approval in March 1946. When they voted for the Bailey Plan, the need for better facilities to house crowded collections was uppermost in the Arboretum staff's mind.

While most who voted in favor of the Bailey Plan agreed on the idea of amalgamating the efforts and the

127. *After ten years in the post, Merrill stepped down as administrator of botanical collections and director of the Arboretum just as Harvard's administration began a reorganization of its botany institutions, largely motivated by the need for modern and expanded facilities for the collections of several of the units. Despite the handicap of cramped conditions, Merrill had strengthened the Arboretum's roles in tropical botany and in North American horticulture. He remained an active researcher and a participant in Arboretum activities until his death in 1956.*

collections, it rather quickly became apparent that the way the reorganization was to be carried out, especially in relation to the staff of the biology department, was unacceptable to many people involved. There began an intricate political tangle and era of controversy and legal debate, especially over the Arboretum's projected role, that would last even beyond 1954, when a new combined herbaria building was constructed and various collections moved in. Before the war, Merrill and his colleagues envisioned a building unifying all of biology, but now only botany was to be remolded.

By the approval of Bailey's report, all botanical activity of the separately endowed institutions was to be divided into two intellectually defined areas, each of which would have its own budgetary and administrative head. One would be the Area of General Morphology, under which would fall the Gray Herbarium and its library, the Farlow Herbarium and library, the library and herbarium of the Arnold Arboretum, the library and dried specimens of the Botanical Museum, and all research pertaining to these preserved collections. The second area, that of Experimental and Applied Botany, would administer the living collections of the Arboretum, the Atkins Institution in Cuba, the Harvard Forest, the Harvard Botanic Garden, the Bussey Institution, the displays of the Botanical Museum, and the ten-year-old Cabot Foundation, an endowment for investigations toward improving the world's timber supply. This area's work would be in genetics, plant physiology, pathology, economic botany, horticulture, and silviculture.

Each of these two areas would be administered by a committee with rotating chairmen: I. W. Bailey would head the general morphology area, while Paul C. Mangelsdorf, geneticist and director of the Botanical Museum, would chair the experimental and applied botany area. The obvious difficulty for the Arboretum was that it was divided between them. To many, it did not look like a workable solution for an institution that had fairly successfully charted its own course for seventy-five years. The idea of its being administered on a rotating basis certainly did not jibe with Sargent's notions of permanence and the importance of a long-term view.

After a few months, Merrill, whose position as administrator of botanical collections was already precluded by official adoption of the Bailey Plan, resigned as director of the Arboretum. He was seventy years old. He retained his Arnold professorship and spent the next ten years in productive research while trying unsuccessfully to keep out of the controversy over the impending move of the herbarium and library collections. By his own account, the most important achievement of his administration was improving the Arboretum's "prestige, not only as a local institution catering to the general public, but also as a national and international one in the research and publication fields" (Merrill, 1946, p. 487). The one item the botanical administrator put forth as the institution's most important need was an endowment restricted to fund the cost of publications. He thought it ironic and absurd that many agencies could be turned to for research grants,

while the cost of publishing scholarly work was rarely subsidized. He thought it paramount that an institution such as the Arnold Arboretum have a guaranteed source of funds to assure broadcast of the results of its investigations.

Karl Sax, geneticist and plant breeder, on the Arboretum staff since 1928, was appointed acting director as of 1 August 1946. A year later he was made director but was not appointed to the Arnold professorship. From the time the Bailey Plan was approved, Paul C. Mangelsdorf, as chairman of the new (and renamed) "Institute" for Research in Experimental and Applied Botany, took some control of activities related to the grounds.

WYMAN CARRIES ON THROUGH WAR AND CHANGE

Despite the turnover in administration and the ensuing inertia of the botanical program, the Arboretum's horticulturist managed to carry on with his agenda for the living collections. Labor shortages during the war years, and the deaths of Schmitt and Judd, who had both served under Sargent, did not make for smooth sailing. Fortunately, the living-collections difficulties were counterbalanced by the addition of a suburban station, accompanied by a generous bequest for its upkeep. As part of the new order of things, an outside consultant was brought in to advise on refurbishing the landscape. Wyman, however, forged ahead to implement a revised collections policy while restoring equilibrium to the style and arrangement of the grounds, which had been upset by Wilson, the hurricane of 1938, World War II, and overcrowding.

During the war there was apparently a period when maintenance of the grounds deteriorated. The temporary loss of several of the grounds staff to the armed services, either by enlistment or the draft, began to have visible effects by about 1943. Both Heman Howard, now Wyman's assistant in charge of mapping and labeling, and Alfred Fordham, assistant in propagation, served in the army for three years. There was no expert pruner left among the grounds staff, and the total number of persons employed full time dropped to eight from an average of

128. Routed direction signs were placed to show visitors the paths to main collections and exit gates, although the use of path names was dropped. Since signs and plant labels were produced by the grounds staff, weather-resistant, low-cost, vandal-proof methods were utilized. This sign indicates that the azaleas on Bussey Hill were considered the main collection before the Meadow Road border was installed in 1948 at the suggestion of Beatrix Farrand.

ten or twelve. In an effort to reduce the amount of grass cutting and hand weeding needed, every other path in the shrub collection was plowed up and kept cultivated with power equipment. The war brought on inconveniences in transportation because of gas rationing: staff members were required to walk between points on the grounds rather than use Arboretum vehicles, and machine operation was curtailed.

Wartime self-sufficiency also prompted new cooperative uses for Arboretum land. Some of the South Street tract was released to the Boston Victory Gardens Committee, and staff members were allowed space for vegetables in various nursery areas. The Massachusetts College of Pharmacy was permitted to install a garden for medicinal plants on the Bussey Institution grounds.

Other events made the war years ones of transition. In late 1944, Louis Victor Schmitt, superintendent of buildings and grounds, who had been associated with the institution since 1905, was taken ill and died. He was succeeded by Robert G. Williams, a 1934 graduate of

Cornell University who had been superintendent of parks in Greensboro, North Carolina, for seven years. Williams reported for duty on 1 January 1945, in time for an unusually busy spring. A heavy, wet snowfall in February caused extensive limb breakage, crushing some shrubs to the ground. Cleanup operations delayed the normal spring rehabilitation, and it suddenly became necessary to move the entire nursery on the Bussey Institution grounds. The state antitoxin laboratory, which had been operating there since the turn of the century, let a contract for a building addition that would use land occupied by Arboretum nurseries. Sudden expansion of the state facility was apparently a result of war demands. Williams and Judd oversaw the hasty preparation of a new nursery site, and the Arboretum's valuable selected stock was transplanted before construction began. Some of the plants were moved to a growing area still maintained on the South Street tract.

There were some good signs for the development of living collections during the early 1940s, the war notwithstanding. One major and two modest land acquisitions took place. With the departure of one of the professors long associated with the Bussey Institution, the mansion house, once Benjamin Bussey's home, was left vacant. For years, exterior maintenance of the large clapboard house had been deferred. Since there was no need for housing Bussey Institution staff now that all its members worked in Cambridge, the mansion's demolition was authorized in 1940. The grounds between it and South Street were thus freed for Arboretum plantings. The area was soon used for some of Sax's hybrid forsythias and crabapples still being tested and as space for extension of the lilac collection.

After the widening and straightening of Centre Street in 1931, a triangular piece of property owned by a quarry company was left abandoned adjacent to the conifer collection. The quarry site was an eyesore, and Oakes Ames once had to protest when a billboard was set up in the space. The spot was located on one of the Arboretum's most heavily traveled automobile approaches, making it a very desirable one to bring under the institution's control. In fiscal 1940–41, visiting-committee member John S. Ames provided the funds to purchase the parcel from the quarry's owners. Immediately, unwanted weeds were cleared out and twenty hybrid flowering crabapples were planted in the new area. An old stone wall marking the earlier boundary between street and Arboretum was left in place for another thirty years, however. Since the city public works department still kept a small building to store equipment on the old Centre Street right-of-way, the adjunct area was referred to by Arboretum employees as "city shack," a name that stuck long after the structure was finally removed in the 1960s.

A welcome potential solution to the question of crowding in the living collections appeared in 1942, with a gift of land and an endowment from Louisa W. Case as a memorial to her father, James B. Case, who had assembled an estate from several farms in Weston, Massachusetts. The fifty-nine-acre parcel was twelve miles from the Arboretum in a suburban community. Right away some of Sax's hybrid cherries and crabapples, and two hundred other trees and shrubs, were established in a nursery there. Two years later, her sister, Marion Roby Case, willed an adjacent share of ninety acres and an even larger endowment to the Arboretum, stating, "It is . . . my earnest hope that the estate be maintained and the bequests utilized to further the development of my land and the adjoining land recently presented by my sister, Miss Louisa W. Case, as an adjunct to the Arnold Arboretum in order that the work of that institution may be amplified in a non-urban center, and yet one that is reasonably close to the City of Boston" (Merrill, 1945, p. 485).

Marion Case had managed her property, Hillcrest Gardens, as a school of farming for boys for several years. The Case "boys" learned by participating in the planting and care of extensive vegetable and fruit gardens, beds of perennial flowers, and a woodland wildflower garden. As part of their training they sold the produce raised, often on special order, to residents of Weston. Marion Case rounded out their education by inviting prominent horticulturists and botanists to lecture at Weston once a week, and each student completed an essay to be published in a yearbook at the end of the growing season. Her interest in perpetuating the use of her land for horticultural education stemmed from this experience. While it would take a few years for the endowment income to become avail-

able, plans were made to use the land for a Cabot Foundation tree-breeding project and Mangelsdorf's experimental maize hybrids, as well as expansion for Arboretum nurseries and displays.

Since the end of the war brought some of the grounds crew back in 1946, and Merrill was able to authorize hiring of others before stepping down as director, Wyman began to rehabilitate the Arboretum after three years of deferred maintenance. He and R. G. Williams oversaw renovation of the overcrowded vine collection, where many of the climbers had become inextricably intermingled. The vines on the wire trellis were repropagated and the old plants removed; this allowed the trellis itself to be entirely restored. In the new arrangement only one vine was planted between each set of posts, so that maintenance and keeping the kinds distinct was made easier. The only plants returned to the trellis were what Wyman termed "truly ornamental" vines; the other species were planted along the boundary walls and fences.

Wyman was pleased with the improvement resulting from machine cultivation between rows in the shrub collection, which allowed efficient application of lime and fertilizers and reduced hand hoeing for weed control. Better soil aeration undoubtedly enhanced growth of these plants as well. An experienced pruner was hired to bring trees and shrubs injured by wind and cold, or overwhelmed with suckers, into presentable condition, although this would take many months of long, painstaking work.

During the years of labor shortage many woody weeds had taken hold in the midst of shrubs and plantings. A vigorous campaign was launched to get rid of these, using

129. The Case Estates in Weston, nearly 150 acres of cultivated ground and wild forest, came to the Arboretum in the early 1940s through bequests of sisters Louisa W. and Marion R. Case. The area was welcomed as a suburban station for Arboretum activities and as expansion room for the woody-plant collections. Other Harvard departments, among them the School of Design and the Cabot Foundation, were welcome to use the site.

the tractor to aid in pulling out unwanted saplings. A program of extensive fertilization was also started after the war, and the use of mulch (spent hops hauled from one of Jamaica Plain's nearby breweries) was tried for the first time.

The horticultural staff was anxious to make the Arboretum look as good as possible for the 1946 growing season in anticipation of visits from two groups, the Garden Club of America in May and the National Shade Tree Conference in August. In one of his last official communications, Merrill prodded Boston Park Commissioner William P. Long to go forward with much-needed resurfacing of roads and repairs to the benches in the Arboretum.

Not long after those attending the national garden club convention were welcomed to view the peak of crabapple and lilac bloom, tragedy struck the Arboretum with the unexpected death of William Judd on 23 May. A widower for many years, the fifty-nine-year-old propagator collapsed from a heart attack after returning to his Jamaica Plain home from a meeting of the Horticultural Club of Boston. Besides Ernest J. Palmer, Judd was the last of the active Arboretum staff members to have worked under Sargent. Untold numbers of plants had been raised for the institution and its correspondents by Judd during his thirty-three-year career there. Like Dawson before him, Judd won the goodwill of many of the Arboretum's associates through his plant-distribution duties, and his good nature and extensive knowledge would be sorely missed. Nearly a year later Richard H. Fillmore, candidate for a master's degree in ornamental horticulture at Cornell, stepped in to head the Arboretum propagation unit. A native of Nova Scotia, Fillmore also had experience working in his family's nursery business.

BEATRIX FARRAND, CONSULTING LANDSCAPE GARDENER

Had Judd lived, he probably would have eased the introductions between Wyman and Beatrix Farrand, who became associated with the Arboretum that same spring. Having asked visiting-committee member Susan Delano McKelvey for suggestions, Paul Mangelsdorf retained Farrand to serve as consulting landscape gardener to the Arnold Arboretum. Since Farrand and Judd were well acquainted, Mangelsdorf hoped Judd would smooth the way for Farrand to work on what he thought were important problems for the Arboretum. Unfortunately, Judd passed away before her first official visit. Farrand had been to the Arboretum before, however; it was where she received initial training for her profession of landscape gardener, a term that she preferred to landscape architect.

In the 1890s, young Beatrix Jones of New York and Bar Harbor, who was greatly interested in plants, had come to live with the Sargents and to study the Arboretum collections under the dendrologist's guidance. After Sargent steered her to her first client, she was off to a career as a designer of private estates and campuses. She often said that the professor's best pieces of advice were to make the plan fit the ground and not twist the ground to fit a plan and to study the tastes of the client. To further her education on lines suggested by Sargent, she traveled widely to study art, architecture, gardens, and plants. She also supplemented her training by being tutored in certain technical subjects, since there were still no schools of landscape architecture in the 1890s. Among Farrand's most notable designs were works on the campuses of Princeton and Yale and the gardens of Dumbarton Oaks in Washington, D.C.

But it was mainly as a former student of Sargent's that Farrand was invited to participate, and made herself welcome, at the Arboretum in 1946.

You and your colleagues have given me the chance to show my gratitude and appreciation of what the Arnold Arboretum has meant to me individually and to my fellow workers in landscape art. More than fifty years ago Professor Sargent began to teach me by precept and example the ideas underlying the Arboretum plantations. They were intended, in his mind, not only to show various trees and shrubs in as complete a collection as might be advisable but he thought the plants should be placed so that their use in the landscape could be understood. (AAA, Farrand to Mangelsdorf, 15 May 1946)

While she and Mangelsdorf agreed her task would be difficult because of the practical, historical, and aesthetic questions involved, Farrand's work for the Arboretum was further hindered by several factors. First, over the years that she acted as consultant the Arboretum was administered by several people: Mangelsdorf, Merrill, Sax, and Wyman. Second, even when she made valuable suggestions and was able to work out plans, there were often no expenditures authorized to carry them out. Then too, she and Wyman, who was the only constant in the chain of command, disagreed as often as they agreed—not only over aesthetics but for practical reasons of managing an arboretum. Nor did Sax and Wyman always agree on the best course of action. Lastly, Farrand's grasp of the raison d'être behind the Arboretum's original layout and subsequent development was somewhat sketchy. She apparently overlooked the concept of the botanical sequence of collections spaced within a backdrop of natural woods and boundary plantations that was devised by Sargent and Olmsted. She tended to look on the Arboretum as a landscape in need of beautification rather than as a series of collections for study. Although Farrand spoke of restoration, many of her suggestions were further departures from the original plan.

While he valued many of her suggestions, Wyman summed up his thoughts years later by saying:

> It was most unfortunate that even with all her extensive experience, she failed to comprehend the practical needs of the Arboretum. . . . It has always been my strong conviction that landscape planning should be left to those on the staff . . . who understand the diverse practical problems of maintaining the plantings. The landscape architect's contribution comes in planning the original lay-out, not in revamping an Arboretum such as this after it is tightly planted. (Wyman, 1970, p. 85)

One of the advantages of setting out the plantings in a botanical sequence, no matter how arbitrary it might have seemed, was that it provided a kind of insurance for long-term continuity of style against the muddle that might otherwise be wrought by the exercise of personal whims of successive keepers of the garden, no matter how tasteful or artistically pleasing any one of them might be. Wyman rarely, if ever, mentioned the overall taxonomic arrangement. Nevertheless, by working with the collec-

tions, he generally grasped their intrinsic unity and maintained it despite change. Although he altered their content, the overall arrangement of the collections was largely undisturbed.

Over the next few years Farrand or her assistant came to the Arboretum often and she wrote letters full of suggestions. The one-year appointment was renewed many times; the period of her most active association with the institution extended until about 1950, with her official retirement coming in 1955. Wyman tended to work as much of Farrand's program into his regular Arboretum operations as his labor and record-keeping force and budget would allow.

COLLECTIONS POLICY SHIFTS

During this period, the alterations to the grounds that did occur were equally influenced by a change in policy that developed to utilize the Case Estates as a supplementary collections area. The one notion on which Wyman and Farrand seemed to agree was that the Jamaica Plain plantings were too crowded in many places. Some plants were duplicated in several locations, and some were less ornamental than (or only slightly different from) other varieties in the same group. There were also overaged and disfigured plants that would be better removed. And "vistas" were needed in certain areas. They agreed that the 265 acres in Jamaica Plain would never gracefully accommodate *all* the hardy trees and shrubs and that, therefore, the tree garden's precious growing space should be used to display the best, most ornamental types in each genus. The idea had been broached in the Bailey Plan, and, with Mangelsdorf's assent, they decided the "less ornamental" plants could be kept, off site, at the Case Estates.

In order to display only the best in Jamaica Plain, the ample new land in Weston would have a dual function. First, new accessions coming to the Arboretum would be placed on trial there to be evaluated for hardiness, correct identification, and superior ornamental qualities before the decision would be made whether to place them in the Arboretum proper. Second, those plants in the crowded

130. One group that had the four-season ornamental qualities Wyman and his fellow horticulturists sought was the genus Viburnum. *These shrubs have conspicuous clusters of flowers in spring, glossy or rugged-textured foliage all summer, bright fruits and colorful foliage in fall, and strong, architectural branching when viewed in winter. Wyman said that if he were to choose only one representative of the group for a garden it would be Siebold viburnum* (Viburnum sieboldii), *pictured here in flower.*

Jamaica Plain collections worth keeping for botanical, historical, or other reasons, but not of interest to most Arboretum visitors, would be placed for the long term in nursery-like rows in one of the big fields, where they could be kept healthy and orderly by machine cultivation. This became known as the "permanent nursery."

The idea goes back to Sargent's early stated need for trial or experimental plantings of a less permanent nature than the main Arboretum arrangement. Wyman made great use of the extra space in the holding areas, moving many plants out to the Case Estates, where he also made his home and kept as watchful an eye on things as in

Boston. It was his opinion that if fewer and better-selected plants were grown under less-crowded conditions, they could be better maintained, and the grounds in Jamaica Plain would be more attractive and instructive for the public. It was an idea he began acting on as soon as the first gift of land from the Case sisters was received, and Farrand backed him once she became consultant. Bailey, too, in his report had recommended "improvement of the demonstrational aspects of the Arboretum [by] elimination of less hardy plants" ([Bentinck-Smith], 1953, p. 583). The notion of a selected display rather than an exhaustive collection also was part of the initial policy of more recently established woody-plant gardens, such as the Scott Foundation and the National Arboretum.

It was up to Wyman to decide what qualities made a plant among the best of its group for ornamental purposes. Although such decisions could be considered arbitrary, and weighted by personal taste, Wyman used a set of standards that can be gleaned from his voluminous writings. He was not alone in his desire to rate the many varieties being selected by breeders and growers in order to aid the gardening public in making sense of all the possibilities. For example, since not everyone would like to look at all five hundred varieties of lilac to decide which one to plant, Wyman attempted to narrow the field for Arboretum users by moving out of the Bussey Hill Road collection those specimens that were unobtainable or "extinct" in the trade, along with some that were just plain inferior in vigor or performance to similar, better types. Many of his colleagues at other botanical gardens and arboreta, in plant societies, and in horticultural organizations were making similar evaluations.

Wyman arrived at his lists of superior ornamentals by making a whole range of observations. Did the plant grow vigorously in the particular area in which it was to be grown? Was it cold hardy? Did it often succumb to a disfiguring or fatal disease? Plants that tolerated extreme conditions—dry soil, poor drainage, salt spray—might be included among the best for this reason. Aside from general health, Wyman preferred plants that had more than one season of interest. Big, colorful, fragrant flowers, that

is, ones that were easily noticed in the landscape, were an important factor. To maintain objectivity he used published charts to compare and describe colors; some representation of the entire range of color in any one species or genus was included. The best possible plant had attractive flowers, attractive fruits, and good autumn color. Those whose leaves and architecture gave interest even when out of flower or fruit were to be preferred. Add highly textured or colored bark or some other winter feature, and it became a four-star plant. Varieties that extended the season of interest for their group by blooming earlier, later, or longer were also favored. Double-flowered varieties were rated highly on account of their general tendency to stay in bloom over a longer period.

Ideally, the horticulturist's evaluations were not based on whim but on years of observation, broad knowledge of the whole group, critical comparison within it, and communication with other experienced plants people. For this reason the process of removing plants from the Arboretum proper to the secondary collection was a slow one. Wyman recognized, too, that his choices were not fixed for all time. He was open to the possibility that at some later date plants from the permanent nursery (or their propagules) might be brought back to Jamaica Plain and other collections moved to Weston. The Case property represented much-needed room to grow, and he considered it fully an extension of the Arboretum.

REARRANGEMENTS ON THE GROUNDS

The transfer of plants to Weston was largely a continuation of what Wyman had started before the war. In the winter of 1946–47 one thousand duplicate or overcrowded plants were removed from the Arboretum. This operation opened up some vistas that greatly pleased Farrand. For example, the remainder of the original euonymus collection was dug up from the west side of Meadow Road and replanted with the secondary euonymus group that had been placed at the foot of the summit section of

Bussey Hill Road in the 1920s. This, combined with the removal of duplicate cork trees, allowed the linden and horse chestnut collections to be seen from Meadow Road.

With judicious thinning in the ashes and elms, a little more of the curving summit of Bussey Hill was visible from the heavily used Forest Hills gate, and from Bussey Hill the Customs House Tower in Boston could be observed. The majority of barberries and many *Ribes* were moved to the Case Estates. The reason given for transporting these two groups was that their roles as alternate hosts in diseases of wheat and white pine, respectively, made them unavailable in the trade and unworthy for view by the public. The *Ribes* bed across Meadow Road from the administration building was completely cleared, as was the row of barberries Wilson had set out along the way to the summit of Bussey Hill. The nearby site of the Bussey mansion was used for the reorganization of the lilac collection, so that only the most ornamental varieties occupied the prime locations next to Bussey Hill Road. One hundred less-important lilac varieties were transferred to Weston for the reasons given above, and Wyman tackled a similar evaluation of the *Weigela, Philadelphus, Deutzia,* and *Spiraea* collections the following spring. Many of these had been spread out under the hickories in Wilson's time. The best were moved into the shrub collection, while the rest went to Weston.

Three areas that Beatrix Farrand especially thought needed improvement were the land around the administration building, the plateau of Bussey Hill, and Peters Hill. She convinced Wyman to augment the young conifer grove that had been planted behind the building after the 1938 blowdown with more pines and hemlocks. She suggested a blitz of Arnold Arboretum introductions in the little valley of Goldsmith Brook to the west of the Jamaica Plain gate and on the hillside beyond the building. Although lists were sent back and forth, apparently no planting plans were drawn up. Some of the last of the old shrubbery that Sargent had first placed between gate and building was cleared. And such Arboretum introductions as hardy cedar of Lebanon, anise magnolia, Arnold Promise witch hazel, and a variety of Amur honeysuckle were arrayed on the slope south of the herbarium wing. Across the road, in order to open a view of the meadow, several

of the old native-shrub masses were removed, and an attempt was made to grow extremely prostrate forms like bearberry, vinca, and memorial rose (*Rosa wichuriana*) in their place. These did not live long in this heavily traveled area, however, and turf eventually replaced all but one or two clumps of fragrant sumac (*Rhus aromatica*) from Sargent's day.

On Bussey Hill, Wyman wished to restore the azaleas to their former glory, since a series of dry summers had weakened the collection. Installation of an irrigation system was one possible remedy for the problem, and it would be valuable for many other places in the Arboretum. The city was approached for aid, but installation of water pipes was not deemed possible, given the state of municipal funds. After ideas on relocating the azaleas were debated, Sax intervened in favor of Farrand's suggestion that a new collection be planted in a border along the east side of Meadow Road. The landscape gardener thought so great an attraction as the azaleas should be located in one of the most heavily traveled sections of the Arboretum. To Farrand, Bussey Hill was too remote; Wilson, on the other hand, had termed it "the heart of the Arboretum" when its Chinese plants, double-flowered cherries, dove tree, and blaze of azaleas brought many visitors up the hill every May. By the mid-1940s the killing winters of 1934–35 and 1942–43 and droughts had apparently taken their toll on the cherries and some of the azaleas. All of the former were removed from the central area of the plateau, leaving the lawn open, as it had been in Sargent's time.

For the new azalea beds, the Arboretum crew began to prepare the ground along Meadow Road by removing more clumps of native shrubs and some of the shadbushes from the existing planting. Several hundred cubic yards of peat were incorporated into the soil. In the fall of 1948, three hundred azaleas and other ericaceous shrubs were planted in long beds sited between the remaining large trees: katsura, silver maples, and cork trees. A few large smoke trees (*Cotinus*) and a fragrant sumac were left in place, remainders of the original sumac collection (Anacardiaceae) in the Bentham and Hooker sequence. Some of the azaleas were moved from the old Bussey Hill plantings, but most were left there, notably the mass of

131. The azalea border on the east side of Meadow Road was created at the suggestion of Beatrix Farrand, consulting landscape gardener, in the late 1940s. Shadbushes, katsuras, cork trees, silver maples, smoke trees, and one clump of fragrant sumac were retained from earlier plantings, and these were interspersed with azaleas arranged in beds by color group.

royal azalea under the oaks near Bussey Hill Road, J. G. Jack's introduction of Poukhan azalea, and some of the *Rhododendron obtusum* varieties. Masses of native American sweet azalea (*Rhododendron arborescens*) and rose-shell azalea (*R. prinophyllum*) were moved from across Meadow Road into the new border. Blueberries, *Enkianthus, Lyonia,* Labrador tea, and *Zenobia* accompanied the deciduous rhododendrons in the new beds. The azaleas were massed according to color of flowers. While the new border was unquestionably attractive, its creation inadvertently contributed to the gradual abandonment of the Bussey Hill plateau as a "must-see" area by the general public.

At the urging of Farrand, the north slope of Peters Hill was rehabilitated. After forty years of undisturbed growth, the hawthorns there had stretched out their branches until they nearly touched. Their low-slung crowns made mowing among them difficult; over the years their spontaneous seedlings, and other weedy shrubs, had sprung up to exacerbate the thorny tangle. Under the trees, tall grass was a fire hazard when it dried out in high summer. Everyone agreed the whole north slope was in need of rehabilitation. Fortunately, John S. Ames was again willing to give financial aid to a project that included costs beyond the normal scope of Arboretum operations: extra soil amendment, heavy-equipment rental, turf improvement, and other items. Palmer's recent taxonomic work on *Crataegus* was undoubtedly helpful in deciding which species and varieties to keep and repropagate for the more modest collection to be returned to the hill slope, once it was cleared and renewed.

The Arboretum staff did most of the labor, beginning in 1948 with removal of dead and surplus trees; more than two thousand were taken out in fiscal 1948–49. Wyman sought the advice of soil specialists to prepare for replanting, a process that took three more years. Alternating strips were plowed following the contour, to which were applied fertilizers and grass seed of a special mixture recommended for its tendency to stay green late into the fall and be less prone to catching fire. Once the new grass rooted to hold the soil, the strips previously untouched were plowed and similarly treated the following year. The process was done gradually to lessen erosion damage and to control weeds while the grass reestablished itself. For the

132. Contour plowing on Peters Hill. The rehabilitation of the north side of the hill took place from 1948 to 1952. The overgrown hawthorn collection was removed, and soil and turf were renewed by cultivating alternate strips over a period of years. While many hawthorns were returned to the slope, crabapples and mountain ashes were added and a large area left unplanted.

summit Farrand suggested a planting of drought- and wind-tolerant New England shrubs (blueberries, sweet fern, wild roses, junipers, and others), but these were never installed. Sax had another idea—he planned to build a branch road from the circuit drive to the top.

On the side where the rows of hawthorns had been, replanting did not begin until 1952, since Wyman was away in Europe for three months in 1951. Eventually groups of crabapples, hawthorns, and mountain ash were placed on the overhauled slope. Each genus was kept in irregular groups, with its individuals well spaced. Some ground was left unplanted to avoid the monotony that had been characteristic of the earlier installation. Pathways along the contours of the ground were left open. Many of the crabapples were results of Sax's breeding program. While this operation vastly improved the ease of maintenance, the renovation of Peters Hill did not greatly change the character of the landscape there from Sargent's original *Crataegus* planting. The groups returned were still small, flowering trees of the rose family; many hawthorns were replaced. The gridlike pattern was gone, but the open, pastoral look remained.

Over the years that Farrand was consultant a few other modifications were made in the general plantings. The shrubby dogwoods were placed along the Bridle Path, among the few shrubs remaining from the earlier Linden Path beds. Many of the tree dogwoods were moved from among the cork trees west of Meadow Road to Bussey Hill and planted in the Azalea Path beds and just below the eastern side of the summit. With some input from Beatrix Farrand, Wyman made changes on the top of the hill, adding a length of chain-link fence to prevent pedestrian shortcuts, which were causing erosion. This was screened with vines, and ground covers were installed on the steep slopes as well.

Next, the horticulturist turned to Centre Street Path, completely renovating the area in 1949 and 1950. Clearing out the weigelas, spiraeas and deutzias in the early 1940s had left an expanse of lawn under the hickories. Nevertheless, the beds, which bordered Centre Street and curved along the edge of Central Woods to a point where the Bridle Path met Valley Road, were still dense with shrubs. Wyman eliminated most of the shrubbery that

133. Through Merrill's contacts in China, in 1948 the Arboretum obtained seed of the "living fossil," dawn redwood (Metasequoia glyptostroboides), soon after it was discovered by Chinese botanists. Seed was sown at the Arboretum and distributed to institutions and individuals around the world. Within a few years, young metasequoias showed a promising growth rate compared with most coniferous trees.

skirted Central Woods, leaving only the part immediately adjacent to Centre Street to be maintained as a mulched bed. The rest of the area, now dotted with scattered trees and shrubs, was kept mown. From the former beds, a collection of boxwoods was relocated in two broad clumps in the space under the hickories recently cleared of deutzias and spiraeas. The new boxwood groups eventually grew up to block the view of the Centre Street beds from Valley Road.

Meanwhile, Beatrix Farrand worked for some time on the idea of completely revamping the shrub collection, although her suggestions never seemed practical to either Wyman or Sax. None of the alternatives provided for displaying the great diversity and numbers of shrubs accommodated in the rows as they existed. However, one of her sketches from this period depicted a new vine trellis paralleling the Arborway wall; this may have been the inspiration for the cedar-wood arbor that was erected for vines in the winter of 1951, at the edge of the shrub collection. On it were planted wisteria, akebia, bittersweet, and others, some sixty taxa of climbing wood plants in all. (The wisterias that had been moved to the Bussey Institution grounds were relocated there.) It was still necessary to grow many other vines on the walls and fences surrounding the Arboretum.

Moving the wisterias back into the Arboretum grounds apparently was a first step in reaction to the recently aired possibility that the Commonwealth of Massachusetts might take the Bussey grounds by eminent domain. Also in anticipation of losing the property, a boundary plantation similar to the original ones of Sargent's day was installed along the top of the Bussey bank in 1953. The screen consisted of native white pine, western American Douglas fir, and the Arboretum introduction, cedar of Lebanon. At the same time, another white pine group was installed on the undeveloped former Adams Nervine parcel to further screen this boundary from the street.

In addition to the supplementary collections and extensive nurseries, two new Arboretum demonstration plantings were installed at the Case Estates in the early 1950s. Wyman believed the use of low and densely growing plants to cover the ground instead of grass could save on labor and maintenance costs in many instances. To show

gardeners and landscape designers the host of choices for this use, he arranged 125 different ground covers in square plots in rows on a piece of level ground. Since people in charge of street and highway planting often turned to the Arboretum for advice, Wyman urged the use of ground covers instead of grass wherever feasible. A couple of years later the collection was expanded to 150 varieties, which included a gift of plants from the Herb Society of America.

The need to use more-appropriate trees for street and home planting where space was limited prompted the creation of a small-tree demonstration area, located near the ground covers. A selection of sixty trees of small stature, and forms with globose or narrow crowns, was lined out as a display of trees suitable for the space between sidewalk and street, under the now ubiquitous power lines. A power company gave financial support for the installation.

Through Merrill's sustained contacts with botanists in China, the Arnold Arboretum was able to introduce one last species of botanical and horticultural significance into cultivation before that country was closed to Westerners for three decades. In late 1946, J. G. Jack's former student, H. H. Hu, then director of the Fan Memorial Institute of Biology in Peking, sent pressed specimens of a coniferous tree to Merrill. The strange tree had been discovered in a remote area of eastern Sichuan Province by foresters and brought to the attention of Hu's colleague at the Department of Forestry of the National Central University, W. C. Cheng. To Cheng, the unknown tree resembled no living species, but it did bear a striking likeness to a genus known from fossils of the Pliocene that had been given the name *Metasequoia* by a Japanese paleontologist only seven years earlier. Hu agreed, and the two subsequently collaborated to formally describe the deciduous conifer and name it *Metasequoia glyptostroboides*.

Quickly grasping the importance of this discovery, Merrill dispatched funds with which Cheng arranged a seed-collecting excursion and transmitted two ample shipments that arrived at the American tree garden in early 1948. The rare find of a living fossil set off a flurry of interest in the United States and plenty of press coverage for the new tree, which was dubbed "dawn redwood."

Even before the kernels sown in the Arboretum green-house germinated, Merrill and the staff had repackaged and mailed out hundreds of shares of seed to organizations and individuals around the globe. Within a few years, as the trees rather quickly reached suitable size, they were planted out in several locations at Jamaica Plain and Weston, where all prospered.

Still actively seeking all the woody plants that would be hardy in Jamaica Plain, Wyman traveled in Europe during the spring of 1951. He visited botanic gardens, nurseries, and private collections to observe trees and shrubs of potential value for the Arboretum and for American gardens and arranged to exchange material with them. While overseas, the horticulturist gave lectures on the work of the Arnold Arboretum, illustrating his talks with color slides. Wyman returned to Boston with propagation wood of hundreds of new plant varieties and promises of more to be sent by colleagues.

During his tenure as director of the Arboretum, Sax began to select and distribute the results of own long-standing breeding program. In 1948 a hardy hybrid cherry, named 'Hally Jolivette' for his wife, was distributed and publicized. A cross between *Prunus subhirtella* and *P. apetala,* 'Hally Jolivette,' although double flowered, is particularly dainty, graceful, and shrubby, unlike any other flowering cherry. Of several *Forsythia* he selected, one was named for Beatrix Farrand. Among the numerous hybrid crabapples resulting from Sax's program, a few found worthy of release to the public were named to honor visiting committee members, among them Blanche Ames, Henrietta Crosby, and Henry F. Du Pont, all worthy advocates of the Arboretum in those years. Sax named an especially large-flowered hybrid star magnolia 'Merrill' for his former director.

COMMUNICATING THE RESULTS

To Wyman, the development of the collections would be meaningless without an equally thorough program for communicating with Arboretum users and horticulturists nationwide. The primary vehicle for education was the institution's nontechnical publication, the *Bulletin of Popular Information,* the name of which was altered in 1941 to

Arnoldia. Merrill advocated changing the names of journals, whenever possible, to one-word titles in the interest of brevity and ease of citation. The new epithet for the Arboretum's publication was chosen to honor James Arnold. Under Wyman's editorship, *Arnoldia* was sent to a wider audience. Its informative articles on various topics in practical horticulture, propagation, and cultivation of woody plants and varietal comparisons were largely from Wyman's pen, but he enlisted other experts to contribute on related subjects.

After the collections survey and detailed mapping were completed, a new guide map was published and made available to visitors in 1940. The new map was a plan view in black and white showing all the major tree and shrub groups; the internal Arboretum roads were not named, nor were the gates. For some reason the naming of roads and most paths was dropped from guides, and apparently from use by the staff, in the 1940s or 1950s. In 1949, Wyman rewrote *Through the Arnold Arboretum,* the guide to the grounds and collections. It did not differ greatly from the 1934 version, but included more explanation of such current horticultural programs as developments in plant breeding, evaluation of hardiness and ornamental attributes of woody plants at the Case Estates, experiments in propagation, the hedge collection, and the Larz Anderson collection of dwarf trees.

As a step toward establishing consistent usage of names for cultivated plants, Rehder and Wyman worked with the American Joint Committee on Horticultural Nomenclature on the preparation of a second edition of *Standardized Plant Names,* which was published in 1942. The committee was made up of landscape architects, nursery owners, park executives, members of plant societies, and others who wished to ease the way for the lay public to utilize plant names accurately. Rehder and Sargent had contributed to the first edition of 1923, when the effort was headed by the landscape architect Frederick Law Olmsted, Jr., the nurseryman Harlan P. Kelsey, and Frederick V. Coville of the United States Bureau of Plant Industry. Their goal was to unify the nomenclature in both scientific and common names.

As stated earlier, the Arboretum was not alone in its desire to educate the horticultural public to discriminate among the many varieties in a given genus of woody

134. Karl Sax's hybridization program, started under the administration of Oakes Ames, began to yield results in the late 1940s. By that time the progeny of breeding experiments with forsythia, crabapples, cherries, magnolias, and others were mature enough for assessment of their vigor and ornamental qualities. Sax named an especially large-flowered magnolia, the result of a cross between two varieties of Kobus magnolia (Magnolia kobus), 'Merrill' in honor of the former director. Merrill magnolia also has the asset of flowering at a younger age than its parent varieties.

plants. To this end the institution cooperated during the 1940s with a new organization in the evaluation and publication of checklists of cultivars in two important genera, lilac and crabapple. *Lilacs for America,* published in 1942, was the result of a survey conducted by the Committee on Horticultural Varieties of the newly organized American Association of Botanical Gardens and Arboretums. Spearheaded by John Wister of the Scott Foundation, the survey attempted to rate the quality and general garden value of the hundreds of varieties being grown. This was distilled in a list of the one hundred most highly recommended varieties of lilacs. In addition the booklet clarified confusion or duplication in use of names, characterized each species or variety by color and season of bloom, gave information on origin and status of availability, and tabulated the botanical relationship of wild species and the parentage of hybrid races. More than forty private collectors, nursery owners, and managers of public collections contributed to the survey; the result was a very useful "who's who" of the lilac world.

Since the Arboretum maintained one of the country's top three lilac collections in numbers of varieties, Wyman's input was important to the survey. Once the listing was compiled, it was used as the basis for expanding and reorganizing the Arboretum's holdings. Since the crabapple collection was also very extensive and there was plenty of expertise on this group between Wyman and Sax, the Arboretum agreed to work on a corresponding *Crabapples for America,* which came out in 1943. Both these handbooks became standard references for professionals and laypersons alike and were updated by their editors at intervals over the years.

Since, by Wyman's calculation, the plants hardy at the Arboretum could be grown in 75 percent of the gardens in the United States and Canada, the knowledge gained from its trials of new woody plants would be valuable to a majority of Americans. To make the Arboretum results more widely known, Wyman worked for some time on two companion books, *Shrubs and Vines for American Gardens* (1949) and *Trees for American Gardens* (1951). Although, like Sargent's *Silva,* they were issued through outside arrangements, they served to interpret the Arboretum's work for a large audience. Both were published by the Macmillan Company, the firm that had produced Rehder's *Manual* and many of Liberty Hyde Bailey's horticultural reference books. The format for these two works was derived from Wyman's earlier book on hedges: a series of introductory chapters including many useful lists, followed by genus-by-genus discussions of recommended species and varieties.

The shrub and tree books also included "secondary lists" of species and varieties that could be grown but were deemed less desirable for any of eleven numbered reasons. In *Shrubs,* 1,100 kinds were described as worthy of planting, while some 1,700 were considered of less interest to most growers. In *Trees,* 750 species and varieties were recommended for continued planting in America, with some 1,600 relegated to the secondary list. As was true of the permanent nursery at the Case Estates, Wyman knew the secondary designation was possibly arbitrary, subject to change with more experience or changing gardening tastes. By including the secondary lists, Wyman followed the encyclopedic tradition of the Arboretum—to deal with all the kinds that can be grown. Nevertheless, having two categories fulfilled his wish to aid and educate his readers by narrowing the field to a manageable amount of information. In *Shrubs,* Wyman included a succinct discussion on the naming of horticultural varieties that could be read with profit by nursery managers and plant breeders today.

The horticulturist informed the homeowners and professionals who made up his readership that the Arboretum collections and the guiding expertise of many of its staff were essential in compiling the volumes: "Particular acknowledgments are due the Arnold Arboretum of Harvard University, and those connected with it, where a majority of these plants have been growing side-by-side for years. Without information from this excellent trial garden and the world-famous research facilities connected with it, the author's suggestions and experiences noted on the following pages would lose much of their value" (Wyman, 1949, p. iv). Alfred Rehder, C. E. Kobuski, R. G. Williams, Heman Howard, and Mrs. Wyman had all helped bring out the books in one way or another. Seventy-five years of experience in cultivating trees and shrubs were, in a sense, summed up in these two volumes—they were catalogs of the collection from a horticultural standpoint. While Rehder's work was the botanically precise and encyclopedic enumeration of all hardy woody plants, Wyman's books were a distillation, written in layman's language. They drew heavily on Rehder's taxonomic investigations, as Wyman readily acknowledged.

BOTANICAL GARDENS AND ARBORETA ORGANIZE

The Arboretum's participation in the production of *Lilacs for America* and *Crabapples for America* was part of a new movement toward greater collaboration among botanical collections throughout the country. One of the prime movers for an organization of gardens was Robert Pyle of West Grove, Pennsylvania. A rose grower and a member of the American Association of Nurserymen's Committee on Botanical Gardens and Arboretums, he had lobbied for the establishment of the National Arboretum in Washington. Pyle rallied the managers of the Arnold Arboretum, the Montreal Botanic Garden, the Cornell Arboretum, the Dominion Arboretum at Ottawa, the Holden Arboretum at Mentor, Ohio, and others to organize as a special chapter of the American Institute of Park Executives in 1940. An association of zoological parks and aquariums had already formed under similar auspices. Donald Wyman served as chairman of the group during its formative years.

At the first meeting various goals for the organization were discussed: to enhance cooperation among member gardens, to assist in the development of arboreta in regions that did not have them, to make gardens' accessions records available to all members, to work with the nursery trade and with government agencies to introduce desirable new ornamental plants to the gardening public, to promote the accurate use of plant names, and so on. The activities of the fifty-member organization were halting at first, owing to the difficulties of travel during the war; the lilac and crabapple surveys were among the few concrete contributions of its first decade. Nonetheless, by 1950, the organization had gained sufficient momentum to sever

135. *Donald Wyman and Brian O. Mulligan, director of the University of Washington Arboretum (now Washington Park Arboretum) in Seattle, 1949. Both were active in the American Association of Botanical Gardens and Arboretums, then a new cooperative venture among North American public gardens and their managers.*

connections with the parent park executives group and become the American Association of Botanical Gardens and Arboreta (AABGA).

Recently founded gardens increased the membership. In the years before and after World War II, more arboreta were established, especially in the western United States. Similar to the Arnold in its dual civic and academic affiliation was the University of Washington Arboretum (now Washington Park Arboretum), founded in 1935 in Seattle. The Olmsted Brothers landscape architecture firm was called on to develop a plan for the grounds; it was to include a woody-plant collection that was comprehensive, yet selective in the larger genera. As in the case of the Cornell Plantations, much of the original construction of trails, roads, and buildings was achieved through federal and state work-relief programs. With its relatively mild climate, the Seattle institution could cultivate plants from a wider geographical range—including representatives from the floras of New Zealand, Australia, South America, and North Africa—than could the Arnold Arboretum.

Equally significant to its urban community was the establishment of the Strybing Arboretum in Golden Gate

Park. It was named for Helene Strybing, who left a bequest for the purpose to the park commissioners of San Francisco in 1939. The nucleus of the new garden was a forty-acre tract to be developed with indigenous California plants, but it subsequently expanded to include exotics, notably those from the Southern Hemisphere. Initiated at about the same time, the Fairchild Tropical Garden extended the geographic and climatic coverage of outdoor plant collections in the United States. Founded through the efforts of Colonel and Mrs. Robert H. Montgomery, who donated land in Coconut Grove, Florida, for the project, this planting featured an extensive palm collection and many South Pacific species introduced by David Fairchild during his long career as a plant explorer.

After the war, West Coast tree collections were joined by the Los Angeles State and County Arboretum in Arcadia. Started in 1947 on 127 acres, its goals were to educate the public and to conduct horticultural research. It was one of the first arboreta to create home demonstration gardens as a means of teaching visitors how to utilize newly introduced plants for their own grounds.

At the opposite end of the public-private spectrum was the Ida Cason Callaway Gardens at Pine Mountain, established in memory of his mother in 1949 by a Georgia industrialist as a sanctuary for native plants. Heavily planted with such woody-plant groups as azaleas, hollies, crabapples, magnolias, and dogwoods native to the southeastern United States, recreation and enjoyment were its chief functions. Although it was a nonprofit organization, the sponsoring foundation planned to sustain the garden by charging admission and operating adjacent resort facilities. While the commercial and recreational aspects of this garden might disqualify it in some minds from the ranks of true arboreta and botanical gardens, Callaway Gardens was intended to teach about plant life. To this end the plants were labeled and their origins recorded, while the foundation developed an extensive educational program in horticulture and conservation. Actually, Callaway anticipated by some thirty years the move to raise funds by providing such visitor amenities as food and shopping that is now quite common in botanical gardens, as well as in most other kinds of museums.

These are but a few examples of the many public

gardens added to the roster during the late 1930s and the 1940s. By 1954 there were sixty-five members of the AABGA. As the diversity in arboreta and botanic gardens grew, the value of experience and information shared through a national organization increased. The Arnold Arboretum moved from its passive role of leadership by example in Sargent's day to actively promoting the AABGA and its collaborative efforts, especially during the long period that Wyman managed its living collections.

SAX AND STATUS QUO

Karl Sax was the first person to serve as director of the Arboretum who was not a taxonomist. A soft-spoken man who preferred advising graduate students and doing experimental work, whether indoors or out, Sax took to his administrative role somewhat reluctantly. Fortunately, Wyman gave much-needed continuity to oversight of the grounds during the uncertain period of controversy.

Like Sargent, Ames, and Merrill before him, Sax occasionally prodded city authorities for improvement or protection of the grounds in Jamaica Plain. In 1951 construction began on the Casey Overpass; this was to carry Route 203, which followed the Arborway, above its junctions with South Street, Washington Street, the Boston and Providence Railroad, and the elevated MTA Forest Hills line. Plans called for a large amount of grading to be done on the Arboretum and Bussey Institution side of the Arborway. Sax intervened to have the high cut-stone retaining wall built to save trees that would be lost if the grading were done. After the wall was erected, a pendant form of forsythia was planted to soften its appearance and to ornament the entrance gate in early spring.

The overpass construction, and the widening of the Arborway that took place two years later, certainly changed the character of the approach to the tree garden on its eastern boundary. Once the overpass was constructed, the pedestrian route from the transportation hub at Forest Hills involved crossing through a constant flow of cars around a rotary and following the now walled-in passage along the Arborway off-ramp. The red oaks lining the Arborway began to suffer from frequent accidental impacts now that cars traveling at high speeds encroached on the planting space originally provided for them. A raised concrete median strip limited access to the Arboretum's main entrance by the southbound vehicle lanes. Like so many improvements in this era, aesthetics and the convenience of foot travelers were compromised for the sake of the automobile.

After Sax was placed in charge of the Arboretum in 1946, several vacancies occurred in the scientific and curatorial staff. Most were not filled. E. J. Palmer, herbarium collector and research assistant, retired in 1947 and returned to his home in Webb City, Missouri, the following year. There he remained an enthusiastic and active botanist working on the flora of his native state until his death in 1962. Two assistants in the herbarium and library, Leon Croizat and Vladimir Asmous, resigned in 1947 and were not replaced. The same year, Hugh M. Raup was transferred to the position of director of the Harvard Forest. In 1948, C. K. Allen, assistant in the herbarium and an authority on the Lauraceae, left because of an illness in her family, and the herbarium's curator, A. C. Smith, resigned to take a post at the U.S. National Museum. Clarence Kobuski was promoted to take his place and given editorship of the *Journal*.

Kobuski did receive some assistance with the addition of Shiu-ying Hu to the herbarium staff in 1950. Hu came to the Arboretum from her home in China in 1946 to study taxonomy with Merrill. She had already begun an intensive examination of the genus *Ilex* (hollies) while in China, and this became the research topic for her doctorate, which she received from the Graduate School of Radcliffe College in 1949. Elmer Merrill and Ivan Johnston sponsored her dissertation, and she had worked closely with all the herbarium staff while studying this woody-plant genus important to the Arboretum living collection. Hu would do much to further the Arboretum's specialty in the flora of China in following decades.

In the late 1940s, both Kobuski and Johnston turned their studies from American to Asiatic members of the plant families in which they were interested. The Arboretum's distinctness from the Gray Herbarium and its identity as an institution for studies in eastern Asian botany were thus strengthened. Although he was retired, Merrill

reactivated his investigations in Asiatic botany once his work on C. S. Rafinesque was completed, beginning research on the history of the plant life of the Pacific islands.

In June 1949, Rehder was pleased to be able to autograph just-released copies of his *Bibliography of Cultivated Trees and Shrubs* for fellow staff members. The book was the result of ten years of sustained endeavor, designed to assemble the accepted names (with synonyms) of all the plants included in his *Manual* and citations of the literature for each. In the days before on-line computerized databases, it was a remarkable resource for those needing to know the correct names for hardy ligneous plants. Rehder died quietly at his home only weeks after the publication of the bibliography; he was nearly eighty-six years of age.

During his industrious and productive career, Alfred Rehder had published more than one thousand papers of value to taxonomic botanists and horticultural science, and he had largely molded the Arboretum's herbarium and research program. Clarence Kobuski wrote of the indefatigable taxonomist:

Rehder's life had been lived just as he wished it. In his later days he once remarked that he was content with what he had accomplished, and that he felt throughout his whole life any decision regarding a contemplated change had been made for the better. . . . Much of his knowledge and experience has been preserved in his writings, but his keen observation and unerring judgment are gone. (Kobuski, 1950, p. 8)

David Fairchild wrote his colleague E. D. Merrill: "I am shocked to hear dear Rehder has joined the fellows on the other side and will never again smile at us and get out of his fertile brain some new facts about Chinese [and] other plants. I have always admired him through these many many years" (AAA, Merrill papers, 9 August 1949).

When, in the space of the next year or so, Oakes Ames and two long-serving botanists with the Gray Herbarium, Charles A. Weatherby and Merritt L. Fernald, died, Reed Rollins termed it the "end of a generation of Harvard Botanists." Rollins, the new head of the Gray Herbarium, observed that "these four men made botanical history at Harvard, each playing a special role within the precincts of the University in developing and maintaining a high

136. *Despite the uncertainty of the projected move of collections to Cambridge, the staff of the library, herbarium, and office posed for Heman Howard's camera in front of the magnolias in 1951. Front row: Miss Eldridge, Shiu-ying Hu (herbarium assistant), and Heriklea Yeranian (herbarium technician). Back row: Ethel Upham (herbarium assistant), Mrs. Craven, Lazella Schwarten (librarian), and Constance M. Gilman (business secretary).*

the opinion of counsel. After two years, the corporation passed a resolution modifying its policy for the Arboretum, essentially revoking the Bailey Plan as it applied to the administration of the institution. Nonetheless, it still determined to move the majority of the herbarium, the library, and some staff to a new building in Cambridge. The opponents of the Bailey Plan still questioned the legality of the move and pressed to have the matter brought before the courts. To effect this they incorporated in July 1953 as a nonprofit organization, the Association for the Arnold Arboretum. Their membership included many Harvard graduates, heads of sister arboreta, longtime supporters of the Arboretum, and some former members

order of research in taxonomic botany" (Rollins, 1951, p. 3). The Arboretum had lost another of its elder statesmen in May 1949 when John George Jack passed away at his farm in East Walpole, Massachusetts; he was eighty-eight.

At this point, with the impending move of herbaria to a proposed new building, Harvard botany faced not only the loss of these pillars of its research reputation but significant alterations of its physical facilities and working relationships. Moreover, as the shape of the new regime under the so-called Bailey Plan took form, many people connected with the Arboretum feared the role of the institution would be compromised. The concern was that with its resources divided, the living collections would become a lesser priority than the scientific program, connected as it was with the herbarium and the library. Some members of the Committee to Visit the Arboretum and some of its interested supporters sincerely questioned whether the proposed changes would constitute a breach of the trust under which the Arboretum was founded and developed. They insisted the matter should be put to a legal test.

Consequently, the Harvard Corporation put off plans for construction of the new building in 1950 and sought

137. *Although Karl Sax relinquished administrative duties in 1954, he remained on the staff until his retirement in 1958. Here he examines the results of experimental grafting techniques aimed at producing dwarf fruit-bearing apple trees, work that had interested him since his student days at the agricultural experiment station at the University of Maine in Orono.*

of Harvard's governing boards. Since the Harvard Corporation declined to seek a judicial decision on its own, the association petitioned the attorney general of Massachusetts to bring a test suit or allow them to do it in his name, since private citizens may not sue a charitable trust. He declined to bring suit, and planning was resumed to locate the new building at 22 Divinity Avenue in Cambridge, near the Farlow Herbarium, the biological laboratories, and the building that housed the Botanical Museum and the Museum of Comparative Zoology.

While this debate was carried on, Karl Sax tried to keep his institution on an even keel despite the tenuous position in which he found himself. With answers to certain questions deferred more or less indefinitely, he continued Arboretum programs as best he could. Nonetheless, it was a period of stasis. Wyman's conservative response to many of Farrand's suggestions was undoubtedly influenced by the attempt to maintain tradition in the face of rumors that horticultural excellence would be abandoned with the change.

The impending move of the herbarium and the library to new quarters in Cambridge, despite the opposition of a cadre of the Arboretum's most influential supporters, wreaked havoc with staff morale. It was difficult for the staff themselves to remain neutral. Although he sympathized with the systematic botanists' plight, Sax was not an "herbarium botanist" himself. Lacking expertise in their field, he felt ill equipped to lead his staff through the transition. More than once he urged that the Arnold professorship be filled and requested to be relieved of Arboretum administrative duties. Once this came to pass in 1954, Sax remained at Harvard as professor of botany on the Arboretum staff, teaching and pursuing his research until his retirement in 1958. However, the institution to which Sax's successor came had made great strides despite its past two decades of disruption and impending change.

CHAPTER EIGHT

SEPARATE BUT EQUAL;
CENTENNIAL AND BEYOND

Although reorganization was an inevitable result of the growth of the herbarium and the library, the Arboretum's new administrator forged a restyled unity in its wake. The division into two locations reinforced the dichotomy between botany and horticulture, yet expansion took place in both and their interface was encouraged at every opportunity. Research in systematic biology, enhanced by cohesion with the Gray Herbarium, extended into the New World with projects in the West Indies, Mexico, and the southeastern United States. The precedent set by Rehder in applying taxonomic principles to cultivated plants was upheld by appointments to horticultural taxonomist positions and participation in a national program to account for cultivars.

As the institution reached its one-hundred-year milestone, several factors created a period of disruption despite the proud retrospect. The retirement of Donald Wyman ended thirty-five years of continuous management of the living collections under a program geared primarily for homeowners and professional growers. At the same time, a much more diverse population found its voice during the social upheaval of the late 1960s and the 1970s, and began to demand more basic information on gardening and the natural environment. Nonetheless, the contemplation of past attainments eventually yielded guidelines for programs relevant to modern needs.

The ensuing reevaluation of the Arboretum's role, combined with historical scrutiny, led to the goals of restoring the living collections and the older arrangement of the grounds. The renewed questioning of humanity's impact on the global ecosystem required strengthening the living collections' representation of wild plant populations. A reaffirmation of its specialty in the vegetation of eastern Asia has likewise infused herbarium research programs. As part of a nationwide reexamination of the value of parks in the life of cities, the Arboretum has been one of many places exploring its original design by Frederick Law Olmsted.

138. Arboretum visitors enjoy the tranquil evergreen beauty of Bussey Brook valley.

RICHARD A. HOWARD
TACKLES THE MOVE

The person who took on the roles of Arnold professor and director in 1954 was Richard A. Howard, a taxonomist who had left Harvard one year before to chair the botany department at the University of Connecticut. Howard attended Miami University, receiving his bachelor's degree in 1936. With an interest in botany already well developed, he worked for a year before entering graduate school at Harvard in the fall of 1938. For four years he worked with both I. W. Bailey and Ivan Johnston on the pantropical family Icacinaceae. Utilizing new techniques to investigate the anatomy of the stem, node, and petiole, Howard combined the results with traditional taxonomic research to elaborate the evolutionary relationships of the group. For this work, Howard received his doctorate in 1942. During his graduate school years he spent summers at the Atkins Institution collecting herbarium specimens in the surrounding Santa Clara province of Cuba for the Arnold Arboretum. This field experience influenced his subsequent research concentration on the plants of the West Indies.

Since the United States had entered into World War II, Howard enlisted in the army after completing graduate work. He was assigned to Randolph Field in Texas, where pilots recently returned from fighting in the Pacific began to ply him with questions about jungle plants and what to eat to stay alive in the remote areas where they often found themselves. Like Elmer D. Merrill, Howard became involved in training and in the development of manuals for survival in tropical regions.

Howard worked at the New York Botanical Garden for a year after the war, and then returned to Harvard as an assistant professor to teach elementary biology. His exceptional teaching abilities and his excellence in taxonomic research were recognized by the staffs of the botanical units and the biology department. After four years, Howard was promoted to associate professor and made an Arboretum staff member; Sax, in particular, thought he had much to offer the institution. Besides his scholarly achievements, Howard was interested in horticultural plants and "gifted in the field of public relations" (AAA,

139. Among Richard A. Howard's first duties as director was to oversee the orderly transfer of more than half the Arboretum's books and pressed specimens to 22 Divinity Avenue. Once the move took place, the new director divided his time between Jamaica Plain and Cambridge, researching the taxonomy of cultivated plants and the flora of the West Indies in addition to governing the institution.

Sax to E. S. Castle, 30 October 1951). Nonetheless, the uncertainty that clouded Arboretum affairs in the early 1950s, and the offer of a tenured position at the University of Connecticut, induced the young taxonomist to move to Storrs. Soon, however, the corporation vote of January 1953 on the Arboretum's future governing policy cleared the way for Howard to accept appointments as Arnold professor, director, and professor of dendrology, effective 1 February 1954.

When he returned, there was no time for Howard to pause and become acquainted; the building at 22 Divinity Avenue was finished, and the Gray Herbarium had already moved in. The unseasoned director of the Arboretum not only had to oversee the transfer of materials and to make their arrangement congruent with the other collections but he had to create a new institutional order in the wake of change. As the one designated to carry out the reorganization, Howard also became the target of the displeasure of those who questioned the legality and wisdom of the move of specimens and books from Jamaica Plain.

As difficult as the Association for the Arnold Arboretum's attacks must have been to bear, they did have a positive impact in that Howard heard the group's point of view. He was repeatedly made aware of the Arboretum's traditional strengths by these prominent people, who cared so strongly about the fate of the institution that they were willing to spend years and a good deal of money fighting its disruption. As if to make the new director's first year even more of a test, two devastating hurricanes struck in early fall, blowing down or severely damaging some four hundred trees. Fortunately, Howard had the energy and perseverance to lead his staff through the change.

It took just three months to formulate specific plans for the reorganization with the advice of the staff and outside consultants. Policies determining what herbarium specimens and library publications were to remain at Jamaica Plain were drawn up and received approval from the Harvard Corporation. The next job was to segregate the collections that would stay in the administration building, not an easy task under crowded conditions. Then the specimens and books to be moved were marked with the seal of the Arboretum, and the books were carefully cataloged. Herbarium specimens in their steel cases and cardboard boxes, and all library materials, were transferred by professional movers in about one month's time. The Art Metal Company, which originally manufactured the herbarium cabinets, was called in to disassemble the cases in Jamaica Plain and reassemble them between the rows of the Gray Herbarium's cabinets on the upper three floors of the new building.

Since the Arnold and Gray herbaria had been kept in the Engler and Prantl sequence, each plant family of the two institutions was placed in close proximity by this initial phase of the move. By June 1955, Howard was pleased to report that for the first time in twenty years all the Arboretum dried-plant specimens were arranged in one sequence and contained in steel cabinets rather than cardboard boxes. Similarly, the library books were moved in on one-half of the new library shelves on the second floor of 22 Divinity Avenue and kept in order. For the next four years the combined herbaria staff worked to fully integrate the two collections to the species level while maintaining all normal herbarium operations. It took three

140. The building located at 22 Divinity Avenue to house the combined libraries and systematic collections of the Arnold Arboretum and Gray Herbarium, as well as facilities for some of the other botanical organizations, was built in 1954. From then on, Arboretum staff worked in Cambridge, as well as in Jamaica Plain. A four-story addition was made to the building in 1979, greatly changing the façade pictured here.

years for the librarian and her staff to completely unite, recatalog, and recondition the books of both institutions in their improved physical environment. Nonetheless, both collections were fully available for use during reorganization.

It was actually not until early in 1966 that the combining of scientific collections was affirmed by a decision of the Supreme Judicial Court of Massachusetts. After the Association for the Arnold Arboretum's first, unsuccessful attempt to convince the commonwealth's attorney general to bring suit, the group renewed its effort when a different attorney general was appointed. Because of this, for more than ten years there was a nagging doubt that the transfer of specimens and books from Jamaica Plain to Cambridge might have to be reversed.

The case was finally heard by the court in 1965. Two of the seven justices, who held positions on Harvard's governing boards, recused themselves. Among the five remaining, the opinion was three to two in favor of legality and the best interests of the Arboretum being served by relocating its collections to the new building in close association with the Gray Herbarium and other botanical collections of the university. After the final dismissal by the supreme court, the name of the Cambridge building was altered to Harvard University Herbaria. Although herbarium operations had already settled to a busy routine, and the research programs had continued and expanded, it was a relief to have the question resolved. With hesitation and uncertainty finally over, a firmer commitment to development of collections and programs for all Arboretum locations could be made.

As a result of the shift of materials and staff, there were now two major divisions of the Arboretum, each with its own focus within the larger picture of research and teaching in plant science. In Cambridge, the herbarium collections, books, and journals were those pertinent to the study of plants in the wild, as components of natural, biological systems. In Jamaica Plain were retained the literature and herbarium specimens devoted to advancing the knowledge of cultivated plants, this subject being considered most relevant to the maintenance of the living collections and to visitor interpretation at the time. This dichotomy notwithstanding, every effort was made to promote useful interaction between the botanical and horticultural units and to maintain a single identity for the institution.

In Jamaica Plain, working conditions were improved once the building was without much of the library and herbarium. Now that the two large rooms on the first floor were free of dried specimens, most of the horticultural library was moved into the western room. New lighting and reading tables were arranged to accommodate outside users (as well as the staff). With administrative offices relocated to the first floor as well, visitors in need of driving permits to view the Arboretum would not have to climb a flight of stairs to obtain them. The opposite hall was altered and redecorated as an area for educational exhibits, classes, and lectures.

The dried specimens of horticultural plants found their home on the first floor of the herbarium wing; its upper floors were devoted to storage of herbarium material in need of processing, much of it collected in the wild in eastern Asia. More staff offices, the library stacks, the photograph collection, and space for future expansion occupied the second floor of the original wing.

HERBARIUM RESEARCH PROGRAMS

From 1954 on, the Arboretum administrator had to be simultaneously occupied with the agendas of the Cambridge and Jamaica Plain locations, as well as with the supplementary living collections in Weston. A scientific program was built upon the expertise of past staff members and that of new appointees when openings occurred. Whenever possible, work that addressed both botanical and horticultural aspects of taxonomic groups was encouraged.

Merrill remained in Jamaica Plain rather than move to the new combined herbaria building. There he wrote up insights gained by examining manuscript records of early expeditions to the islands of the Pacific; these were published as "The Botany of Cook's Voyages, and Its Unexpected Significance in Relation to Anthropology, Biogeography and History" in *Chronica Botanica,* a periodical devoted to the history of biology, in December 1954. Little more than a year later he died. Howard summarized Merrill's career and accomplishments in an obituary for the *Journal of the Arnold Arboretum:* "Our path has been made clearer by his keen mind and his extensive record of publications. . . . In retrospect only one word can satisfactorily describe Elmer Drew Merrill and his contributions to the knowledge of botany. He was a builder" (Howard, 1956, pp. 212, 213). In the Philippines, at the University of California, the New York Botanical Garden, and the Harvard botanical units, Howard estimated that Merrill had been responsible for amassing botanical collections totaling more than one million sheets of herbarium specimens. These merely symbolized the extent of institutional and human resources he had encouraged and inspired.

Ivan Johnston, too, declined new quarters in the herbaria building, although he maintained the schedule of offering an upper-level course to Harvard students. Possibly the staff member most adamant in his opposition to the move of Arboretum specimens and books to Cambridge, Johnston kept somewhat aloof from the institution's affairs for the remainder of his career. Nevertheless, he persisted with taxonomic research in the borage family. Contract work with the U.S. Army took him to Panama several times during 1955 and 1956, after which Johnston resumed field and herbarium work on the flora of Texas and adjacent Mexico. He had nearly completed a manuscript treatment of the Boraginaceae of Texas when he died, suddenly, at his home in Jamaica Plain in May 1960.

I. W. Bailey officially retired only one year after Howard took charge, although the veteran plant anatomist remained active for another twelve years until his death in 1967. The Arboretum staff honored Bailey on his retirement by combining two numbers of the *Journal* in a special issue dedicated to him. For many years after his retirement, Bailey reorganized and curated the Arboretum wood collection. The wood anatomist turned his researches to a new subject—the comparative anatomy of the leaf-bearing members of the cactus family.

In 1959, Karl Sax became professor of botany, emeritus, having reached the age considered normal for retirement at that time, sixty-six. For the spring semester, Sax took sabbatical leave to accept an appointment as visiting professor of forestry at the University of Florida. For the 1959–60 academic year he was given a visiting professorship at Yale; following that, he received a Guggenheim Fellowship to continue his research at Oxford for a year. He eventually moved his residence to Pennsylvania, accepted other visiting professorships, and continued his research in genetics, with a few papers listed in the bibliography of the published writings of the Arboretum staff annually until a few years before his death in 1973.

Richard A. Howard's research took a formative step in his first year of Arboretum administration when the Institute of Jamaica appointed him senior botanist on a project to survey native vegetation on bauxite (aluminum-rich) soils. One goal of the survey was to determine what plants of economic value could be grown on areas where aluminum had been mined out. In the mid-1960s Howard turned to a critical study of the environment, composition, and biology of an unusual forest zone, the "elfin forest," on the mountains of Puerto Rico. He also extensively studied *Coccoloba,* a genus of the buckwheat family with many species occurring in the Caribbean as well as in Central and South America. His growing knowledge of West Indian plants was eventually focused on production of a flora of the Lesser Antilles, which involved many years of field and herbarium work.

Having trained under Bailey, Howard also continued the Arboretum's expertise in anatomical studies. The major program was a comparative survey of the anatomy of the node and petiole in dicotyledonous plant families. In the course of this work, as well as his systematic and ecological studies, Howard trained many students and professional assistants.

Not long after the start of the new director's term, in the midst of hurricane debris, moves, and renovations, Arboretum employees welcomed a new staff member, Carroll E. Wood, Jr., in September 1954. Wood, a Virginia native, entered graduate school at Harvard in 1947, after serving in the army during the war, and received his Ph.D. in 1949. He was an instructor at Harvard when Howard was teaching introductory biology, then joined the staff of the University of North Carolina. Personable and a good teacher, Carroll Wood was a taxonomist with considerable field experience in the southeastern United States. Once at the Arboretum, he was asked to oversee an adult education program in addition to conducting research based on the living and preserved collections.

Wood's interest in plants of the southeastern United States led to formulation of a plan, initiated in 1956 in conjunction with the Gray Herbarium, to compile a biologically oriented generic flora of this vegetatively diverse area. The project was initially supported in part by a private donor, then received successive grants from the National Science Foundation starting in 1960. The grant funds permitted the hiring of postdoctoral scholars who contributed to the project. Wood also took over editorship of the *Journal of the Arnold Arboretum* in January 1958, relieving Clarence Kobuski of a duty that, combined with his responsibilities for managing the herbarium, had kept the chief curator from his research on the camellia family.

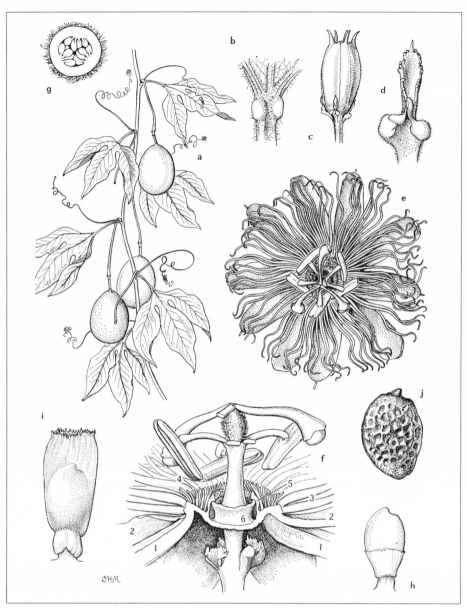

141. Passiflora incarnata, *plate to accompany one of a series of original taxonomic treatments of the genera of seed plants of the southeastern United States published in the* Journal of the Arnold Arboretum. *The generic flora was started not long after Carroll Wood joined the staff as taxonomist. The project aimed to combine modern biological information with traditional taxonomy for each genus occurring in this region of great diversity. Essential to the papers were illustrations prepared to document taxonomic characters and biological peculiarities. The legend for* Passiflora, *drawn by Dorothy Marsh, the project's first artist, is as follows: a, portion of stem with fruits, × 1/4; b, summit of petiole with glands, × 3; c, bud with bractlets, × 1; d, bractlet with glands, × 5; e, flower, color pattern omitted, × 1; f, flower, central portion in partial vertical section, × 2, 1–6 flower parts; g, ovary, diagrammatic cross section, × 6; h, very young seed with developing aril, × 3; i, older seed with aril, × 3; and j, seed with sarcotesta removed, × 4.*

For the generic flora of the southeastern United States, Wood presided over the production of precise illustrations and written treatments that were published in the *Journal*. Over the years, many capable taxonomists participated as full-time staff members or visiting fellows or as scientists with other institutions. Wood prepared treatments himself, contributed to others, edited all manuscripts, and maintained a bibliographic index. In this work, he and his collaborators broke with tradition to meld information from all fields of biology into a taxonomically accurate framework.

Another new appointee to the scientific staff was Lorin I. Nevling, an expert on the daphne family (Thymelaeaceae), who arrived to be assistant curator in the fall of 1959. His graduate work on this family was completed at Washington University in Saint Louis, where he also utilized the prominent herbarium collections of the Missouri Botanical Garden and contributed to its *Flora of Panama*. Despite the Arnold's well-known Asiatic interest, its holdings in Mexican and Central American plants (which started with Sargent's *Silva* investigations and continued through the work of Johnston) were substantial, especially when taken together with the Gray Herbarium material. Consequently, in 1966 Nevling joined in collaboration with the University of Mexico to investigate the flora of the state of Veracruz. One of the Mexican botanists involved was a former visiting fellow at the Arboretum. Nevling traveled to collect in the field and worked to identify many specimens sent by the Institution of Biology when he was in Cambridge. An innovation on this project was the use of computers to manage collection data and produce specimen labels.

The Arboretum scientific corps was expanded once again in January 1962 with the appointment of Bernice G. Schubert as associate curator. She had been working as a taxonomist with the Department of Agriculture in Beltsville, Maryland, on a project to determine the alkaloid content of as many species of plants as could be tested. She received a doctorate from Radcliffe College and had worked for the Gray Herbarium some years earlier, assisting M. L. Fernald in his work on an eighth edition of *Gray's Manual*. Her specialties were the difficult worldwide legume genus, *Desmodium,* and American species of

142. Bernice G. Schubert was appointed to the staff in 1962. A taxonomist specializing in the genera Dioscorea, Desmodium, *and* Begonia, *she also edited the* Journal *from 1963 to 1983. A frequent visitor to the Institute of Biology at the University of Mexico in Mexico City, Schubert collaborated there with botanists studying the legume family, of which* Desmodium *is a member.*

Dioscorea (yams), as well as *Begonia,* a genus with many cultivated taxa. After five years of editing the *Journal,* Wood turned this duty over to Schubert a year after her arrival at the Harvard herbaria.

Schubert rigorously edited the *Journal of the Arnold Arboretum,* the medium through which much of the institution's research results were communicated. She maintained its high standard for twenty years until her retirement, and she forged ahead with contributions to the knowledge of her genera of specialty.

In May 1963 the staff of the herbaria and Arboretum were saddened by the death of Clarence Kobuski, who had been with the Arboretum since 1927. He had worked closely with Alfred Rehder, and he performed the lion's share of herbarium management during the administrations of Ames, Merrill, Sax, and Howard. His dedication and good humor had been an invaluable asset during the integration of the Gray and Arnold herbaria and in the creation of an orderly management and record system for the distinct but intertwined collections that resulted. More than once, Kobuski's kind personality and sense of fair play smoothed over inevitable conflicts that arose over details of carrying out the reorganization.

While the scientific supervision of the combined collection was assumed by Nevling (whose appointment then became a joint one with the Gray's), the direct management of herbarium operations was assigned to Michael A. Canoso, as senior curatorial assistant and later manager of systematic collections. Canoso had worked at the Gray Herbarium since before the move to Divinity Avenue. Like Kobuski, he had a healthy view of life and was quite able to continue the diplomatic role, balancing the requirements of two directors and staffs.

Arboretum expertise in eastern Asian flora was kept up by two of Merrill's former protégées and a new appointment. Lily M. Perry finished identifying and labeling the New Guinean collections resulting from the fourth Archibold expedition of L. J. Brass, then published her continued findings on plants of that region. Although Perry retired in 1964, she carried on the Arboretum's Far Eastern involvement by starting a study of the medicinal uses of plants as reported on herbarium specimen data. *The Medicinal Plants of East and Southeast Asia* was published in 1980 by the MIT Press.

For some years, Shiu-ying Hu prepared treatments for a flora of China and kept up a card file of species of that vast region. Some of the groups she investigated were the genera *Philadelphus, Hemerocallis, Paulownia,* and *Clethra,* along with the families Malvaceae and Compositae in China. When her hopes to enumerate a flora of China were put off by the Communist revolution there, she turned to work on Hong Kong and the New Territories while the mainland was inaccessible. The new botanist who worked on plants of Asia was Thomas G. Hartley, a University of Iowa graduate with field experience in Papua New Guinea. He came to the Arboretum in 1965 and investigated members of the citrus family native to the Pacific region, as well as other problems of Papuasian vegetation.

Teaching students and training other botanists was an important component of the scientific program. In addition to their research, Howard and Wood regularly offered courses of instruction for Harvard students and supervised the work of graduate students. From the bequest of Martha Dana Mercer (received in 1962), some funds were used each year to subsidize scholars pursuing research related to the Arboretum's collections—living, herbarium, and library. Called Mercer fellows, their terms were anywhere from one to three years. For many it was a period of advanced training with Arboretum staff. Some came with topics of their own to pursue, whereas others worked on such Arboretum projects as the generic flora, investigation of wood anatomy, or the taxonomy of horticultural plants, horticultural practices, and garden management.

LIVING COLLECTIONS REFINED

Once the replanting of Peters Hill was complete in the early 1950s, the Arboretum concentrated on improving the content of the collections and coordinating plant introduction programs with sister institutions. The Arboretum horticulturist diligently scanned the literature, botanic garden seed lists, and published catalogs of selected nursery firms for notice of species and varieties not included in the collection; he then made every attempt to obtain plants, cuttings, or seeds. The cooperating-nursery program, started in 1940, served not only to get some of the Arboretum's best finds into circulation but also generated much goodwill on the part of these growers. Many companies in America and abroad were quite willing to donate samples, when requested, to the Arboretum.

Wyman's trip to Europe in 1951 had netted several hundred new clones for trial at the Arboretum. This material had to come as living specimens, since seed would not yield plants true to the characters for which the clones were selected. Yet, bringing many of them in from overseas was no simple matter. Certain groups that had long been completely forbidden entry into the United States by the Department of Agriculture were now allowed in, but only after being held for varying lengths of time in quarantine. After a period of observation of their health, lack of disease organisms, and noninvasive character, the plants would be released to the Arboretum. The institution joined in a cooperative effort with other American botanical gardens and arboreta to obtain plants from overseas through the USDA's own plant introduction department.

Many lots of seed still came in from botanical exploration in wild areas through the continued policy of supporting reliable collectors in the field. In 1955 seed from northern stations and high mountains in Japan arrived from one institution, while extreme southern Argentina was the source of another shipment in 1958. From a four-thousand-mile trip to work on the flora of the southeastern United States, Carroll Wood and his assistant brought back seed of *Liriodendron, Calycanthus, Philadelphus, Malus, Amelanchier, Robinia, Lonicera,* and *Diervilla* for the Arboretum.

All new accessions were considered to be on trial no matter what their source. The tests were for cold hardiness, correct identity, and, in the case of cultivars, distinctness or ornamental attributes superior to others of their kind. For seedlings the test could last from five to ten years, depending on their growth rate. From its start at the greenhouses, nearly every plant was moved out to a protected nursery for its first year or two, then to the Case Estates. There it was held without winter protection until it reached flowering age.

If a taxon failed to withstand the Weston climate, it might be tried again from another source where colder temperatures prevailed. While the number of kinds of plants in the collection averaged six thousand in these years, Wyman often stated that there were another two thousand, for trial or for replacement of aging plants, in the nursery system. He summed up this aspect of his work as follows: "If a plant does not prove hardy here, or if a new variety does not demonstrate clearly that it has outstanding ornamental qualities, it may never reach the Arboretum plantings and only necessary observations and notes will be made about it." The records of failed plants, and of those tried but not kept, were a valuable source of information in their own right.

Record keeping was key to the horticulture staff's ability to manage the thousands of plants, and for some thirty years Heman Howard kept the records. His participation in the surveys between 1938 and 1940, and in the subsequent revisions of each of the maps, built his experience with the collections to the point where he field-checked and updated one-third of the 108 records maps each year. He made sure that every plant was tagged with both record and display labels, replacing an average of one to two thousand per year. In addition, Howard and Wyman systematically documented Arboretum plants with photographs of flowers, fruit, and habit. The photograph collection became another extensively used resource for publication and teaching.

In 1957 the horticultural staff gave special attention to plants introduced into cultivation for the first time by the Arboretum during its eighty-five-year history. A survey and notation of the records was made to identify them or their descendants, and special yellow-painted labels were attached to the plants on the grounds. Once these specimens were marked by the unique tags, visitors were interested to find that some of the most common and sought-after woody plants had been brought to American gardens by the Arboretum. It was now easier for the staff to keep track of these significant plants as well.

Wyman traveled to Europe again during the spring and summer of 1965 "to select from many sources, including botanic gardens, private gardens, nurseries, and the wild, plants worthy of introduction into the United States" (Howard, 1966, p. 329). A glimpse of how this modern plant-hunting journey proceeded may be gained through an excerpt from a letter written by Wyman from Boskoop, the Netherlands:

Spent three full days with Harold Hillier [nursery owner of Winchester, England]. One of these included a 12 hour session going through everything he grew with Rehder['s *Manual*] perched on my knee and his catalogue and notes on his. That was a most exhausting deal . . . all in his office with 30 minutes out for lunch. It resulted in noting 280 species that he can eventually supply us. Some will turn out as synonyms although we tried to exclude them as we went along, and some he may not have. I am convinced he is the most knowledgeable nurseryman in the world as far as woody plants are concerned. He grows about 13,000 different things. His knowledge, help and friendship are worth much to us. We

143. Hamamelis intermedia *'Arnold Promise.' This witch hazel was given the name Arnold Promise in compliance with the new International Code of Nomenclature for Cultivated Plants of 1958. The original plant was one of several started from seed collected from a Chinese witch hazel (H. mollis) in the Arboretum in 1928. In 1945, Alfred Rehder noticed that many of these original seedlings combined the attributes of Chinese witch hazel and Japanese witch hazel (H. japonica), which grew near the seed parent. Realizing that they were hybrids, he gave them the name Hamamelis intermedia. By the late 1950s, the staff had observed that one* Hamamelis intermedia *planted adjacent to the administration building consistently bloomed for a longer period and bore larger flowers than others of its kind, thus meriting its distinction as a named cultivar.*

also promised him some material and Kew asked for about 100 things. This give and take makes for good relationships. (AAA, Wyman to Howard, 28 April 1965)

The trip yielded plants or propagules of some 930 species and varieties in 119 genera to be grown and evaluated at the Arboretum.

Karl Sax remained active in experimental horticulture, utilizing Arboretum collections for about five years after he stepped down as director. The progeny of crosses made years before were still reaching maturity, and each year a few were found worthy of naming and distribution. Sax's hybrids utilizing *Malus sargentii*, for example, were favored for their dwarf habit and brilliant foliage, which added interest even when not in bloom. With the retirement of Sax the comprehensive program of woody-plant breeding was dropped from Arboretum activities, however. Only very rarely after that was any hybridization undertaken by the staff in the woody-plant collections, although the topic was often reiterated as worthy of the institution's collections and human resources. In the main, other organizations and agencies outstripped the Arboretum in attempts to breed woody plants for ornament or reforestation; accordingly, the Arnold became a leader in establishing systems for accurately describing and naming new selections.

HORTICULTURAL TAXONOMY AND NOMENCLATURE

That every Arboretum plant be accurately identified and correctly named meant keeping up with the latest procedures in nomenclature. To this end, Richard Howard attended the nomenclature sessions of the Fourteenth International Horticultural Congress in the Netherlands in August 1956. There, experts from around the world grappled with the difficulty of applying the code of botanical nomenclature to ornamental and crop plants. It was not the first such meeting to discuss the problems of applying the rules worked out for wild plants to forms selected under cultivation. The Arboretum and similar organiza-

tions had long needed uniform rules, together with the means of recording the name, origin, and distinguishing features of cultivated plants.

Especially troublesome were the numerous varieties selected for particular traits that made them worthy of cultivation by nursery operators and other growers. No uniform standards had been used in classifying and naming these cultivated entities. Only recently had the term "cultivar" been coined to refer to artificially selected varieties. The nomenclature committee began work on a revised International Code of Nomenclature for Cultivated Plants. At the Netherlands conference, Howard agreed on behalf of the Arboretum to compile a directory of listings of cultivars. *Crabapples for America* and *Lilacs for America* were examples of the kind of roster needed for groups in which breeders and growers were continually selecting additional varieties.

With the new commitment to documentation of cultivated plants, one of the herbarium assistants was assigned the task of locating the type specimens of horticultural taxa described by Rehder and Sargent in the herbarium. Likewise, the decision was made to update the card index of references to original publications of cultivated plants that Rehder had compiled for his *Bibliography*.

In 1958, Kobuski wrote a paper on the horticultural herbarium, urging the development of such collections and the importance of using pressed specimens to document the work of horticulturists, nurserymen, cytologists, and geneticists. To further the Arboretum's role in this field, Burdette L. Wagenknecht was appointed the Arboretum's first horticultural taxonomist in July 1958. He was given responsibility for the herbarium collections in Jamaica Plain and urged to carry on studies of cultivated woody plants. Specifically, Wagenknecht was to see that every living plant in the Arboretum was represented by flowering and fruiting stages in the herbarium. Emphasis was to be given to all the accessions added since 1949 (by which time Palmer and Rehder were gone), and their names were to be checked with Rehder's *Bibliography* and the updated index.

A revised code for horticultural nomenclature was adopted by the next International Horticultural Congress, held in Nice in 1958, and several American organizations made a concerted effort to put the code's recommendations into effect. To avoid duplication and disorder in the naming of new cultivars, these organizations decided to act as registrars, or registration authorities, for ornamental plants. The effort was coordinated by the American Horticultural Society and the American Association of Botanical Gardens and Arboreta. The AHS enlisted many plant-fanciers groups, such as the American Begonia Society and the National Chrysanthemum Society, to tackle cultivar registration in their respective specialties. Member institutions within the AABGA took responsibility for woody genera for which they held extensive collections or had particular expertise.

The Arnold Arboretum agreed to be registrar for nine genera (*Chaenomeles, Cornus, Fagus, Forsythia, Gleditsia, Malus, Philadelphus, Pieris,* and *Ulmus*) and for all woody taxa not yet assigned to any organization or institution. This arrangement was to be tried for two years; it was subsequently renewed, and the Arboretum fulfilled this role for more than two decades. In acting as registrar, the Arboretum also accepted the responsibility of publishing lists of cultivars for which it received registrations.

While many of the institution's botanical staff, including Howard, applied their expertise to the identification and correct naming of Arboretum accessions, new appointees to horticultural taxonomist positions had this work as a priority. Peter S. Green, former scientific officer at the Royal Botanic Garden, Edinburgh, joined the staff in January 1961. His specialty was the olive family (Oleaceae), which includes such woody genera of importance to the Arboretum living collection as lilac, forsythia, privet, ash, and fringe tree. While Green addressed many problems in the Arboretum, and handled the registration of cultivars after Wagenknecht resigned in June 1961, he also worked on such Asian and African tropical genera as *Osmanthus* and *Jasminum,* for which the Arboretum collections in Cambridge had ample material and library resources.

In 1963, Theodore R. Dudley, a recent graduate of the University of Edinburgh, joined the staff. He continued research on the nonwoody but ornamental genus *Alyssum,* but also tackled problems in two families of importance

in the Arboretum: the Aquifoliaceae, the holly family, and the Caprifoliaceae, which includes honeysuckle, viburnum, elderberry, and weigela among its extensively cultivated genera.

Some visiting fellows also concentrated on certain genera to produce much-needed checklists of cultivars. During the 1960s the Arboretum brought out comprehensive surveys for flowering quince (*Chaenomeles*), elm (*Ulmus*), *Weigela, Lantana,* and dogwood (*Cornus*), in addition to publishing the registrations of scores of cultivars of other woody plants. Along with this work on improving the documentation of plants in cultivation, the herbarium in Jamaica Plain was built up as a repository for all cultivated plants according to the plan devised in 1954. By 1962 the collection of pressed horticultural plants and the staff working with them had grown to the point where expansion was necessary. New steel cabinets were installed on the second floor of the herbarium wing in Jamaica Plain, doubling the specimen capacity and work space.

The two horticultural taxonomists resigned in fiscal year 1965, however. Green joined the staff of Kew, and Dudley left to accept a position at the United States National Arboretum. Gordon P. DeWolf of Georgia Southern College was the next horticultural taxonomist, engaged two years later. In addition to master of science degrees from two institutions and a doctorate from Cambridge University, DeWolf had accumulated varied work experience (at an agricultural institute in Costa Rica, in the Bailey Hortorium at Cornell, for the Colonial Office in London, and at Kew) before going to Georgia. For the Arboretum he was to work on problems of horticultural taxonomy and nomenclature and to continue research on his specialty, the pantropical genus *Ficus,* of which one species is the edible fig and many are houseplants and ornamentals of warm climates.

PROPAGATION SCIENCE, NEW FACILITY

The propagation of woody plants was another topic of investigation given continued emphasis from Merrill's administration on. Before the propagator Richard Fillmore left the Arboretum in 1952 to work in the commercial side of growing plants, he published several papers on new methods applied to difficult-to-propagate plants. For two years, the propagating unit was headed by Lewis Lipp, whose lasting contribution was initiating the use of polyethylene film in various propagating techniques.

Roger G. Coggeshall, who had joined the Arboretum staff in 1950 as assistant propagator, took over from Lipp. This graduate of the Stockbridge School of Agriculture at Amherst continued to experiment with various propagation techniques. Among many procedures tried and documented in Arboretum records were improving or hastening seed germination by treatment with varying periods of cold or heat, using hormones to induce the rooting of cuttings, and comparing rooting success when such factors as time of collection of scion, holding period in cold storage, or the rooting medium were varied. The chief propagator presented his findings at professional meetings and published his results. When he, too, resigned in 1958 to go into business, his place was taken by Alfred J. Fordham, an Arboretum staff member since 1929.

Fordham had begun work as an assistant to Judd in 1930; in 1936, he took a one-year leave to work at the Royal Botanic Gardens, Kew, to further his training in horticulture and his knowledge of plants. After World War II, Fordham became assistant superintendent of grounds under the recently arrived R. G. Williams. He welcomed the opportunity to work in propagation once more in 1958, and he headed this aspect of the Arboretum's work for the next twenty years, carrying on extensive experimentation and publication on woody-plant propagation.

The need for more-modern propagation, laboratory, and classroom facilities had been suggested by the horticultural staff since the early 1950s. As a result of two events that occurred not long after Fordham took up the duties of chief propagator, construction of a new greenhouse complex became feasible. In 1959 the Commonwealth of Massachusetts started eminent domain proceedings to obtain most of the Bussey grounds for its department of health. When the possibility of losing the Bussey land became a certainty, planning for a new propagation unit began. A site on the parcel abutting Centre Street, purchased from the Adams Nervine Asylum in the 1920s, was chosen, since its use would not disturb existing collections. While some native trees had seeded in, few accessioned plantings had been made there. Because the land was not included in the agreement with the City of Boston, it could be kept closed to the public for the sake of security; there would be ready access for deliveries, and for visitors to its proposed classroom, from Centre Street.

Fortuitously, in the spring of 1960 the Arboretum re-

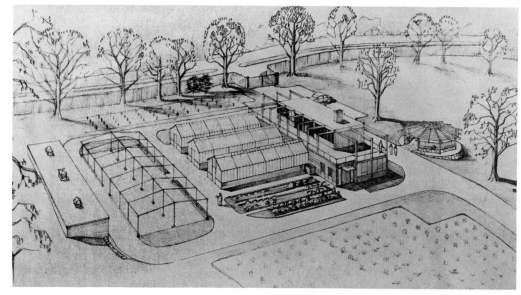

144. Architects' proposal for the Charles Stratton Dana Greenhouses by the firm of Griswold, Boyden, Wylde and Ames, 1960. A gift by Martha Dana Mercer as a memorial to her father made construction of these new facilities for propagation, laboratory work, teaching, and display of the bonsai possible. The greenhouses, cold-storage unit, nurseries, and special growing areas were located on land that had been purchased from the Adams Nervine Asylum in 1927.

ceived notice that it had been designated one of three beneficiaries of a perpetual charitable trust established by the will of Mrs. Martha Dana Mercer. A native of Boston, Mercer had long been a friend of the Arboretum until her death in February 1960. Since use of income from the fund was unrestricted, it was applied to much-needed improvement of physical facilities and increases in research staff.

The architectural firm of Griswold, Boyden, Wylde and Ames was selected to provide plans for a new propagation unit. Since Wyman, Williams, and Fordham worked closely with the architects, the resulting design was quite satisfactory. A ground-breaking ceremony was held on 12 May 1961 and construction completed by March of the following year. Named the Charles Stratton Dana Greenhouses in honor of Mercer's father, who also had a lifelong interest in the Arboretum, the unit's main building consisted of a head house finished with redwood panels and exterior steel beam construction, and three glasshouses (with room for a fourth when the need arose). In addition to ample work space for propagating activities, it included walk-in cold rooms, a laboratory, a classroom, and plenty of storage space. It was designed for easy accessibility, with well-lit, comfortable working conditions, and flexibility of use. Most operations were on one floor, with basement storage readily available by a motorized dumbwaiter. The power plant was designed to heat twice its original capacity should the glasshouses be expanded on modular units. An emergency generator and automatic controls for heat and ventilation were trouble-saving innovations.

A cinder-block cold-storage unit, set into a hillside, provided further controlled growing space and an area for wintering the bonsai. The dwarf trees were also provided a hexagonal, cedar slat-house for display during the growing season. Nursery areas and display beds were prepared on adjacent ground.

With much of the preparation having been made during the previous fall and winter, hundreds of plants, pieces of equipment, people, records, and supplies were moved into the new facility in the spring of 1962, in time for an open house held on Lilac Sunday. The hedge collection was lifted and relocated to a level site near the Adams

145. *Alfred J. Fordham, who started working at the Arboretum in 1929, was chief propagator from 1958 until his retirement in 1977. Among the many studies he conducted were those on seed dispersal, seed dormancy, and obtaining dwarf conifers from seed or cuttings of abnormal growths known as witches'-brooms. Fordham, a founder of the International Plant Propagators Society, readily shared his knowledge through lectures, workshops, and publications.*

Nervine boundary, down a pine-clad slope from the greenhouse complex. A new access road to the nursery area from the Arboretum's Bussey Hill Road was built between the *Zelkova* and *Viburnum* collections. The entire four-and-a-half-acre parcel was fenced and could be locked up to protect the bonsai and small plants, which had not been possible before. Howard summarized the prospects for the new greenhouses: "This development has met a long-standing need of the Arboretum for modern greenhouses with experimental facilities. Their completion and occupancy make possible continued contributions by the staff to the study of ornamental plants hardy in New England" (Howard, 1962, p. 443). Experiments in plant propagation and research in anatomy were henceforth conducted at the new facility.

MANAGING THE GROUNDS

While efforts aimed at growing a complete, accurately named collection, leading the horticultural community in applying scientific principles to naming plants, and conducting experimental propagation were carried on, Wyman also had to maintain the health and beauty of the Arboretum collections.

Cleaning up after the destructive forces of New England weather periodically occupied Arboretum caretakers, and sometimes the damage wrought significant changes to the landscape. Although two 1954 hurricanes ("Carol" on 31 August and "Edna" in late September) caused nowhere near the demolition of the 1938 storm, their impact was great enough to impress the new director with the uncertainties of growing a garden of trees. Within days, however, Wyman and the grounds force had pulled 25 percent of the blowdowns back upright and staked them. All the plants in danger of dying were repropagated. Cleanup operations were boosted by the purchase of a brush chipper and two new chain saws.

In the fall of 1955 heavy rains caused some of the worst flooding ever experienced at the Arboretum. Although a hurricane hit, it caused less damage than the 1954 storms. One notable loss was a large old white oak, under which the memorial service for Charles Sargent had been held three decades earlier, that leaned toward Bussey Brook. The brook channel was so seriously eroded by this, and by more flooding two years later, that in 1958 the Boston Parks Department reset the large stones that lined the streambed at its entrance to the Arboretum near Walter Street and along the section through the rhododendrons.

For Wyman and Williams, the methods of maintaining the grounds and the health of the plants were subjects of experimentation and topics for publication. Reducing costs by eliminating hand operations whenever possible was a constant consideration in choosing methods to control weeds and pests. In Sargent's and Ames's day, weeds were kept in check by hand hoeing. Rising costs, as laborers sought a decent standard of living after World War II, meant that institutions with a limited budget like the Arboretum had to find "labor-saving" practices. From the time that Wyman first initiated the use of mulch to reduce weed growth at the close of the war, he and his superintendent experimented with different materials. The ideal mulch was one that improved soil texture, yielded nutrients, subdued weeds, retained soil moisture, and was fire resistant. Each new substance was tested in selected areas before being generally applied. After the double hurricanes of 1954, the purchase of a chipper yielded woodchips that were tentatively applied as mulch; spent hops, however, which were available free for the hauling from Jamaica Plain's breweries, were preferred. But by 1957 the closing of these local businesses meant hops were no longer readily available. Wyman then turned to another local industrial by-product, cocoa shells from a chocolate factory in Dorchester Lower Mills. After only a few years of testing, the Arboretum staff found cocoa shells so satisfactory that their recommendations created a commercial demand; once more, free supplies were no longer available. Woodchips ultimately came to be the ubiquitous material applied to the Arboretum's beds and to the ground immediately surrounding young plants, since they could be obtained relatively cheaply as a by-product of the institution's own pruning and tree-removal operations. Only very recently has the aesthetic and nutritional impact of this practice been questioned.

In the 1950s, chemicals to control weeds and to destroy insect and fungal pests were widely used throughout the country, and the grounds management staff readily adopted them. Using the latest spraying equipment and knowledge of the life cycle of the pest involved, new chemicals were applied at the season of greatest vulnerability of the pest. If effective controls were not known, the Arboretum sometimes tested new materials. When fire blight threatened the crabapple collection, a new antibiotic, donated by the manufacturer, was used to curb this disfiguring fungal disease for a few years. Resistance to such diseases became an increasingly important criterion in choosing the best cultivars in a particular group.

As soon as a "biological control" for Japanese beetles became available, it was tried on the grounds in the vicinity of plant groups usually hardest hit by the pest. Fertilizers and such nutrient additives as lime, cottonseed meal, and castor pomace were rigorously applied on a rotating basis. The superintendent and his assistant kept

146. The shrub collection. To reduce the cost of maintenance, Wyman tried a variety of organic materials as mulches. In the shrub collection, weeds were kept in check by the use of mulches from World War II onward. In the 1960s this was still an area of great diversity and interest among the collections, despite the somewhat formal arrangement of its rows.

records of what materials were used where and when each year so as to evaluate their efficacy over the long term.

Fortunately, the City of Boston Parks Department, under commissioner Frank Kelley, took a renewed interest in the upkeep of Arboretum roads starting about a year after Howard's arrival. Each summer for four years a section of the cobblestone gutters was cleaned of weeds and debris by city workers; once cleaned, the gutters were kept free of weeds by the chemical controls then in use. The extra work on this seeming detail markedly improved overall appearance of the grounds. The parks department also repaired many benches and patched some sections of the asphalt drives and walks during the 1950s.

Between 1960 and 1964, all the Arboretum drives were resurfaced under authority of the city parks department. In 1963, construction also began on an all-weather road to the top of Peters Hill, something that Wyman had desired for many years. The dirt track established in Sax's day was inadequate, muddy in wet weather and impossible to plow when covered with snow. In the interest of

visitors' ability to partake of the view from the now ubiquitous automobiles and tour buses, the parks department executed a well-drained, paved road that curved slightly as it approached a broad "turnaround" at the top. The grade was steeper than Olmsted and Company would have allowed in planning for horse-drawn vehicles of their day. Except for this, and the bareness of the broad, paved circle on the summit, the new approach road agreed in style and detail with the rest of the tree garden. On both sides of Bussey Street the Arboretum added new gates to enhance the approach to the Peters Hill area.

Wyman continued to adjust certain areas in the late 1950s and in the 1960s. The rhododendron collection was expanded onto the lower slope of Hemlock Hill across Bussey Brook where hemlocks had been destroyed in 1938. The hickory (Carya) collection and nearby edge of Central Woods came under revision during the 1958 season. More of Sargent's original group plantings of Carya were removed. To Wyman, who was used to specimen trees, they were still too densely planted and represented more than enough duplication. Where many shrubs had been taken out eight years earlier, he installed a planting of *Ilex pedunculosa* (long-stalk holly), a serviceable broad-leaved evergreen introduced when Sargent journeyed to Japan in 1892. A few other rare and unusual shrubs remained from the earlier bed, and more were added.

With so much spring bloom to see in the main Arboretum—along Meadow Road, Bussey Hill Road, and Valley Road—a recurring problem was to entice visitors to venture over onto Peters Hill when the crabapples, pears, hawthorns, and mountain ashes created their own spectacular displays. The subdued green of the conifer collection and of Hemlock Hill disrupted the momentum of those seeking spring color, and Bussey Street was a physical and psychological barrier. Although Wyman and other Arboretum staff always took tour groups to Peters Hill when bloom was at its height, the "casual visitor" apparently often overlooked the area. When Oakes Ames was supervisor, a few crabapples were planted along Hemlock Hill Road near the Bussey Street entrance in an attempt to lure visitors toward the gate and across the street to see more. In 1958 and 1959, the approach to the Peters Hill circuit road, opposite the Bussey Street gate, was regraded so that more of the crabapple-covered hillside could be seen from the public street. Selected crabapples were planted on the newly graded land, extending the collection closer to the most frequently used entrance.

After the relocation of propagation operations to the Dana Greenhouse complex, various improvements were made in its immediate vicinity and elsewhere on the grounds. In 1962, a collection of American hollies was donated by the children of Wilfred Wheeler, a recently deceased breeder from Falmouth, on Cape Cod. They were planted next to the approach to the greenhouses from the Arboretum.

A few years later, Wyman and Williams decided to try their hand at the ancient art of espalier (training a plant to grow against a flat surface) by starting with ten plants trained against the blank walls of the new cold-storage house. They used several different plants, such as yew, pyracantha, and flowering dogwood, and varied the forms to demonstrate the range of possibilities of this specialized pruning technique. Because of their success and the interest shown by visitors, some sixteen more plants were soon espaliered on the cedar fence that separated the nursery area from Centre Street. In front of the cold-storage unit and flanking the head house, beds were planted with the kind of plants that had once been grown in the rock garden at the Orchard and Prince Street property: naturally low-growing shrubs of alpine regions.

ANOTHER LANDSCAPE CONSULTANT

In the spring of 1965, Seth Kelsey was retained as a "horticultural consultant" to look over the Arboretum and make recommendations for certain problem areas. Kelsey was a nursery owner and the son of Wilson's colleague Harlan P. Kelsey; he also served on the Arboretum visiting committee. His twenty-four-page report, based on his observation and discussions with staff members, was submitted to the director in May. Kelsey considered the document to be preliminary, the basis for continued thought and planning, rather than a final set of directives. Like Beatrix Farrand's many suggestions, some of Kelsey's were appropriate and some were not.

In the report Kelsey showed little understanding of the integrity of the original, sequenced plant groups and the native character of supporting plantings, and he tended to suggest the liberal use of broad-leaved evergreens and ground covers. The goal was beautifying and "adding interest" rather than developing the collections. For example, he would have added no less than thirteen masses of about eleven kinds of broad-leaved evergreens along both sides of Meadow Road between the tulip trees and the ponds. With all this "evergeen color and interest added," who would ever look at the katsuras, the lindens, the azaleas, or the maples? Besides, what makes the *Kalmia* and *Rhododendron* collections at the base of Hemlock Hill so dramatic and pleasing is that very few broad-leaved evergreens are seen until that part of the Arboretum is reached by following the planned drive.

Nonetheless, a few of Kelsey's recommendations were put into effect. Some could well have been those suggested for his report by the Arboretum staff. The dwarf

147. *A terraced dwarf conifer garden was established on the slope below the Dana Greenhouses in 1967. Laid out on the recommendations of the horticultural consultant Seth Kelsey, the collection was also the result of Al Fordham's interest in compact and slow-growing forms of conifers. Although the plants looked stark at first in their beds covered with crushed stone, they soon filled in to form a rich interplay of color, texture, and shape.*

conifer garden, for example, had been under discussion prior to 1965. With refinement the report could have served as the basis for fund-raising. Whatever their stimulus, several projects on the grounds were started the following year.

Over the fifty-odd years since its formation, Azalea Path's once level, grassy expanse had slumped sideways downhill, making for an unkempt appearance and an unsafe surface for walking. To remedy this, the pathway was leveled and a fieldstone retaining wall set into its upper side. The unobtrusive wall, two feet high, was made up of single stones for the most part. It edged the planting bed and would prevent slumping of the path in the future. Ever since the establishment of the Meadow Road azalea border, the yellow-flowered Scotch brooms (*Cytisus*) added by Wilson in the 1920s outshone the few remaining azaleas. The epithet Azalea Path had been dropped, and the area was referred to as the *Cytisus* beds in 1965, even though many unusual Asiatic plants still thrived there.

Careful scrutiny was given the *Viburnum* collection in the mid-1960s by the horticultural taxonomist Theodore Dudley. He worked at verifying the identity of Arboretum plants in Jamaica Plain and Weston, while Wyman and Kelsey troubled over the health of the ones in the main collection on the north side of the junction of Bussey Hill and Valley Roads. Five years of drought between 1962 and 1967 caused poor performance and disease susceptibility in many Arboretum plantings. Because the *Viburnum* collection was apparently underlaid by ledge, the plants there were more vulnerable to drought. As a result, many had weakened and succumbed to disease. From the several alternatives suggested by Kelsey, Wyman chose to start a supplementary *Viburnum* collection under the walnuts and oaks, not far away, across the road. Several masses, each an assortment of species and varieties, were placed among the large trees. The effect was not unlike the way azaleas had long been arranged under the oaks, further south on the same side of Bussey Hill.

A major innovation was the terraced garden of dwarf and slow-growing conifers constructed on the slope below the new bonsai house. Plants that remain small and well-shaped without high maintenance were of prime interest to the postwar generation of homeowners. A start had been made in expanding the dwarf conifer collection into a cleared area of Central Woods near the original collection, where many of the oldest plants had actually grown quite large. More recent selections of spruce, fir, pine, hemlock, arborvitae, and other evergreen cone bearers promised a greater variety in color and shape (and more compact dimensions) in the group. In 1963 William Gotelli of New Jersey gave the Arboretum propagating material for some eighty-four cultivars of dwarf conifers. A few years later, when more were obtained, it was evident that the Central Woods site would not be large enough for the collection. A protected location within the fenced greenhouse complex was chosen for a new garden of diminutive evergreens.

The terraces were planned by Kelsey and installed by the Arboretum staff, using a small front-end loader to place stones that came from the Case Estates. Among the wide, stepped planting areas, walks were set in and stairs of railroad ties were formed. In 1967, 125 new dwarf conifer varieties were planted in the beds and mulched with crushed gray stone. A few years later, two more levels were added at the lower portion of the slope to accommodate another 25 specimens. Although the interesting little plants looked stark in their new surroundings, they soon grew in. Fortunately, a grove of white pines planted in 1951 blocked out the view of Centre Street. More large white pines were dug from the woods at the Case Estates and set on the other side of the garden, between it and the fence separating greenhouse area and Arboretum.

For a collection of plants with such varying colors and odd forms, many of which resemble cartoon caricatures, it is hard to imagine a more compatible setting than that created in 1966. Within the limits of space, materials, and cost, the installation's style was naturalistic enough to suit the plant materials. Its proportions, too, were suited to the plants and to its surroundings. Framed by masses of pine, subtly ornamented by a small selection of flowering and deciduous materials, the collection's eccentricity was neatly tucked into the hillside out of view from the rest of the Arboretum landscape. Nevertheless, it delighted all who entered it.

Some of the far reaches of the property received attention in the late 1960s. These three areas had not been part of the original acreage, and each contained ambiguous mixtures of plants. Nor were they accessible by obvious paths. Although Harvard owned the triangular piece of the land left when Centre Street was straightened and widened in 1931, Boston's department of public works had kept the right-of-way. A building used for equipment storage, a stone wall, and abandoned pavement interfered with Arboretum use of its land beyond. In 1966 the public works department agreed to turn the right-of-way over to the purposes of the Arboretum. Following this, the pavement was torn up, the roadbed excavated, and the area was filled with topsoil and compost. After two years of preparation, the former right-of-way was seeded to grass, making the space ready for expansion of the conifer collection. The old stone wall was not removed, however. More crabapples and other spring-flowering species were planted near the street to enhance the display along this approach to the Arboretum.

Another small piece of land was officially added in 1966, although it had been used for Arboretum purposes for some time. After the commonwealth took most of the Bussey grounds, a five-acre piece that included the site of the mansion was transferred from the Bussey Institution to the Arboretum. Existing plantings consisted of trees and shrubs from the landscape of Bussey and his heirs: clumps of rhododendron, several grand old European beech varieties, a pin oak, a tulip tree, and a wide-spreading American elm whose canopy filled the lawn area formerly fronting the house. Near South Street were many plants from Karl Sax's experimental breeding: some forsythia and crabapples, as well as hybrids between the American white pine (*Pinus strobus*) and its Asiatic counterpart (*P. armandii*). Some of Wilson's introductions of spruce had been planted there, and in the 1940s Wyman developed extensions of the lilac and mountain ash collections.

On the site of Bussey's mansion, a rectangular four-foot-high mound of earth with cut-granite slab steps, a few granite fence posts, a millstone that covered an old well, and the crumbling walls of a stone outbuilding overgrown with weeds were still in evidence. Each spring thousands of snowdrops popped up on the sides of the

148. One of the last of the original American elms to grace the Arboretum property stood in front of the site of Benjamin Bussey's mansion. Only a few years after this photograph was taken, in the mid-1970s, the tree was so stricken with Dutch elm disease that it had to be cut down. Although a mixture of plantings has been added in this vicinity, large European beeches, rhododendrons, a tulip tree, and a few other plants remain from the mansion's landscape.

earth foundation. For twenty years little would be changed, however, once the far side of Bussey Hill was officially transferred to the Arboretum. Not accessible by the paved drive, which had become by far the most popular route for pedestrians as well as drivers, the "back side" of the hill was rarely visited, and the spot near the outbuilding was used to stockpile manure, compost, and logs.

Because there was pressure from the community to utilize the seemingly idle land, development of the Weld-Walter tract began in 1968. Legislation that proposed to create a ski tow on Peters Hill had recently been defeated by enlisting many Arboretum advocates to oppose it, but only by fully putting its land to use could the Arboretum successfully defeat motions to employ it for organized recreational activities. This kind of proposal became more frequent in the 1960s: parts of the Arboretum were suggested as sites for picnic grounds, swimming pools, or playing fields. The 1956 sale of the adjacent city woodland, known as Joyce Kilmer Park, and subsequent development as a facility for care of the elderly had already caused consternation in the neighborhood. In several of the city's parklands, the original design and intent was usurped in favor of "development" or facilities for active recreation in the 1950s and 1960s.

Beginning in the winter of 1968, the Weld-Walter land was cleared of brush in some spots. Its surrounding wall was repaired, and a chain-link fence was placed on top. A roadway was graded to sweep gracefully from Walter Street to the highest point, where a turnaround like the ones at the tops of Bussey and Peters Hills was to be developed. Junipers and masses of other ground-hugging, drought-tolerant plants were installed along the bank above Centre Street and just below the turnaround area. The plan was to open the site officially in 1972, the centennial year of the Arboretum, but this was not realized for lack of funds. For the most part, these peripheral, later additions to the grounds, together with the South Street tract, were still not integrated with the original Arboretum as space for well-defined collections.

The vicinity of the administration building and Arborway gate underwent several transformations after Kelsey's report. His many ideas for changing the plantings were aimed at improving the visitor's first impression on entering. This was an area that had seemed problematic to Beatrix Farrand, as well. Following her suggestions, some large rhododendron masses immediately west of the gate were removed in 1959. That same year, the curving streambed was straightened, moved further from the Arborway, and deepened, all in an attempt to prevent flooding. Both Farrand and Kelsey wanted to develop garden-like plantings of mixed shrubs in this dell; instead, it was used to extend the magnolia collection. Across the Arboretum road to the east of the gate, modifications were made in 1965. In hopes of better drainage, the meandering streambed was straightened and moved toward the Arborway. A large grove of that celebrated Arboretum introduction, dawn redwood, was planted over the old stream course directly opposite the building; the "living fossil" tree had been found to thrive on moist soil conditions. However, further modifications in the plantings near the building were put off until after the addition of a garage in the vicinity.

For sixty years all the equipment and supplies for maintenance of the grounds had been kept in the basement of the herbarium wing of the administration building. In early days the equipment was simple, but, as more motorized devices came into use, all had to be driven down a ramp and made to fit like pieces of a puzzle between pillars in the cramped space. After two years of planning and preparation, a garage for storage and repair of equipment was constructed to the north of the administration building. The one-story, brick-faced structure was built into a bank. The plans included a paved area for staff cars to eliminate parking in front of the building.

While Kelsey suggested a split-cedar fence and plantings to screen off the area, the Arboretum staff decided on more secure and durable chain-link fencing; its transparency was actually less imposing than a solid wall of cedar posts would have been. Nearly twenty kinds of vines were started on the fence; three venerable witch hazels were carefully preserved to mask the front side of the fence. A young *Cedrus libani* variety *stenocoma* was added, in keeping with the use of Arnold Arboretum introductions in the vicinity of the building and Arborway entrance. One specimen, the original *Malus* 'Bob White,' which came from a plant on Francis Parkman's estate, was saved to stand in an island in the middle of the parking lot.

Once the garage was in place, other improvements near the building could be made. In the late winter of 1969 an ice storm, followed by two heavy snows in February and March, caused destruction that Howard likened to the damage wrought by some of the recent hurricanes. Most conspicuously ruined were several star magnolia trees in front of the administration building and a cluster of Sargent cherries that Wilson had placed on the slope beyond. That year's growing season was spent cleaning up broken limbs (and even a few toppled trees) throughout the grounds. The magnolias were so badly affected that in the early spring of 1970 a new "foundation planting" was installed in front of the building.

Several large long-stalk hollies were moved by a contractor from the recent planting near Central Woods and planted among the remaining magnolias, adding another Arboretum introduction to the surroundings. Umbrella pines (*Sciadopitys verticillata*), cherry laurels (*Prunus laurocerasus*), and Dahurian rhododendron (*Rhododendron dauricum* variety *sempervirens*) were other evergreens added. To further the evergreen effect another long-stalk holly and a mass of mountain andromeda (*Pieris floribunda*) were planted immediately to the west of the Arborway gate and a mass of *Rhododendron carolinianum* to the east. The lawn southwest of the building was regraded and reseeded in 1970 to improve its ease of maintenance. The low expanse was apparently made more level, creating an abrupt rise at the base of the slope. Beyond the lawn, under the grove of tulip trees, a mass planting of inkberry (*Ilex glabra*), yet another broad-leaved evergreen, was installed. The lost Sargent cherries were replaced with young plants, and many more magnolias were added to the slope of the hill above the lawn.

EDUCATION AND DEMONSTRATION

While the programs of research and curatorial activity in both herbarium and living collections were strengthened, so too were the educational offerings. The botanical staff continued to teach courses and advise graduate students within Harvard, drawing on all the collections for inspiration and examples. The comprehensive assemblage of living trees was visited regularly by classes from many other New England colleges and universities, as well. The audience for *Arnoldia,* education courses, and lectures was still made up largely of homeowners and professionals with an interest in planting trees, most of whom resided in the suburbs, not in the city. Observing the Arboretum by car or from tour buses became a more frequent mode of visitation. Driving permits were issued on request on weekdays, but they were never given out on weekends.

Taxonomist Carroll Wood initiated expansion of the adult education program soon after his arrival. Almost continuously since 1891, the Arboretum had offered one educational course to the public, the field classes conducted on the grounds each spring and fall. These were taught first by John G. Jack, and then by Donald Wyman, who continued until his retirement in 1970. To this, Wood's education program added Plant Propagation, Basic Botany for the Home Gardener, and Tropical Botany, taught by Arboretum staff members Coggeshall, Wood, and Howard, respectively. Some four such short courses for adults were taught each spring and fall thereafter. Although the subjects varied from year to year, this level of course offering continued for the next two decades, with propagation methods, the field class, and some botany course nearly constant. Plant identification was another popular topic. Subjects of courses were invariably those in which the staff had plenty of expertise and interest.

Other aspects of education and demonstration were expanded as well. Once the herbarium no longer filled the first floor of the administration building, the original museum function was revived, although some of the space for displays also served as a lecture hall. The exhibits were generally changed two or three times a year. The first show in the new lecture hall was a group of paintings of Caribbean trees that had been commissioned to illustrate a book on the subject. This was followed by Christmas plant material and by an exhibit of photographs of plants on the grounds. Sometimes material that had been used in the Arboretum's exhibit at the Massachusetts Horticultural Society's flower show was set up in the lecture hall: a mulch demonstration and a pruning exhibit, for example.

Color photographs of some of the Arboretum's important introductions, a glass case filled with the Native American artifacts found by Ernest J. Palmer, and a topographic model of the Arboretum became permanent exhibits in the first-floor hallway. Although the library had always been available to interested parties for reference, moving a large portion of it down to the first floor was a new gesture aimed at encouraging Arboretum friends and the public to use it.

Open House day at the Case Estates in May was initiated in 1955, starting a tradition that swelled almost to the popularity of Lilac Sunday for many years. The Case Estates became the site for more demonstration plantings that could not be accommodated in Jamaica Plain but that were appreciated by the homeowner audience: ground covers, small trees, various mulches, the effects of different pruning methods, and herbaceous perennial collections installed in cooperation with plant societies.

The staff guided tours for groups of twenty-five or more who registered in advance, and presented numerous off-site illustrated lectures to garden clubs and civic and other organizations that greatly increased awareness of the Arboretum's value to the community. On the lecture circuit and on the grounds, Wyman shared the lead role with Arboretum staff members who interpreted other dimensions of the institution. "A Botanist in the Grocery Store," topics in tropical botany, and plants of the Caribbean were subjects of lectures and classes given by Richard Howard. The Arboretum continued to host national conventions of professional and amateur gardening organizations. In the mid-1950s, publicity for the landscape and collections, as well as for the contributions of its research programs, was frequently and actively sought in the papers and on Boston's newly organized educational television station.

The perennial demand for certain back issues of *Arnoldia* stimulated the publication of *The Arnold Arboretum Garden Book* by D. Van Nostrand Company of New York in 1954. The volume consisted of Wyman's selection of forty-two *Arnoldia* articles, all but six of which he had authored. To accompany them, Wyman wrote an introductory chapter on the past history and current work of the Arboretum.

Arnoldia continued to publish information directly useful to gardeners and horticulturists and attempted to provide leadership in the correct application of plant names. Additionally, such articles as Richard Howard's "The Meadow," Peter Green's "Herbaceous Aliens in the Arnold Arboretum," and Lorin Nevling's "Some Ways Plants Climb" carried on the kind of botanical interpretation that had been the province of John G. Jack, Ernest J. Palmer, and Hugh M. Raup in earlier years.

In January 1965, Secretary of the Interior Stewart L. Udall announced that the Arnold Arboretum was desig-

149. On the steps of the administration building in May 1966 members of the Arboretum visiting committee, the director, and city officials accept the certificate designating the Arnold Arboretum a National Historic Landmark. Among those present were (left to right) Eleanor C. Bradley (visiting-committee member, seated), Henry Scagnoli from the mayor's office, L. Gard Wiggins (representing Harvard's president, Nathan Pusey, and the corporation), A. H. Parker (chairman of the visiting committee), Edwin Small of the National Park Service (representing the secretary of the interior), and Richard A. Howard.

150. Donald Wyman talks about the virtues of the cork trees along Meadow Road to his Friday morning field class. Wyman was widely known among professional and amateur horticulturists for his interpretation of the Arboretum collections through public speaking and publications.

nated a National Historic Landmark, joining five hundred other institutions and localities so designated throughout the country. With combined responsibility for the Arboretum, Harvard University and the City of Boston agreed to preserve its historical integrity as far as possible and to continue to use the property only for purposes consistent with its historical character. The following year a bronze plaque was placed on one stone pillar of the Arborway gate, and a certificate was handed over from the Interior Department to Arboretum and city officials at a meeting of the Arboretum visiting committee.

WYMAN RETIRES

As the year slated for Donald Wyman's retirement, 1970, approached, Richard Howard began preparations for the transition. He urged Wyman to write up his many operations, as records of the past and guidelines for those who followed; his independent operation of the horticulture department was an asset the director regretted to lose. With no one on the horizon in 1969, the possibility of extending Wyman's employment until the centennial year was considered briefly. Nonetheless, Wyman stuck to the scheduled retirement date of 31 August 1970. To pay tribute to his thirty-five years of service, the staff feted the horticulturist at a luncheon held on the day of his last field class. Harvard authorities gave Wyman the title of horticulturist emeritus, an honor rarely bestowed on administrative officers—his exceptional publications, teaching, and lecturing made him worthy of the distinction. The staff prepared and dedicated to him a cumulative index to *Arnoldia,* the publication that was so much the result of his efforts over the years. Howard commented in his annual report that Wyman's contributions had helped define and establish the role of the Arboretum.

While not swerving greatly from the intent of the Arboretum's originators, Wyman built programs of significance to mid-twentieth-century students and growers of plants. He concentrated his powers on making sense of the institution's encyclopedic collection for the general and professional public. Out of the thousands of species and varieties grown by the Arboretum, Wyman continually refined selections of the most promising materials for the needs of those who sought information. This he communicated through his writings, lectures, participation in professional gatherings, the cooperating nurserymen's program, and his popular field classes on the grounds.

Like E. H. Wilson, Wyman was primarily a plantsman, a curator of collections from which the landscape designer or garden maker could draw the elements for beautifying, ameliorating, and creating environments. And while he had definite ideas about what makes a pleasing design or a beautiful garden, he rarely imposed his taste on the Arboretum's overall design. The plan of sequenced collections set in the background of native woods (with numerous exceptions to the sequence) was complicated enough without further cluttering it with elements added just for their beauty or "landscape effect." Whether Wyman was conscious of the details of Sargent and Olmsted's original intent or not, he tended to develop the existing design with the best of modern plant material.

When trends toward formality or reduced maintenance in gardens demanded demonstrations not readily made in the Arboretum's naturalistically styled, parklike dimensions, special plantings were created outside of the original layout. The hedge demonstration plot was located first on the Bussey Institution grounds and later within the Dana Greenhouse complex; the bonsai, the dwarf conifer collection, and the espaliered plants were also placed on the latter site; the small-tree plot, ground cover demonstrations, and mulch and pruning exhibits all were sited at the Case Estates. With their educational mission uppermost in his mind, Wyman let the demonstrational aspects dominate the arrangement of the special collections. The hedges were set in rows and uniformly clipped, varying only in height appropriate to scale of plant used.

Wyman had tried adding ground covers in Jamaica Plain with mixed results. Diminutive and slow-to-start plants like bearberry, *Pachistima canbyi,* and even native lowbush blueberries failed to take hold among the Meadow Road azalea border. Various ground covers were used to stabilize soil on the steep slope of the summit of Bussey Hill. Nonetheless, Wyman settled on rows of square plots at the Case Estates as the best way to compare the qualities and merits of plants for this purpose. There was no charming design or added elements to distract from the study of ground covers in this particular area—the plants themselves were free to tell their story. Moreover, every planting had to come within his small staff's ability to maintain it at high standards, and in this regard simplicity was usually the best choice.

Although, to a large public, Wyman's horticultural role alone represented the Arboretum, he worked in tandem with its scientific developments. Supplying plant material to scientists around the world for study was an obligation always fulfilled, just as the herbarium staff answered numerous requests for the loan of specimens each year. The taxonomic work of Alfred Rehder and such succeeding botanists as Carroll Wood, Peter Green, Richard Howard, and others was the basis for accurate classification and naming of plants under Wyman's care. With the emphasis given to solving problems germane to naming garden forms, Wyman became an interpreter and spokesman for more precise use of names by commercial and private growers.

For the Arboretum's first fifty years, botanical and horticultural roles were often combined in one person. Sargent and Wilson could talk up the best plant for a particular horticultural purpose and make sound recommendations for its care; they could explain its origin in nature and argue its place in a classification scheme with equal alacrity. Although the department of "horticulture" was sifted out during Merrill's regime, Arboretum administrators through Howard tended to favor the interaction and overlap between botany and horticulture.

By the close of his thirty-five years at the Arboretum, Wyman had published more than one thousand articles, along with five books. After retirement, the horticulturist emeritus continued his writings, most notably with the popular compilation *Wyman's Gardening Encyclopedia,* published in 1970 and subsequently revised. American horticulture lost one of its leaders when Wyman died in September 1993.

Wyman's assistant in mapping and plant records, Heman Howard, with longer incumbency at the Arboretum than his boss by a few years, also chose retirement in 1970. He accepted a post as horticulturist at the Heritage Plantation of Sandwich, Massachusetts, a museum of early Americana located on the former grounds of the rhododendron hybridizer Charles O. Dexter. For the Arboretum, Gordon P. DeWolf was appointed horticulturist; he was assisted by Robert S. Hebb, a Rhode Island native who had recently graduated from the gardening program at Kew.

CENTENNIAL BENCHMARK

When the Arboretum's ninety-fifth anniversary was attained in 1967, Richard A. Howard began looking ahead to the institution's centennial. A committee of supporters and visiting-committee members was formed to plan and raise funds for appropriate activities. Howard and an administrative assistant conceived the idea of investigating the early history of the Arboretum, which led to the decision to produce a biography of Charles S. Sargent. The assistant, Stephanne B. Sutton, was appointed research fellow to pursue this work.

Sutton's *Charles Sprague Sargent and the Arnold Arboretum* was published by Harvard University Press in 1970. The book concentrated on the founding years of the Arboretum, especially on Sargent's scientific work, his advocacy of forest conservation, and his oversight of the explorations of Wilson and Rock.

Other steps were taken in anticipation of the centennial to interpret the institution and its history for the public. The Arboretum produced a new guide to the grounds in 1968. Although it, too, was titled *Through the Arnold Arboretum,* the forty-eight-page booklet was not a revision of Wyman's 1949 version of the guide, but a photo-essay with vignettes highlighting the tree garden's main features and natural charms.

Sutton and others at the Arboretum compiled a more informative book for publication in 1971, *The Arnold Arboretum: The First Century.* Generously illustrated with color and black-and-white photographs, this work related the institution's history, described collections, and presented current research by using a series of quotes from publications of the staff. A foreword by the Boston historian Walter Muir Whitehill added flavor to the interpretation of the Arboretum's role in the city. Howard's afterword spoke of the institution's strengths and the challenges it faced for the future.

In the fall of 1970 the Arboretum commissioned Peter Chvany, instructor in filmmaking at Boston University, to create a documentary on the Arboretum to premiere during 1972. In anticipation of an increased demand for group tours of the grounds, a corps of volunteers was given an intense seven-month training program coordinated by a volunteer working closely with every department of the Arboretum. The recent historical publications added to the basis for interpreting the institution. This was the first organized attempt to enlist volunteers. Before their training was complete, Robert Hebb convinced the already enthusiastic group to plant daffodil bulbs for naturalizing on the slope south of the administration building. E. H. Wilson had probably been the last to oversee bulb planting here, and apparently the daffodils were in need of renewal after forty-five years. Meanwhile, the centennial committee also worked with staff members to organize a week of concentrated celebrations, as well as yearlong observances.

The week of 21 to 28 May 1972 was, fortunately, the sunniest one in a cold, wet spring. Nearly seven hundred persons accepted an invitation to attend one or more events or the whole week of activities. The program was headquartered at a downtown hotel, where an opening banquet featured an address by William T. Stearn, Senior

Principal Scientific Officer of the British Museum (Natural History), an esteemed taxonomist and botanical historian. His speech explained "how the modern arboretum, so well exemplified by the Arnold Arboretum, has developed from the medieval English deer park, by way of the 18th-century landscaped park, through the anxiety over timber shortage and feeling for landscape, which together led to an interest in trees for their aesthetic and scientific value as well as for their economic use" (Stearn, 1972, pp. 173–74).

For a daylong symposium—Potential of Arboreta and Botanical Gardens—twelve papers were presented by Arboretum staff and by colleagues from such varied institutions as the Gray Herbarium, the Smithsonian, the USDA, the Hunt Institute for Botanical Documentation, and the New York Botanical Garden. Addressing concurrent botanical and horticultural sessions, the speakers covered a broad range of topics pertaining to the role of institutions with living and preserved plant collections. Among horticultural subjects considered were plant introduction, propagation, hardiness evaluation, plant ailments, and hybridization. Discussion of the capabilities in botanical research touched on using computers for taxonomy, the value of collections for anatomical study, and cytogenetics. Some of the talks had historical themes, and one explored the importance of the public education function at botanical gardens and arboreta.

Three public lectures were sponsored during the centennial week. Representative of the times was the presentation by Paul R. Ehrlich, a biologist from Stanford University, "The Population-Environment Crisis—Where Do We Go from Here?" For three days, program participants could either take daylong tours in eastern Massachusetts or stay in Boston and Cambridge to study in the collections and hear the lectures. The tour program offered a choice between sites of botanical interest or private and public gardens. The Arnold Arboretum film premiered on an evening in which participants had enjoyed walks through the Case Estates and a picnic supper.

An exhibit of old and rare books from the Arboretum library was organized at the Houghton Library, Harvard's rare-book collection, and opened with an evening reception during centennial week. An illustrated catalog, including complete bibliographic descriptions of the sixty items on display, was compiled by the Arboretum staff. The five active days came to a close with a final outdoor buffet at the Arboretum on Friday night. Representatives from fellow institutions around the world presented official greetings, which were gratefully accepted. The Arboretum facilities and staff remained available to visitors all weekend. The guests represented twenty-seven states, ten countries, sixty-nine institutions, and forty-six organizations. It was a fine mingling of growers and students of trees, colleagues in research and collections management, garden administrators and garden lovers. The week had been marvelously orchestrated by the thirty-two-person committee, in concert with the staff.

Special, though less intense, commemorative activity continued all year. In the best of long-standing tradition, the horticultural staff made several distributions of plant

151. Botanists and horticulturists came from 27 states, 10 countries overseas, and 115 institutions and organizations to participate in the weeklong celebration of the Arboretum's centennial in May 1972. Tours of the grounds and many activities provided the opportunity for advocates of arboreta and botanic gardens to exchange ideas and information.

152. *Among centennial activities was the move of a forty-two-foot specimen of giant sequoia (Sequoiadendron giganteum) in 1971 from a Chestnut Hill estate owned by Boston College. The property's former owner, Chandler Hovey, had acquired the tree in the 1940s, but had died before acting on his intention to donate it to the Arboretum for the centennial. When Boston College acquired the estate, it permitted Hovey's wish to be carried out. Here the commercial truck loaded with giant sequoia makes its way down Valley Road toward its destination in the main conifer collection above Bussey Brook.*

material. A young plant of the katsura tree (*Cercidiphyllum japonicum*) was mailed out to every member of the Friends of the Arboretum. Every botanical garden and arboretum listed in the International Directory of Botanical Gardens was offered propagating material from a list of type plants and original introductions in the Arboretum living collections—plants that were named and introduced by Sargent, Rehder, Wilson, and others. A large shrub or tree was given to each of 153 garden clubs throughout New England for planting in a public spot. One hundred and twelve colleges and universities accepted gifts of an Arboretum plant for teaching collections on campus.

Because of the publicity for the weeklong celebration—the plant distributions, film showings, and traveling exhibits of photographs of the Arboretum—visitation to the grounds greatly increased in the centennial year. The new corps of volunteer guides aided greatly by leading tours for organized groups. Several articles on the Arboretum's accomplishments and history were featured in *Arnoldia*: Howard wrote on scientists and scientific contributions of the institution; Sutton contributed "The Arboretum Administrators: An Opinionated History"; and Wyman compiled a list of all the Arboretum plant introductions for the second fifty years, 1923 to 1972. The majority of these were the results of Wyman's efforts, although Sax's hybrids and Merrill's *Metasequoia* haul were notable additions. The texts of the symposium talks and centennial lectures also found their way into the publication.

Although the anniversary was an appropriate time to pause and look at the past and contemplate the future, the retirement of Wyman was a significant break in continuity for the institution. His management of the grounds extended back to a time not far removed from Sargent's and Wilson's. In 1970, Wyman said that when he arrived "plantings made under the direction of these men were considered sacrosanct and were not changed or removed" (Wyman, 1970, p. 81). For thirty-five years, he concentrated on utilizing the collections for horticultural evaluation of woody plants and communication of the results. The arrangement of plantings he found when he arrived was generally suited to this agenda, especially once the

Case gifts provided additional land and the funds to manage it.

Over years of daily observation of the grounds and the plants, through their worst losses to hurricanes, and through two changes of administration, Wyman and his assistants developed a horticultural program that stood firmly on its own. Nonetheless, he and the directors maintained its link with the Arboretum's botanical investigations. By well-thought-out plant distributions, publications, and public relations, the Arboretum offered its resources to the people, fellow botanical gardens, the nursery trade, investigators in many fields of plant science, and to patrons and supporters. Equally utilizing the institution's strengths at the century mark, the botanical staff interpreted their critical plant studies for students at Harvard, their colleagues worldwide, and for the public. Director Howard strove to build cohesion between these two phases of the institution. The Arboretum now thought of itself in terms of the two subjects of botany and horticulture, yet the cross-connections between them were as numerous as ever.

The research that had gone into centennial-year interpretation was to have important long-term effects on the institution, although it was some time before they were felt. Sutton's biography of Sargent, Chvany's film (which included a historical sequence), the *Arnold Arboretum* illustrated history, the many in-house and traveling photograph exhibits that were mounted, and the preparation necessary to teach volunteers to give tours—all contributed to a new institutional memory that would influence subsequent program development, especially once a nationwide revival of interest in Frederick Law Olmsted gained momentum in the 1980s. The fascination with E. H. Wilson, for example, led to more thorough investigation of his contribution and to several articles for *Arnoldia* later in the 1970s. This in turn was a great stimulus to pursue fieldwork in the Far East, once China reopened its doors. The continued organization of scattered files, correspondence, old papers, notebooks, scrapbooks, and photographs into an archive provided a focal point for understanding present conditions in the light of the past. Before many centennial-stimulated initiatives and restorations could occur, however, the Arboretum had to weather the 1970s, a period of great questioning and, for the Arboretum, one of substantial change of staff for the living collections.

THE SEVENTIES: IDENTITY CRISIS

Besides the departure of Wyman, the year 1970 marked the height of a social crisis that touched nearly every institution in the country, including the Arboretum. Howard wrote: "Many adjectives have been used to describe the events of the past year. It has been a disturbing year—a year of student unrest and rebellion, a year of inflation, a year that could not fail to provoke a reappraisal of values" (Howard, 1970, p. 201). The sometimes violent turmoil of student protest interrupted staff in Cambridge, and lawlessness among the urban population threatened destruction or disfigurement of plants and property in Jamaica Plain. The unrest had been building since the late 1960s, and, while it caused very immediate problems of vandalism in the living collections, this period of questioning and sudden awareness of "the environment" by the populace ultimately brought about great shifts in policy for institutions charged with the study of natural science. The process of change took place in fits and starts, however, like so many other revolutions that began during this time.

In regard to the effect of many traditional practices on environmental quality, the Arboretum had already made specific adjustments. In 1967, adoption of air pollution legislation prohibiting open fires in Boston meant the Arboretum had to chip more of its brush for composting and mulch or else pay to have debris removed. In 1969, the staff stopped using DDT and some other insecticides and chemical weed killers. Concern about the environmental impact of certain horticultural procedures ultimately led to the adoption of many aspects of the idea of "integrated pest management" and a commitment to prevent plant disease by maintaining the health of individual specimens rather than relying on the wide use of toxic chemicals.

Ironically, at the same time that many people became suddenly concerned about the health and protection of the living world, there was an upsurge of random destruction of property and disfigurement of parks. Large groups of young people began to congregate in the Arboretum and leave litter strewn behind them. Many parks came to be considered abandoned territory into which stolen cars could be taken and set afire. For the first half of the twentieth century the large gates at every entrance to the Arboretum had always remained open, although only the staff and those who applied for a permit were allowed to drive on the grounds. By the mid-1960s, unauthorized vehicles and vandalism prompted Boston's parks department to close all the large gates on weekends. By 1972 it was necessary to keep all but two of the vehicle gates shut at all times. The pedestrian gates flanking these remained open, but the expansive, inviting views from outside the grounds were lost.

The roots of unrest lay in the social and economic realms, and there seemed little a botanical department of a major university could do to make change. Nonetheless, the Arboretum was rather suddenly overwhelmed with requests for information and plant materials by a host of community groups, government agencies, and organizations determined to beautify, replant, and regreen urban locations or to preserve natural sites. In response, "community and professional service" became a part of the institution's programs. While many staff members gave their expertise to several projects, one energetic recipient of a Mercer Fellowship was retained on the staff for a few years to coordinate requests from the community, especially with regard to the distribution of excess plant material. A large part of the job was to teach enthusiastic, but inexperienced, city residents how to dig, plant, and provide care for trees.

Up to this time the Arboretum's main audience had been people who already knew the basics of gardening. Much of its educational effort had instructed the advanced grower in the fine points of woody-plant identification and selection. Now, a different group of enthusiasts, who had little knowledge of plant material or horticultural techniques, desired information. The Arboretum was not the only organization in Boston besieged with requests from a new generation of city gardeners and beautifiers. The Massachusetts Horticultural Society, the Suffolk County Extension Service, the Waltham Field Station of

153. *The Arboretum community coordinator, Nancy Page (behind tree), joined citizens in planting ornamental trees at the Worcester/Claremont garden during a Boston Urban Gardeners harvest festival. The 1970s marked the beginning of a great revival of interest in improving and appreciating the environment of cities.*

the University of Massachusetts, and a new organization that arose out of the victory-garden revival, Boston Urban Gardeners, all provided what guidance they could.

As of 1970, *Arnoldia* put on a new face by adding an illustrated cover and altering its frequency of publication from twelve to six times a year. Although the number of issues was reduced, each contained more articles. Its authorship and subjects became more diverse, the latter expanding to include horticulture, history of botany and gardening, herbaceous plants, and book reviews. Sometimes a whole issue was given over to a single topic; "A Guide to City Trees in the Boston Area," "Wild Plants

in the City," and "Poisonous Plants" were three numbers of value for the urban audience.

The core of volunteers trained for the centennial year was subsequently expanded, and this dedicated group added greatly to the Arboretum's ability to interpret its collections to garden clubs and to social and civic groups. The volunteers contributed to many other facets of the work in Jamaica Plain as well, from reviewing books and writing articles to washing pots at the greenhouse. Nonetheless, the Arboretum was seriously hampered in its relations with the general public in the 1970s as greater numbers of people quite suddenly began to value the experience provided by "oases of green" in their cities. The general increase in visitors created unexpected demands for amenities like rest rooms, water, food, and a place to picnic, as well as for information about the plants and the landscape. The small staff and modest facilities were often overtaxed—answering the telephone, for example, became more than a full-time job. Rotating duty to keep the administration building and the greenhouse area open on weekends was tried for a number of years, but it was not a satisfactory solution. The lack of resources to handle visitors successfully and the cumulative effect of vandalism in provoking further vandalism were exacerbated in this decade by a decline in Boston's ability to repair benches, paths, roads, fences, and gates.

Given the distractions, disruption of continuity in horticultural staff, and constraints of inflation on finances, it is not surprising that few developments took place on the grounds or in the collections. One exception was the rehabilitation of the top of Bussey Hill in 1974, a project funded in part by an anonymous gift. The landscape-design firm of Shurcliff, Merrill and Footit drew up plans and supervised the major construction; the Arboretum staff collaborated in selecting and acquiring plant materials.

The summit was originally designed by Olmsted as part of the City of Boston reservation, a spot where visitors in carriages would converge to take in the view. Once the Arboretum took it over in 1895, a mixture of plant material was added, and in the twentieth century the concourse was paved. The steep slope between Bussey Hill Road and the summit suffered from erosion where walkers had cut unplanned paths through the shrubberies. The overriding goal of the renovation was to eliminate the unwelcoming expanse of pavement from the very top, creating a pleasant place to sit and take in the view southward to the Blue Hills. Although a space to turn autos and buses around would still be needed, it could be moved away from the best viewing area. A further aim was to manage pedestrian access to prevent erosion and shortcuts. One technical difficulty was the existence of an unused underground reservoir, left from the time of Bussey's or Motley's occupancy.

The objectives were met by the creation of a sitting area raised above the level of the road surface, defined by steps and a wall made of cut puddingstones that were "recycled" from the old wall along the former Centre Street right-of-way above the conifer collection. On the south side, the wall was of a suitable height to sit and view the Blue Hills. Simple, sturdy granite-and-wood benches provided additional resting spots, and a low fence of granite uprights and wood cross-members separated a circumferential path from areas densely planted with shrubs. Four gradually rising asphalt paths provided choice of access for walkers. Plant material for the steepest faces of the slope was chosen to be impenetrable and to hold the bank. Plantings around the top emphasized summer bloom: butterfly bush (*Buddleia davidii*), *Elsholtzia,* shrubby cinquefoil (*Potentilla fruticosa*), and Russian sage (*Perovskia atriplicifolia*). A grove of white pines on the east slope was preserved and balanced by the addition of a few Japanese white pines (*Pinus parviflora*) to the raised area. These hilltop trees echoed the ancient pines that remained at the far side of the plateau at the end of Azalea Path, as well as the more recently planted grove immediately to the east.

An important curatorial development that took place just as the new horticulturists stepped in for Wyman was participation in a national movement to create a computerized listing of garden accessions. The idea had been under discussion for a number of years by members of the American Association of Botanical Gardens and Arboreta and the American Horticultural Society, and the latter organization received grant funds to carry it out. After a year of preparation by Arboretum staff, all of the cards in the living-collections record file were microfilmed by the staff of the Plant Records Center of the American Horticultural Society in 1972. The records center then processed the data on its computer in Virginia and made printouts of all taxa listed alphabetically available to the Arboretum. A system was worked out for additions, deletions, and corrections to be submitted twice yearly. Updated printouts were issued on request, and the data could be manipulated to some extent to create listings by family or by location, for example. The Arboretum not only obtained useful catalogs of its holdings but shared its inventory with all of the gardens participating in the database. The center eventually became the Plant Sciences Data Center and issued the master inventory on microfiche. Participating institutions use this inventory to locate sources of living plants throughout North America.

The appointment of Constance Derderian as honorary curator of the bonsai collection in 1969 brought new life to this intriguing group. Derderian had considerable experience with the horticultural art of bonsai, having studied for ten years with experts at the Brooklyn Botanic Garden and in Japan; she was also a director of the American Bonsai Society. For the Larz Anderson collection she initiated a regular program of repotting and careful pruning. The honorary curator was instrumental in adding new specimens to the collection and taught classes in the technique for beginners and advanced students. This revival of interest inspired greater understanding of the miniature trees within the staff and better efforts to interpret them for the public.

154. *The Arboretum's exploration of eastern Asia underwent a revival starting in 1977, when taxonomists Stephen A. Spongberg and Richard E. Weaver traveled in northern Japan and the Republic of Korea. Spongberg, Weaver, and the recently appointed propagator Jack Alexander examined and cataloged the seeds collected on the trip at the Dana Greenhouses soon after their return.*

NEW DIRECTIONS

In the 1970s, the program of research remained relatively constant in its scope and subjects despite some personnel changes. With the resignation of Lorin Nevling in 1973, Arboretum participation in compiling the flora of Veracruz was dropped, although cooperation continued with the institutions subsequently involved. The study of ligneous genera of eastern North America and eastern Asia was revived with the addition of taxonomists Stephen A. Spongberg and Richard E. Weaver, Jr., to the staff in 1970. Peter F. Stevens carried on the enumeration of plants of Papuasia, starting with his appointment in 1973. Additional botanists joined Carroll Wood in his work on the generic flora of the southeastern United States for varying periods.

While botanical institutions like the Arboretum had long been stores of knowledge on the changes in natural range, the increasing rarity, or the extinction of plants, this expertise was given greater emphasis starting in the mid-1970s with development of federal and local legislation to protect threatened and endangered species. Once the movement to pinpoint species and habitats in need of protection was under way, Harvard's combined herbaria and the botanical staff were consulted more frequently by those making inventories of various areas.

The first volume of Richard Howard's projected *Flora of the Lesser Antilles* was published by the Arboretum with a grant from the Stanley Smith Horticultural Trust in 1974. It included the treatment of the orchid family by Leslie A. Garay and Herman Sweet of the Oakes Ames Orchid Herbarium of the Botanical Museum. Howard provided an introduction to the project and an extensive discussion of the phytogeographic and floristic relationships of the islands. The second volume was published in 1977; it included treatments of the ferns and fern allies (Pteridophyta) by Howard's longtime colleague George R. Proctor of the Institute of Jamaica.

Although the orchid and fern volumes were an auspicious start to publishing the flora project, Howard still had the remainder of the flowering-plant families to write up. Intending to complete the flora and to continue teaching, Howard requested to be relieved of administrative duties, effective June 1978. He had served as director for twenty-four years, and he looked forward to a sabbatical year as a Guggenheim fellow to renew his concentration on research. Howard's final year as Arboretum administrator was nonetheless momentous for two events, which reestablished ties with eastern Asian botany and set the stage for his successor.

Parallel to the reconsideration of values going on in many other parts of society, the botanical community began to appraise its methods and collections in the light of questions about the relevance of its research to problems in the natural environment. A related issue was whether living collections of plants were adequately serving *botanical* research. Occasional findings—that misnamed botanical-garden plants were actually the progeny of spontaneous hybrids that had occurred in the botanical garden or nursery where seed had been collected—led to the questioning of the propriety of these sources for collections to be utilized in research.

Seed collected from naturally occurring, wild populations was deemed a more reliable source of plants for scientific gardens. From the kind of interaction that produced the Plant Sciences Data Center's master inventory of botanical-garden and arboretum holdings, it became apparent that many of the plants introduced from overseas were represented in North America by a very small sampling of genetic diversity; of these, many had originated from only one or a few collections and then had been continually repropagated from the original, relatively scant stock. In order to strengthen the role of botanic gardens as conservators of rare species and providers of material for a host of biological investigations, new efforts were launched toward obtaining plant material from wild sources. One of the Arnold Arboretum's notable steps in this direction was a field trip to northern Japan and Korea in the fall of 1977.

The two Arboretum botanists keen on woody plants of the north temperate zone, Stephen Spongberg and Richard Weaver, undertook the excursion. It was the first expedition sent out from the institution to temperate Asia since Rock returned from China in 1927. Their goals were to extend the diversity of Japanese and Korean plants in Western horticulture; to obtain propagation material of plants previously introduced that had failed or been lost over the years; to collect from the northernmost stations many plants that had been obtained from southern localities in an attempt to introduce hardier strains; and to collect herbarium specimens to document each accession from the wild. Korea was of special interest. Few American plant explorers had been there, and its flora was closely allied with that of China, which was, in 1977, still inaccessible.

The investigators spent a month in Japan, concentrating on Hokkaido and the northern district of Honshu, where colleagues from Hiroshima, Hokkaido, and Tohoku Universities provided welcome guidance and support. Two weeks were spent in South Korea traveling with Carl Ferris Miller, proprietor of Chollipo Arboretum, located on the western coast. In contrast to Japan, where temperatures are everywhere influenced by the sea, the climate of Korea is continental, with hot summers and very cold winters, making it more like that of New England. Spongberg and Weaver had a chance to collect from several localities in Korea: the northeastern mountains, the western coast, and the southwestern part of Chŏlla-namdo Province. Over the six-week tour they made 505 collections of 327 taxa in 69 families. As soon as many of the resulting seedlings were mature enough, they were distributed to some thirty-three institutions that responded to a notice in the newsletter of the American Association of Botanical Gardens and Arboreta. Most of their finds were added to the living collections in Jamaica Plain, where their evaluation and study continue.

When Weaver and Spongberg returned from Korea and Japan, they had little inkling that within seven months the Arboretum's director would be on his way to the People's Republic of China. Selected upon application as a member of a delegation representing the Botanical Society of America, Howard participated in a tightly scheduled twenty-eight-day tour of botanical and forestry institutes, universities, academies, botanical gardens, communes, parks, temples, and pagodas in the spring of 1978. While the late-spring journey was not a collecting trip, it was a valuable opportunity to reforge old ties and establish new ones with the younger generation of botanists. Howard had alternate presentations to offer at each institution if requested: a lecture on the vascular patterns of stem, node, and petiole of the dicotyledons, or the centennial film on the Arnold Arboretum. Since the Harvard tree museum was still widely known by reputation among Chinese botanists, the latter was frequently chosen.

The interest and goodwill generated by the monthlong trip prompted the Botanical Society of America to extend

155. *Peter S. Ashton's interpretation of the Arboretum's role included a strengthened commitment to applying botanical research to problems in the conservation of biological diversity. He also initiated many programs to broaden the Arboretum's constituency. Here, Ashton wields the microphone at a rare-plant auction held annually at the Case Estates to raise funds.*

an invitation for a return visit by a delegation of Chinese botanists, which took place in May 1979. While at Harvard, the delegation visited the Arboretum, where it pleased them to see many Chinese plants thriving. At the end of their tour of the United States, a meeting was held at Berkeley to discuss future collaboration; this proved to be fruitful for the Arboretum within a year.

RESTORATION, REAFFIRMATION

When Peter Shaw Ashton arrived from Aberdeen University, Scotland, in late December 1978 to take up duties as the sixth person to head the Arboretum, almost every aspect of the institution was in a state of change. The scientific program had recently reaffirmed its emphasis on Asiatic plants, although the facilities for the combined herbaria in Cambridge were in the midst of construction of a major addition, the preparation for which disrupted routine work for at least two years. A reevaluation of living-collections policy had made tentative strides with

the recent growth of interest in wild material, yet the staff members charged with care and management of the living collections were all new to their jobs by Arboretum standards. Within two years of the change of directorship, all of the senior posts in living-collections management had also changed hands. Alfred Fordham, propagator, and Robert Williams, superintendent, both retired; Gordon DeWolf, horticulturist, and Robert Hebb, assistant horticulturist, resigned.

A revival of interest in the life and work of Frederick Law Olmsted presented a challenge to institutions that were marked by his legacy. There was only a modest staff assigned to manage the production of *Arnoldia,* the adult education courses, the friends' membership organization, public relations, and the myriad requests for data from professionals and the public. The attempts to provide information and education to all of the Arboretum's constituencies, including Boston residents and the Harvard community, taxed every department, yet many felt its role in these areas should be enlarged.

Ashton's area of expertise was tropical forest botany. After receiving a doctorate from Cambridge University in England in 1962, he spent many years during his association with Aberdeen University in the field in Brunei, Sarawak, Sri Lanka, and the Malay Archipelago, studying the ecology of forests. Nonetheless, once at Harvard he became directly involved with the problems and promises of a north-temperate garden of trees. Many of the programs initiated during his eight years of oversight were intended to be restorations, or to forge new links between the Arboretum and the other biological departments of the university. Ashton tended to look back to Sargent's time for the living collections, and to Merrill's for the herbarium, but his sights were likewise on the present and the future.

Building on the base of past and present taxonomic inventory in Asia and the Americas, Arboretum botanists, like their colleagues throughout the world in the 1980s, came to a firmer and more outspoken conviction that their work must contribute to the preservation of the planet's organisms and the natural systems they form. It was beginning to be apparent that conservation of particular habitats or species is not enough, that diversity itself

156. Korean plum yew (Cephalotaxus koreana) *is one of many plants maturing on the grounds from seed gathered on the 1977 expedition to Japan and Korea. Recently, Arboretum collections policy has emphasized expanded representation of botanical species and genera, and plants of documented wild origin over those of garden provenance. The shining, dark green foliage of this shrub promises bolder texture than any of its more commonly cultivated relatives, the yews* (Taxus *species*). *The Korean plum yew may find its place in American gardens.*

must be maintained so that the global ecosystem can continue to evolve and adapt. The value of studying evolutionary biology became strikingly evident for its potential role in the inventory and preservation of the diversity of organisms.

New initiatives in the study of temperate Asiatic flora proceeded from the breakthroughs in relations with China. Organization of the 1980 Sino-American Botanical Expedition to western Hubei was the result of discus-

sions at the Berkeley meeting that concluded the Chinese delegation's visit to America in 1979. Fortunately, Stephen Spongberg received an invitation from the Chinese Academy of Sciences to participate on the three-month excursion along with four other American botanists. The two regions visited were the *Metasequoia* valley in Lichuan Xian and the mountainous Shennongjia Forest District. In both locations the Americans were impressed to see plants, known to them only from pressed or cultivated specimens, growing in their natural environment. All the botanists were concerned to see that, although the mature *Metasequoia* trees were being preserved, there was little chance for regeneration of the species and its natural forest association because of encroaching agriculture; subsequently, the Chinese and American scientists made joint recommendations for habitat preservation. By the end of its excursion the team had pressed more than twenty thousand specimens and made five hundred collections of seed or living plants. The material was shared by all institutions participating in the fieldwork.

In contrast to the decades when American botanists could rely only on dried specimens for their knowledge of Chinese plants, they could now use field experience as an aid to understanding the flora most like our own in many respects. The importance of this development was expressed by the team members David Boufford and Bruce Bartholomew: "When examining herbarium specimens from China, we now can recall the kinds of situations under which the plants may have grown in the field, and can consider the various species that might have grown with it. The observations that are only available through fieldwork are most important in providing a clearer understanding of many aspects of biology, plant geography, taxonomy, and evolution that would otherwise either be speculative, or remain completely unknown" (Boufford and Bartholomew, 1986, p. 20). Boufford, who represented the Carnegie Museum of Natural History on the 1980 field trip, joined the Arboretum staff in 1983. He returned to China in 1984 for another Sino-American expedition, this one to Yunnan, a region floristically quite different from western Hubei. In Yunnan the vegetation is more closely related to that of the Himalayan area and of northern Thailand and Burma. Although living plants and seed were not collected on this mission, it yielded another twenty thousand specimens, which would be of value in China for botanists actively working on a *Flora of Yunnan,* as well as a compendium for all of the People's Republic. The American sets of specimens were again shared among the participating institutions, and extra duplicates of certain genera were sent to experts around the world.

The China connection has continued for the Arbore-

tum, with many visits by that country's botanists for fieldwork and examination of the living and herbarium collections. Both Boufford and Spongberg, as well as other Arboretum botanists, have returned to north-temperate Asia for field study and other cooperative work. Since 1988, a joint project to produce an English-language version of the *Flora Republicae Popularis Sinicae,* under supervision of a Sino-American editorial committee and coordinated by the Missouri Botanical Garden, has meant even closer collaboration for the Arboretum. The projected twenty-five-volume series will be published over a fifteen-year period.

Peter Ashton's own expertise in the Old World tropical family Dipterocarpaceae, whose large trees are major components of many tropical rain forests of Southeast Asia, brought new interest to the tradition of research in this area. In addition to straightforward taxonomic investigations, Ashton, his graduate students, and his research associates have studied ecosystem dynamics in the forests of Malaysia, Indonesia, and Sri Lanka. Recent projects include biological assessment combined with socioeconomic research to provide information to be used for long-term management of forests, sustainable use of natural products, and conservation of biodiversity. Peter Stevens's work toward monographs of woody genera of the mangosteen family (Clusiaceae) and additions to the knowledge of biodiversity in New Guinea continue to be part of the research and teaching of the Arboretum. A recent Far Eastern initiative has been collaboration with the National Cancer Institute to survey the plants of Indonesia for biochemical agents of potential use in the medical treatment of cancer and AIDS.

LIVING-COLLECTIONS POLICY SHIFTS ONCE MORE

Although the idea of returning to better representation of species naturally occurring in the wild had been broached, the notion was formalized as a new, or "restored," collections policy in the early 1980s. It was actually the first time the words "collections policy" were used to describe the sets of decisions made in the course of obtaining plants for the Arboretum. Up to that time the only written doctrine was the indenture of 1872, which stated that the Arboretum should "contain, as far as is practicable, all the trees, shrubs, and herbaceous plants, either indigenous or exotic, which can be raised in the open air." Within this broad guideline, the administrators and keepers have had the latitude to work with shifting tastes and changing scientific and educational goals. On account of its limited budget, the Arboretum has always had to be "opportunistic" in its modes of securing plants for the collection. Charles S. Sargent obtained the majority of his specimens from the wild because that was the only source for them. He did not shun plants under cultivation or those coming from nurseries, however; when they became available, he collected cultivars. While Donald Wyman never set down an accessions policy as such, he definitely picked and chose certain items to the potential exclusion of others. These men worked within their times. In Wyman's case it was an era when the belief in man's ability to improve on nature, to conquer the world through experimentation and technology, was at its height. Collaboration with commercial nurseries and fellow arboreta brought many cultivated forms into the inventory at Jamaica Plain.

When Ashton became director, the time was ripe for a new look at the content of the collection. After discussion among the living-collections staff, a policy was precisely written up by Spongberg and published in *Arnoldia*. In order to strengthen the botanical role, *species* of documented wild origin are to be preferred over those of nursery or garden origin. Taxa of infraspecific rank (i.e., naturally occurring varieties, subspecies, and forms) are to be included to the extent that they demonstrate the range of characters of the species to which they belong or if the variant is the only hardy representative of that species. Cultivars now have to meet even more stringent criteria to be retained. These human-selected clones shall be included only if their origin can be fully documented and if their distinctness is unlikely to be challenged for many years. Each cultivar accessioned is to be grown for a definite purpose, which is to be stated in its record: either it must represent a hybrid of botanical interest or it must be an unusual individual that exhibits one end of the range of variation in characters of the species.

Naturally, the new policy has not been implemented overnight. It is a gradual process, like the growth of trees and the succession of forests in nature. The beauty of the fully planted landscape is to be preserved, and even enhanced, through the change. Plants that do not meet rigorous standards will be replaced in time with carefully documented and botanically scrutinized individuals. The initial step—evaluating existing collections for accuracy of documentation and identity and for health and suitability—has been going on since 1980. Essential to implementation of the new policy has been the updating and modernizing of the plant records by the introduction of electronic data-management systems. From the 1980 Sino-American expedition, from subsequent seed exchanges with botanical-garden staffs in Eurasia, and from several North American collecting forays, material from the wild with accompanying documentation has been received. The attempt to represent every hardy North American tree and shrub has been reinstated in the light of the discovery that many such plants had been lost from the collections and not replaced.

Related to the reappraisal of the provenance of their accessions by many botanical gardens and arboreta was the desire to strengthen their role as safe harbors for rare and endangered species. In 1984 a new organization was formed to address this issue nationwide. Named the Center for Plant Conservation, it coordinates the attempt to maintain, in botanical gardens, North American species threatened with extinction. Although habitat preservation is recognized as equally important to the survival of species, the added insurance of controlled cultivation is the center's particular mission. While many gardens keep these plants in their collections for education and research, the Center for Plant Conservation ensures that too much overlap in effort does not occur and that the less "glamorous" plants are not overlooked.

The center organized a network of institutions, each of which is responsible for certain geographic areas or taxa. Within its area of specialty each participating garden works with the center to identify the plants in need of cultivation. Care is taken in the collection process not to further the wild population's demise but to obtain as wide a sample as possible. Some progress has been made through the member organizations on reintroducing rare plants to the wild after their numbers have been increased under cultivation. The center maintains a database on the status of every species and records of garden holdings; this model effort has been consulted by many countries and by conservation organizations throughout the world. The Arboretum was a participating member of the program from its start. In fact, the center had its offices at the Arboretum for its first six years; it is now housed at the Missouri Botanical Garden. Trips to the field and propagation of the plants, some of which have rarely been artificially increased, have added many endangered native American plants to the Arboretum collections. Finding sites on the grounds with compatible growing conditions has presented a challenge.

RESTORATION OF THE SARGENT-OLMSTED PLAN

Although the renewed interest in Frederick Law Olmsted coincided with the 150th anniversary of his birth, in 1972, it was the relevance of Olmsted's work to the social and ecological crises of the 1960s and 1970s that sparked the rediscovery of the parkmaker as a major figure in the American experience. Within the decade surrounding 1972, many of his important publications were reissued, two full-length biographies were published, his life and work were the subject of doctoral dissertations, and several other titles were brought out, not the least of which was the first volume of his papers, edited by Charles C. McLaughlin and Charles E. Beveridge. In New York, Boston, and other cities throughout the country, park advocates urged a return to Olmsted's farsighted vision for scenic communal spaces as important civilizing forces in urban society.

Coming to Boston with the British viewpoint on the influence of the eighteenth- and nineteenth-century naturalistic landscapes on Downing and Olmsted and their compeers, Ashton suggested that the Arboretum landscape might benefit from a closer look at the origins of its design. From this evolved the idea of reviving the style and intent of Olmsted and Sargent, if not the original layout of the tree garden.

The examination of the original sketches, drawings, and plans for the Arboretum, still housed at the Olmsted Associates offices in Brookline in the late 1970s, greatly stimulated moves "to restore the Olmsted-Sargent plan, both as a work of art and, to the extent feasible, as a scientific collection" (Ashton, 1979, p. 343). In the course of preparing a feasibility study for the proposed restoration, horticultural taxonomist Richard Weaver and librarian Sheila Connor discovered that the Arboretum had originally been arranged according to the botanical sequence devised by George Bentham and Joseph Hooker, one of many facts that had been lost with time. For decades it was stated that the Arboretum plantings were scientifically arranged, but by the 1970s some of the staff had begun to wonder what this meant. It was obvious that a large number of the plants were grouped by genus, but the ordering of genera did not seem to follow the widely used sequence of Adolf Engler and Karl von Prantl. Because of the many alterations to the original plan by Sargent and others, discovering the Bentham and Hooker sequence took considerable botanical sleuthing. A prospectus for gradually returning the collections to this original sequence was developed, to be implemented as appropriate plants reached the right age and size.

As might be expected, many of the same stumbling blocks encountered by Sargent were met in the attempt to fit all the presently known entities into the century-old botanical system and to have at least some representative of every family along the Arboretum's main drive. What planners in the 1980s did not fully appreciate was that the "Sargent-Olmsted plan" that they relied on as the goal for restoration was a working document—it was a *plan,* not a depiction of the collections at the conclusion of planting. As has been mentioned, maps of the collections as planted and as they were altered for the remainder of Sargent's term and for Ames's have not been found. That the plan was made with relatively few non-American species in mind, and that it could not have included the measures Sargent did take to accommodate subsequent growth of the collections, also were not fully understood when the restoration objectives were formulated.

157. The restoration of Willow Path during the 1980s was inspired by investigations into the history of the grounds and the roles of both Olmsted and Sargent in creating its initial layout. Because of intervening plantings and changes in the course of Goldsmith Brook, the new Willow Path leaves Meadow Road at a point farther north than the original. A grove of dawn redwood, seen through the magnolia on the left in this picture, was planted at the site of the earlier path.

One of the goals of the restoration process was to reinstate at least some of the mown-grass paths developed at the turn of the century and to reintroduce native, herbaceous woodland and meadow vegetation along them. Looking back to the 1885 projection, the living-collections staff decided to go ahead with the long-since-abandoned idea of locating the rose family in the area where the shrub collection existed. All nonrosaceous shrubs were to be moved to sites near their tree relatives or elsewhere.

One of the best guides for restoring the paths was the photograph collection, especially those images captured on film early in the twentieth century by T. E. Marr. The decision was made to start with Oak Path, a route that had fallen into disuse once its lower section, which left Meadow Road at the junction with Valley Road, was obliterated by all-season mowing. The section that climbed the western slope of Bussey Hill to meet the lower end of Azalea Path required regrading since it had slumped over the years. Along this part of the path, white oaks and hemlocks that antedate Bussey's ownership still stand. Several mass plantings of native wildflowers were installed, and the existing naturalized flame azaleas were weeded and pruned to restore their health and abundance. Beds of sweet fern and bayberry were added on either side to define the path entrance from Bussey Hill Road and to serve as the main collection of the bayberry family (Myricaceae). The plan was to cease mowing all ground in the area, except for the path and the margins of the main drive.

Willow Path was also given a major facelift in the early 1980s. It was the route that left Meadow Road across from the administration building, paralleled the Arborway wall, and (by skirting the North Meadow and the maple collection) reached the Forest Hills gate. Swampy at its margin, it needed building up to provide a pleasant walking experience. A wooden bridge was erected to cross Goldsmith Brook. It was necessary to plant only a few new herbaceous wildflowers, since the meadow itself was rich with them. Along the wettest margins of Willow Path a supplementary alder collection was added. Both paths were readily adopted by visitors and staff.

Their apparent popularity inspired the living-collections

supervisor, Gary Koller, to improve, extend, and connect other paths to create a circumferential passage, named Sargent Trail. This was not a restoration, but a novel treatment of visitor circulation to areas not planted with major collections. Construction was completed in 1986, and the new trail opened some formerly less-frequented areas and provided a circular tour for pedestrians. The bulldozer on hand to create a section of Sargent Trail on the south side of Bussey Hill was also used to level the earthen mound that had been the foundation of the Bussey mansion. Not long after, a portion of the old lilac hedge that had lined Bussey's path to the top of the hill was removed, and the old pathway was extensively graded. These changes have opened new areas for collections development on the southeast side of the hill.

The greatest change to the Arboretum landscape resulting from the restoration idea was the creation of the Eleanor Cabot Bradley Collection of Rosaceous Plants on the site of the shrub collection. Bradley, long a member of the Arboretum visiting committee and an avid gardener, offered to provide financial assistance for the project and, most important, an endowment for the collection's future maintenance. As a tribute to her generosity, it was decided to name the collection for this dedicated lover of plants. The staff developed plans for displaying rose-family plants in the area, while the long process of removing the great mixture of shrubs from the rows and relocating or repropagating them began in 1981.

Once the area was cleared of all but a few specimen plants, large planting beds, interspersed with wide grassy lanes and openings, were created. Turf installation and soil improvement necessitated periods of rest before the beds were filled with the diverse woody relatives of rose, apple, and cherry. The rose family is one of the largest in numbers of species, especially among those families that contain hardy woody plants. Botanists and horticulturists generally divide it into four natural subfamilies, most easily distinguished by fruit type, and the arrangement of plants in the Bradley collection follows the subfamilial groupings. Thus, it continues the tradition of meshing scientific arrangement with graceful planting design. A granite bench was sited on the knoll overlooking the collection, and the area was dedicated in 1985, when planting was

158. *The Bradley Collection of Rosaceous Plants was started in 1985, inspired by the notions that the shrub collection was intended to be temporary and that Sargent and Olmsted meant to locate the rose family at this point in the Bentham and Hooker sequence. The new garden's free-form beds of trees and shrubs, interspersed with grassy paths and openings, are organized by genera within the botanically recognized subfamilies of the Rosaceae. Most of the nonrosaceous plants of the former shrub collection have been dispersed throughout the grounds.*

under way. Subsequently, the bench was visually anchored to the hillside by adding native-stone retaining walls, and granite cobblestones were laid flush with the ground to line the edges of the planted beds. These last give the area greater architectural definition than most other spaces in the Arboretum, except perhaps the summit of Bussey Hill.

Some reorganization of the lilac collection was initiated in the late 1980s in an attempt to improve the ease of maintenance and renew the vigor of some of the older plants. It was not a restoration, and the style of planting used was more gardenlike than that of any other place in the Arboretum. Over several years many varieties were propagated; when new plants attained suitable size, they were clustered in beds on the slope and underplanted with pachysandra, bleeding hearts, Spanish bluebells, and other spring-flowering bulbs. In creating a few large planting beds instead of siting each shrub individually on the steeply sloped lawn, the hope was to reduce the amount of mowing and the consequent potential of injury to the plants. The dense herbaceous layer would keep grass and people's feet away from the base of the specimens. The plants grouped in each bed were selected to illustrate certain aspects of lilac classification and the history of lilac breeding.

Also in the restorative mode (and a pleasing visual improvement to the grounds) was the complete repair and painting of the iron gates at all the entrances. The refurbished gates were protected against the worst of vandalism by the installation of heavy ornamental bollards, painted and styled to match. The bollards were sited to prevent vehicles from hitting the gates and to keep motorcycles from entering through the pedestrian openings. The work was made possible by grants from two City of Boston funds.

A significant change in the landscape and collections that took place in this decade was wholesale destruction of the elm collection by disease. While all the magnificent vase-shaped American elms that studded the landscape when Sargent laid out the collections had died by the early 1970s, many other species and varieties remained healthy and covered a large portion of the upper slopes of Bussey Hill. One or two would succumb every few years; the horticultural staff usually watched them closely. New kinds, some of them disease-resistant clones, were added over time; more recently, the new inject-method treatments had been tried on ones that showed signs of failure. Suddenly, most of the rest of them died over a couple of years. One summer, the upper hillside looked like a winter scene as scores of large trees stood dead. After their removal, the open grassy area looked desolate only to those who once knew its shades.

PUBLIC PROGRAMS

The Arboretum's resources for communicating with the populace were greatly increased during the 1980s. While for most of its existence the public relations and education function had been a responsibility shared among the horticultural and botanical staff, the 1970s brought the first hiring of public relations professionals and a fuller commitment to adult education classes. Public programs took on a life of their own in the decade that followed. Adult class offerings were greatly expanded to cover many topics

in horticulture that had no relation to the Arboretum's collections or fields of specialty. Instructors were no longer drawn solely from the institution's staff. For some years, however, an effort was made to interpret the Arboretum's scientific contributions and those of allied biological institutions at Harvard through a newsletter, *plantSciences*. A shop for the sale of botanical and horticultural books, Arboretum publications, maps, postcards, souvenirs, and items related to nature study was opened. The building was altered to better accommodate visitors and kept open on weekends, when visitation is greatest. Similar steps were being taken in gardens and museums everywhere.

A further departure from the traditional, relatively passive educational role was the initiation of children's programs; these were made available for a modest fee to public and private schools for pupils in grades four through six. Halting at first, the program gradually developed two components. One was the offering of different "field study experiences" at the Arboretum conducted by volunteers, with preparatory and follow-up materials for the teacher's use in the classroom. The other was an intensive summer training program for science teachers, presented in conjunction with several Boston area museums.

Relations with residents of the immediate neighborhood were enhanced by the spontaneous organization of a group known as the Arboretum Committee. Its birth came when citizens rallied to support the deployment of urban park rangers in 1983. The concept went back to Olmsted, who believed that parks ought to have keepers of the peace who would also act as teachers for visitors. As of the 1980s, New York's Central Park had created a ranger force as part of its restoration, and the claim was that the saving in litter cleanup and reduced vandalism covered the cost. The Arboretum Committee formed, with the assent of Ashton, to raise money to fund park rangers when Boston officials were still skeptical of the idea. Since the first year, Boston Park Rangers on horseback, on foot, or in vehicles have been assigned to nearly every park, including the Arboretum. They serve as guides, as liaison officers with the police when needed, and as educators, offering lectures, classroom programs, and tours when not on patrol. The committee has continued to support the ranger programs, as well as organizing cleanup days for the periphery of the grounds and promoting care of the Arboretum as one of the city's parks.

These education and public relations efforts notwithstanding, interpreting the collections and the landscape for the visitor who is not part of an organized group remains a challenge. For many decades administrators and staff believed that the novelty of the collection itself and the brief labels attatched to each plant were sufficient to attract the interest of those seeking knowledge. More recently, however, many kinds of museums have come to realize that although their objects may speak for themselves, not all visitors come equipped to understand their language. People will get much more out of their observations if they receive some guidance. Explanatory labels or signs can do much to inform museum viewers, but their use may diminish the visual experience. For some kinds of objects it is possible to rearrange collections or to select a few examples to better demonstrate their place in human culture or role in the natural world.

In a landscaped garden such as the Arboretum, however, special exhibits drawn from the living collections are not feasible. As plans were made to restore both its scientific and artistic aspects in the mid-1980s, it was agreed that anything more than minimal signs to aid visitor orientation would be inappropriate to preservation of its naturalistic beauty. One step taken to convey the Arboretum's value and purpose to the public was the publication of sourcebooks utilizing information from the humanities to explain the institution and its collections; the present volume and its companions are the result. But before an accompanying interpretation program could be implemented, Arboretum administration changed hands once more.

ANOTHER HORIZON

In 1987 Peter Ashton relinquished his administrative duties, intent on continuing research and teaching in tropical forest ecology and biodiversity conservation. The post of director of the Arboretum was assumed in 1989 by Robert E. Cook, a specialist in plant population biology. Having received his doctorate from Yale in 1973, Cook was a Cabot fellow at Harvard from 1974 to 1975, then served on the faculty of the Department of Biology for seven years before going to Ithaca to head the Cornell Plantations. His appointment marks a change in the administrative relationship between the Arboretum and the university: the Arboretum now is overseen by the vice president for administration rather than by the faculty of arts and sciences.

The most recent Arboretum director has addressed problems in physical facilities resulting from deferred maintenance and has initiated long-range planning to carry the institution beyond the year 2000. Most notable with regard to facilities has been the rehabilitation of the "Hunnewell Building," which still serves as both visitor center and administrative headquarters. Improving structural soundness, providing modern security and fire safety measures, upgrading utilities to meet current code requirements, installing climate control, and rendering the entire structure accessible to people with disabilities were the major goals of the renovation. To provide access for those in wheelchairs, the landscape in front of the building was altered to form a path that rises gradually to the main entrance. The new plantings here illustrate the floristic relationship between eastern North America and eastern Asia with several examples.

There have been further innovations involving the living collections since 1989. On the model of the endowed rose-family collection, another area of the Arboretum has benefited from landscape improvement and the provision of long-term maintenance, in this case as a result of a memorial trust. Under the guidance of Julie Meservy, a landscape architect and author from Wellesley, Massachusetts, a path through the rhododendron collection along the east side of Bussey Brook was rehabilitated in 1990 and 1991. Native stones were used subtly to protect areas from erosion, to guide pedestrians, and to provide seating among the evergreen shrubs and their complementary plantings; a rustic bridge was constructed to cross the brook. The refurbished walkway was named the Linda J. Davison Rhododendron Path in memory of the wife of the project's financial supporter.

An equally creative collaboration for the Arboretum

159. *Both Arboretum taxonomists Carroll Wood and Richard Howard retired from the faculty in June 1988. They have, however, continued their research: the final issue of Howard's six-volume* Flora of the Lesser Antilles *was published in 1989; nearly 80 percent of the papers projected to complete the generic flora of the southeastern United States have been published under Wood's guidance.*

living collections in the early 1990s was the permanent loan of several large trees to a new park, known as Post Office Square, in downtown Boston. In planning the park, which occupies space over an underground parking garage, the landscape designers and sponsoring Friends of Post Office Square wished to use a wide range of plant material and mature specimens if possible. When they turned to the Arboretum for advice, the idea of utilizing

160. Robert Cook, director of the Arboretum since January 1989, adds a symbolic shovelful of soil around one of six Arboretum trees on loan to the park at Post Office Square in downtown Boston at the dedication in early April 1991. With him are Amy Cheung, winner of a special poetry contest run by the Arboretum Children's Program for the event, and Boston's mayor, Ray Flynn. Labels on the trees and photographs in the entry to the parking garage that lies below the park remind visitors of the Arboretum's presence in the city.

some plants slated for removal from the collections took the form of a permanent loan as a means to acknowledge the origin of the trees once planted in the financial district. Six trees, no longer needed at the Arboretum because they do not meet present accessions policy and are well repre-

sented by other specimens, were moved to the park in the spring of 1991. With labels identifying their status, the trees extend Arboretum presence into the heart of the city.

At the herbarium, work is aimed at enhanced utilization of the collections through development of a national electronic database for type specimens, even as the newly named Center for Asian Botany builds on programs of the 1980s and earlier decades. Since giving up his administrative duties, Richard A. Howard has continued to work on the *Flora of the Lesser Antilles* and related topics in the history of botany in the West Indies. The third volume of the *Flora,* which treated the monocots, was published in 1979. The dicotyledonous families were enumerated in three final volumes; two of them came out in 1988 (the year of Howard's official retirement) and the last in 1989. Likewise, Carroll Wood has continued to oversee publication of papers for the generic flora of the southeastern United States project after retiring in 1988. With the suspension of the *Journal of the Arnold Arboretum* in 1990, subsequent contributions to the Generic Flora papers are published in a series supplementary to the *Journal,* the first of which was issued in 1991. In all, some 140 papers of the anticipated 184 installments have contributed to our knowledge of the plants of the Southeast, the center of tree species diversity in North America.

Planning for the development and curation of living collections to meet the next century's research and educational needs is the challenge to be met by the Arboretum staff. While looking back to the original design may provide inspiration for future management, it is just as valuable to realize that the tree garden's designers intended the Arboretum to be a viable resource for at least a thousand years. Its collections were meant to grow and change to serve educational roles about which the founders could only speculate.

Recently, interpretation for Arboretum visitors has been the subject of experimentation in collaboration with the National Park Service through the staff of Fairsted, the Frederick Law Olmsted National Historic Site in Brookline. While the Arboretum affords one of the best-preserved examples of Olmsted's artistry in Boston, its status as a combination of park and scientific collection makes it an exception within the emerald necklace. Nonetheless, the idea of an arboretum was one in which Olmsted took exceptional interest, as evidenced by his intense involvement toward the end of his career with the creation of a tree collection and experimental forest at George Vanderbilt's Biltmore. And, too, Boston's Arnold Arboretum must be understood equally as the creation of Sargent and succeeding curators and as a part of Cambridge's Harvard University.

A WORK OF MANY MINDS

From the institution's beginning there has been a constant interplay between science and art. The links between investigations of the natural world and the creative process of garden building began to play a role in the Arboretum's development at an early date. To bring together in one place all the trees and shrubs known was itself an experiment in a time when Darwin's theory had barely taken hold. The men who founded the Arboretum (Bussey, Arnold, Emerson, President Eliot) lived with the notion that "one of each kind" was sufficient to demonstrate the entirety of arboreal creation. The few historic precedents for arboreta employed a style of design suitable to the size, proportions, and longevity of trees. Although the concept of the public arboretum evolved along with the naturalistic landscape style for parks, estates, and rural cemeteries, it was the Arboretum's creators, Sargent and Olmsted, who most effectively commingled scientific order and serene beauty.

The emergence of biological science out of natural philosophy necessitated the formation of a library and herbarium; it also required the extension of the subjects of research and publication far beyond the Arboretum walls. In the forests of North America and the Far East, the institution's agents have surveyed the flora and returned with both information and plants. When raised at the Arboretum, the plants continued to be the subject of scrutiny after they joined their cohorts to improve upon the landscape. The tree garden became a popular resort, a refuge of beauty and tranquillity for Boston's citizens, yet its plants continued to be utilized for any number of research projects at the Arboretum and elsewhere through the institution's constant sharing of its material. As the science of taxonomy matured to the point of recognizing species as dynamic populations of mutable individuals, its conclusions came to be based on numerous examples taken from as many localities as possible. Observations of plants in their wild habitat, carried back in the form of

161. Surrounded by early green, and the glorious flowering of forsythia and cherries, children contemplate the nitty-gritty—tadpoles, fish, mud, water—at the edge of Dawson Pond. It is an Arboretum event as perennial as spring itself.

pressed specimens, outdistanced observations of plants in the garden as the basis for studying classification and relationships. Despite the diversification in plant science and its greater reliance on fieldwork and laboratory experimentation, the Arboretum living collection could still be used for surveys preliminary to field and laboratory studies. Although Bentham and Hooker's nineteenth-century model of family relationships was outdated by others, this classification-based arrangement was still considered the most manageable for the Arboretum's myriad of uses. Even for students of horticulture and landscape design, the most useful comparisons were facilitated by botanical arrangement of the collections. For those who sought contact with nature and spiritual restoration, the Arboretum landscape provided a maximum of diversity moderated by the harmony of its generic groupings.

When Sargent died, when Merrill arrived to unify the "botanical orphans," when the Bailey Plan was voted in, when the social unrest and intellectual questioning of the late 1960s was at its height—at each of these times of change the fear was voiced that the Arboretum would become "a mere park," that science would desert the pleasure ground, and that the Arboretum would somehow be diminished. However, there is nothing simple or inferior about a park, as the story of Olmsted and his antecedents has shown. Since the sesquicentennial of Olmsted's birth, there has been a slowly but surely growing recognition that urban parks are the result of highly creative processes aimed at fulfilling human intellectual and social needs. Park landscapes, especially an arboretum landscape, are works of the mind, and they should be understood and protected as such.

For Arboretum scientists, the age of environmental awareness brought a greater demand for information relevant to the conservation of individual species and to the study of world biodiversity. The conservation and management of forest resources were subjects just as important to the Arboretum's founders more than a century ago. To contribute its expertise in temperate and tropical forest floras to solving problems of global ecosystem management is the next challenge for the botanical program.

In the early 1970s the Arboretum came under great public pressure as an "oasis of green" and one of the city's suddenly valued "open spaces." From the horticulturally sophisticated homeowners of the 1950s and 1960s, who could appreciate the fine points of distinction among cultivars of woody plants, the Arboretum audience has expanded to include urban dwellers and others who have no gardening experience but want to "regreen" the city. To meet this demand for its horticultural expertise is another challenge for the Arboretum. Likewise, it must more actively interpret itself as a creation of some of our best parkmakers and plant scientists. It must reveal its "green" as multihued, and its "open space" as truly full with the plants and the stories of the people who got them and nurtured them on its undulating acres.

BIBLIOGRAPHY

In quoting from the correspondence of the staff of the Arnold Arboretum, I have used the abbreviation AAA in the text to indicate that the original letters or copies thereof are housed in the Archives of the Arnold Arboretum in Jamaica Plain, Massachusetts.

Abbott, K. M. 1903. *Old Paths and Legends of New England.* New York: G. P. Putnam's Sons.

Allen, C. K. 1964. An appreciation—Clarence Emmeren Kobuski (1900–1963). *Bulletin of the Torrey Botanical Club* 91:331–34.

Ames, J. S. 1954. Position of the Association for the Arnold Arboretum in the controversy with Harvard College. *AABGA Newsletter* 18:16–19.

Ames, O. 1933. The Arnold Arboretum [report for the year ending 30 June 1932]. In Report of the President of Harvard College and reports of the departments 1931–32, *Official Register of Harvard University* 30(1):249–58.

————. 1936. The Arnold Arboretum [report for the year ending 30 June 1935]. In Report of the President of Harvard College and reports of the departments 1934–35, *Official Register of Harvard University* 33(4):275–84.

Anderson, E. 1934. Hardy forsythias. *Bulletin of Popular Information,* 4th ser., 2:9–14.

Anderson, E., and D. Brockhoff. 1967. Henry Shaw. A pictorial biography. *Missouri Botanical Garden Bulletin* 55(6):1–16.

Anderson, I. 1926. *Under the Black Horse Flag.* Boston: Houghton Mifflin.

Anonymous. 1847. Reviews. A Report on the Trees and Shrubs of Massachusetts. *Horticulturist* 1:565–67.

————. 1848a. The Massachusetts horticultural show. *Horticulturist* 3:190–91.

————. 1848b. The horticultural festival at Faneuil Hall, Boston. *Horticulturist* 3:224–38.

————. 1872. Untitled editorial notice. *Gardener's Chronicle* 1872:1522, 16 November.

————. 1881. George B. Emerson. *Proceedings of the American Academy* 16:427–29.

————. 1927a. *The Arnold Arboretum and Its Future.* Jamaica Plain, Mass.: Harvard University.

————. 1927b. Charles S. Sargent memorial service. *Horticulture* 5:235–36.

————. 1927c. Prof. Sargent's funeral is held. Many prominent men attend Brookline services. *Boston Herald* 26 March.

————. 1927d. Sargent memorial. Arnold Arboretum holds memorial to Professor Sargent. *Boston Evening Transcript* 9 June.

————. 1928. New Arboretum bridle path affords fine canter. *Boston Herald* 8 April.

————. 1932. Mr. William H. Judd. *Gardener's Chronicle,* 3d ser., 91:232.

————. 1933. A new arboretum. *National Horticultural Magazine* 12:268–70.

———. 1934. Arboretum issues new guide for visitors. *Boston Herald* 5 August.

———. 1944. Harvard's botanical empire. *Harvard Alumni Bulletin* 46(9):269–72.

———. 1946. Blueprint for botany. *Harvard Alumni Bulletin* 48(11):443–47.

———. 1964. Introduction to an arboretum. *Bulletin of the Old Dartmouth Historical Society and Whaling Museum* Winter:1–4.

———. 1967. The builders of Cornell Plantations. *Cornell Plantations* 23(2):25–28.

Ashton, P. S. 1979. The director's report. The Arnold Arboretum during the fiscal year ended June 30, 1979. *Arnoldia* 39:330–68.

———. 1984. Botanic gardens and experimental grounds. In V. H. Heywood and D. M. Moore, eds., *Systematics Association Special Volume No. 25, Current Concepts in Plant Taxonomy*. London: Academic Press.

Avery, G. S. 1957. Botanic gardens—What role today? *American Journal of Botany* 44:268–71.

Bacon, E. M. 1897. *Walks and Rides in the Country Round About Boston*. Boston: Houghton, Mifflin [for the Appalachian Mountain Club].

Barber, L. 1980. *The Heyday of Natural History, 1820–1870*. Garden City, N.Y.: Doubleday.

Barker, J. G. 1885. Report of the committee on gardens. *Transactions of the Massachusetts Horticultural Society* 1885, part 2, pp. 323–42.

Baron, R. C. 1987. *The Garden and Farm Books of Thomas Jefferson*. Golden, Colo.: Fulcrum.

[Barry, P.]. 1850. Association for collecting rare trees and plants. *Horticulturist* 4:483–84.

Baxter, S. 1909. The remarkable work of the Arnold Arboretum. *World's Work* September, pp. 1182–94.

Bean, W. J. 1910. A visit to the Arnold Arboretum. *Kew Bulletin of Miscellaneous Information* 1910(8):261–69.

Bender, T. 1975. *Toward an Urban Vision*. Baltimore: Johns Hopkins University Press.

Benson, A. E. 1929. *History of the Massachusetts Horticultural Society*. Boston: the Society.

———. 1942. *History of the Massachusetts Society for Promoting Agriculture, 1892–1942*. Boston: Meador.

[Bentinck-Smith, W.] 1953. Storm over botany. *Harvard Alumni Bulletin*. 55(14):583–585, 589.

Bigelow, J. 1987. [excerpt from] A report on Mount Auburn Cemetery, 1831. *Sweet Auburn* Fall/Winter.

Boston Landmarks Commission. 1983. *1982 Survey and Planning Grant, Part 1—Jamaica Plain Project Completion Report*. Unpublished report, library of the Arnold Arboretum.

Boston Society of Natural History. 1930. *The Boston Society of Natural History, 1830–1930*. Boston: the Society.

Boufford, D. E., and B. Bartholomew. 1986. The 1984 Sino-American botanical expedition to Yunnan, China [includes discussion of 1980 expedition]. *Arnoldia* 46:15–36.

Boufford, D. E., and S. A. Spongberg. 1983. Eastern Asian–eastern North American phytogeographical relationships. *Annals of the Missouri Botanical Garden* 70:423–39.

Bouvé, T. T. 1880. Historical sketch of the Boston Society of Natural History with a notice of the Linnaean Society which preceded it. *Anniversary Memoirs of the Boston Society of Natural History, 1830–1880*. Boston: the Society.

Brent, H. J. 1868. *Was It a Ghost? The Murders in Bussey's Wood: An Extraordinary Narrative*. Boston: Loring.

Britton, N. L. 1898. Botanical gardens. Origin and development. *Science*, n.s., 4:284–93.

Brooks, G. 1962. *Boston and Return*. New York: Atheneum.

Brooks, Van Wyck. 1940a. *The Flowering of New England, 1815–1865*. rev. ed. New York: E. P. Dutton.

———. 1940b. *New England: Indian Summer, 1865–1915*. New York: E. P. Dutton.

Bullard, J. M. 1947. *The Rotches*. New Bedford, Mass.: privately published.

Burkill, J. H. 1956. Prof. E. D. Merrill. *Nature* 177:687–88.

Bussey, B. 1841. *Will of Benjamin Bussey of Roxbury*. Privately printed.

———. 1899. Benjamin Bussey. *Dedham Historical Register* 10:70–76 [posthumously published autobiography].

Chock, A. K. 1963. J. F. Rock, 1884–1962. *Taxon* 12:89–102.

Cleveland, H. W. S. 1869. *The Public Grounds of Chicago: How to Give Them Character and Expression*. Chicago: Charles D. Lakey.

———. 1873. *Landscape Architecture as Applied to the Wants of the West, with an Essay on Forest Planting on the Great Plains*. Chicago: Jansen, McClurg.

Clifford, D. 1966. *A History of Garden Design*. London: Faber and Faber.

Constance, L. 1964. Systematic botany—An unending synthesis. *Taxon* 13:257–73.

Copeland, R. M., and H. W. S. Cleveland. 1856. *A Few Words on Central Park*. Boston: n.p.

Coville, F. V. 1925a. Bill introduced for a national arboretum. *American Nurseryman* 41:10.

———. 1925b. Practical value of big national arboretum. *American Nurseryman* 41:44.

Cranz, G. 1982. *The Politics of Park Design*. Cambridge, Mass.: MIT Press.

Crapo, H. H. 1917. Banks of Old Dartmouth. *Old Dartmouth Historical Sketches* 46:23–51.

Croizat, L. 1938. Mapping the Arboretum. *Bulletin of Popular Information*, 4th ser., 6:23–26.

Curtis, J., W. Curtis, and F. Lieberman. 1982. *The World of*

George Perkins Marsh, America's First Conservationist and Environmentalist. Woodstock, Vt.: Woodstock Foundation.

Curtis, R. W. 1945. He speaks for himself. *Cornell Plantations* 1(4):59–60.

Darwin, F., ed. 1887. *The Life and Letters of Charles Darwin.* 2 vols. New York: D. Appleton.

Davis, P. H., and V. H. Heywood. 1965. *Principles of Angiosperm Taxonomy.* Princeton, N.J.: D. Van Nostrand.

Derderian, C. E. 1980. Bonsai at the Arnold Arboretum. *Bonsai Journal* 13(4):75–78.

DeVos, F. 1979. The nature of present day collections. *Bulletin AABGA* 13:12–15.

DeWolf, G. P., Jr. 1969. Dr. Donald Wyman. *Arnoldia* 29: supplement pages not numbered.

Downing, A. J. 1844. *Treatise on the Theory and Practice of Landscape Gardening.* 2d ed. New York: Wiley and Putnam.

[Downing, A. J.]. 1849. Public cemeteries and public gardens. *Horticulturist* 4:9–12.

Downing, A. J. 1850a. The Derby Arboretum. *Horticulturist* 5:266–68.

———. 1850b. Mr. Downing's letters from England. *Horticulturist* 5:217–23.

Drake, S. A. 1900. *Old Landmarks and Historic Personages of Boston.* Boston: Little, Brown.

Dumbarton Oaks Colloquium on the History of Landscape Architecture, eds. 1980. *John Claudius Loudon and the Early Nineteenth Century in Great Britain.* Washington: Dumbarton Oaks Trustees, for Harvard University.

———. 1982. *Beatrix Jones Farrand (1872–1959). Fifty Years of American Landscape Architecture.* Washington: Dumbarton Oaks Trustees, for Harvard University.

Dunbar, J. 1897. Has no equal in the world. The Arnold Arboretum of Boston and how it was established. *Rochester Democrat* 22 December.

Duncan, F. 1905. Professor Charles Sprague Sargent and the Arnold Arboretum. *Critic and Literary World* 47:115–19.

Dupree, A. H. 1959. *Asa Gray, 1810–1888.* Cambridge, Mass.: Harvard University Press, Belknap Press.

Eliot, C. W. 1872. *Annual Reports of the President and Treasurer of Harvard College, 1871–72.* Cambridge, Mass.: Harvard University.

———. 1895. President's report for 1893–94. In *Annual Reports of the President and Treasurer of Harvard College, 1893–94,* pp. 3–46. Cambridge, Mass.: Harvard University.

———. 1902. *Charles Eliot, Landscape Architect.* Boston: Houghton, Mifflin.

Ellis, G. E. 1880. *Memoir of Jacob Bigelow, M.D., LL.D.* Cambridge, Mass.: John Wilson and Son.

Ellis, L. B. 1892. *History of New Bedford and Its Vicinity, 1602–1892.* Syracuse, N.Y.: D. Mason.

Emerson, G. B. 1846. *Report on the Trees and Shrubs Growing Naturally in the Forests of Massachusetts.* Boston: Commonwealth of Massachusetts.

———. 1859. *Forest Trees* [reprint from the Transactions of the Norfolk Agricultural Society for 1859].

———. 1878. *Reminiscences of an Old School Teacher.* Boston: A. Mudge and Son.

———. 1887. *Report on the Trees and Shrubs Growing Naturally in the Forests of Massachusetts.* 4th ed. Boston: Little, Brown.

Emerson, G. B., S. May, and T. J. Mumford. 1876. *Memoir of Samuel Joseph May.* Boston: American Unitarian Society.

Emerson, R. W., W. H. Channing, and J. F. Clarke, eds. 1852. *Memoirs of Margaret Fuller Ossoli.* 2 vols. Boston: Phillips, Samson.

Fairchild, D. 1944. *The World Was My Garden.* New York: Charles Scribner's Sons.

Farrand, B. 1946. Contemplated landscape changes at the Arnold Arboretum. *Arnoldia* 6:45–48.

———. 1949a. The azalea border. *Arnoldia* 9:6–7.

———. 1949b. Peters Hill. *Arnoldia* 9:38–43.

———. 1959. Beatrix Farrand, 1872–1959. *Reef Point Gardens Bulletin* 1(17): pages not numbered.

Faull, A. F. 1962. Joseph Horace Faull, 1870–1961. *Journal of the Arnold Arboretum* 43:223–33.

Fein, A. 1972. *Frederick Law Olmsted and the American Environmental Tradition.* New York: George Braziller.

Finan, J. J. 1972. Edgar Anderson, 1897–1969. *Annals of the Missouri Botanical Garden* 59:325–45.

Foley, D. J., ed. 1969. *The Flowering World of "Chinese" Wilson.* London: Macmillan.

Fuller, H. W. 1873. Report of the committee on ornamental gardening. *Transactions of the Massachusetts Horticultural Society for the year 1873,* pp. 146–56.

Gambee, E. B. 1974. The founding fathers. *Garden Journal* 24(2):34–37.

Geary, S. C., and B. J. Hutchinson. 1980. Mr. Dawson, plantsman. *Arnoldia* 40:51–75.

———. 1981. The original design and permanent arrangement of the Arnold Arboretum as determined by FLO and CSS—A chronology. Unpublished manuscript, Archives of the Arnold Arboretum.

Goode, P., and M. Lancaster, eds. 1986. *The Oxford Companion to Gardens.* Oxford: Oxford University Press.

Graustein, J. E. 1967. *Thomas Nuttall, Naturalist. Explorations in America, 1808–1841.* Cambridge, Mass.: Harvard University Press.

Gray, T. 1842. *A Tribute to the Memory of Benjamin Bussey, Esq.* Boston: I. R. Butts.

Greene, J. C. 1984. *American Science in the Age of Jefferson.* Ames: Iowa State University Press.

Haas, W. J. 1988. Transplanting botany to China: The cross-cultural experience of Chen Huanyong. *Arnoldia* 48:9–25.

Hadfield, M., R. Harling, and L. Highton. 1980. *British Gardeners. A Biographical Dictionary.* London: A. Zwemmer.

Hall, M. T. 1967. The arboretum as a cultural institution. *Morton Arboretum Quarterly* 3:49–52.

Hammond, C. A. 1982. "Where the arts and virtues unite": Country life near Boston, 1637–1864. Unpublished Ph.D. dissertation, Boston University.

Hedrick, U. P. 1950. *A History of Horticulture in America to 1860.* New York: Oxford University Press.

Higginbotham, J. S. 1990. Four centuries of planting and progress. *American Nurseryman* 171(12):36–59.

Higginson, T. W. 1884. *American Men of Letters.* Boston: Houghton, Mifflin.

Hosmer, R. S. 1947. *The Cornell Plantations. A History.* Ithaca, N.Y.: Administrative Committee of the Cornell Plantations.

Hovey, C. M. 1840. Notes on gardens and gardening in New Bedford, Massachusetts. *Magazine of Horticulture* 6:361–66.

[Hovey, C. M.]. 1855. Suburban visits, residence of H. H. Hunnewell, Esq., West Needham. *Magazine of Horticulture* 21:378–81.

Hovey, C. M. 1868. The progress of horticulture. *Magazine of Horticulture* 34:1–14.

Howard, R. A. 1954. Progress at the Arnold Arboretum. *AABGA Newsletter* 18:13–15.

———. 1955. The Arnold Arboretum and the Supreme Court. *AABGA Newsletter* 22:1–6.

———. 1956. Elmer Drew Merrill, 1876–1956. *Journal of the Arnold Arboretum* 37:197–216.

———. 1961. Ivan Murray Johnston, 1898–1960. *Journal of the Arnold Arboretum* 42:1–9.

———. 1962. The director's report. The Arnold Arboretum during the fiscal year ended June 30, 1962. *Journal of the Arnold Arboretum* 43:439–60.

———. 1963. Clarence Emmeren Kobuski, 1900–1963. *Journal of the Arnold Arboretum* 44:411–20.

———. 1965. Susan Delano McKelvey, 1883–1964. *Journal of the Arnold Arboretum* 46:45–47.

———. 1966. The director's report. The Arnold Arboretum during the fiscal year ended June 30, 1966. *Journal of the Arnold Arboretum* 47:323–46.

———. 1968. Irving Widmer Bailey, 1884–1967. *Journal of the Arnold Arboretum* 49:1–15.

———. 1970. The director's report. The Arnold Arboretum during the fiscal year ended June 30, 1970. *Arnoldia* 30:201–50.

———. 1972. Scientists and scientific contributions of the Arnold Arboretum. *Arnoldia* 32:49–58.

———. 1973. Lazella Harenberg Schwarten, 1900–1973. *Journal of the Arnold Arboretum* 54:419–21.

———. 1974. Karl Sax, 1892–1973. *Journal of the Arnold Arboretum* 55:333–43.

———. 1978. Botanical impressions of the People's Republic of China. *Arnoldia* 38:218–37.

———. 1980. E. H. Wilson as a botanist. *Arnoldia* 40:102–38, 154–93.

———. 1993. Lily May Perry, 1895–1992. *Harvard Papers in Botany* 4:86–87.

Hu, S. 1949. The genus *Ilex* in China. *Journal of the Arnold Arboretum* 30:231–344.

Hubbard, T. K. 1930. H. W. S. Cleveland. *Landscape Architecture* 20:92–111.

Humphrey, J. E. 1896. Botany and botanists in New England. *New England Magazine* 14:27–44.

Hunnewell, H. H., ed. 1906. *Life, Letters, and Diary of Horatio Hollis Hunnewell.* 3 vols. Boston: privately printed.

Hyams, E. 1971. *Capability Brown and Humphrey Repton.* New York: Charles Scribner's Sons.

Irwin, H. S. 1971. The role of the botanical garden in the modern world. *Garden Journal* 21:167–70.

———. 1972. Short history of arboretums. *Garden Journal* 22:6–10.

Jack, J. G. 1904. The Arnold Arboretum. *Transactions of the Massachusetts Horticultural Society* for the year 1904, part 1, pp. 59–75.

———. 1949. The Arnold Arboretum. Some personal notes. *Chronica Botanica* 12:185–200.

Judd, W. H. 1931. Dr. E. H. Wilson, V.M.H. *Journal of the Kew Guild* 5(38):79–81.

Kammerer, E. L. 1963. The Morton Arboretum. *American Horticultural Magazine* 42(2):101–5.

Kobuski, C. E. 1950. Alfred Rehder, 1863–1949. *Journal of the Arnold Arboretum* 31:1–38.

———. 1962. Ernest Jesse Palmer, 1875–1962. *Journal of the Arnold Arboretum* 43:351–58.

Lawton, B. 1968. George Engelmann, 1809–1884. Scientific father of the garden. *Missouri Botanical Garden Bulletin* 56(6):10–17.

Lee, E. B. 1849. *Memoirs of Rev. Joseph Buckminster D.D. and His*

Son *Rev. Joseph Stevens Buckminster.* Boston: Wm. Crosby and H. P. Nichols.

Leet, J. 1990. He dared. Dr. Donald Wyman. *American Nurseryman* 171(12):118–19.

Lehmer, M. 1961. The Walter Street "berrying" ground. *Arnoldia* 21:75–82.

Lewis, M. D. 1958. *Lumberman from Flint. The Michigan Career of Henry H. Crapo, 1855–1869.* Detroit: Wayne State University Press.

Lombardi, W. L. 1967. Arnold Arboretum—Scene of famous murders. *Prudential Center News* 1 April.

Loudon, J. C. 1827. On the natural arrangement of plants [untitled note by the conductor]. *Gardener's Magazine* 2:300–303.

———. 1828. Foreign notices—Germany. *Gardener's Magazine* 4:497–98.

———. 1829. Arboretum in the garden of the Horticultural Society. *Gardener's Magazine* 5:344–48.

———. 1832. Birmingham Botanical Garden plan. *Gardener's Magazine* 8:407–28.

———. 1835. Remarks on laying out public gardens and promenades. *Gardener's Magazine* 11:644–69.

———. 1840. *The Derby Arboretum.* London: Longman, Orme, Brown, Green and Longmans.

Mann, A. 1954. *Yankee Reformers in the Urban Age.* Cambridge, Mass.: Harvard University Press, Belknap Press.

Manning, R. 1880. *History of the Massachusetts Horticultural Society, 1829–1878.* Boston: the Society.

Marsh, G. P. 1874. *The Earth as Modified by Human Action.* New York: Scribner, Armstrong.

Mayr, E. 1982. *The Growth of Biological Thought.* Cambridge, Mass.: Harvard University Press, Belknap Press.

McFarland, J. H. 1905. A tree garden to last a thousand years. *Country Calendar* 1:232–36.

McKelvey, S. D. 1936. The Arnold Arboretum. *Harvard Alumni Bulletin* 38:464–72.

McLaughlin, C. C., and C. E. Beveridge, eds. 1977. *The Papers of Frederick Law Olmsted.* vol. 1. *The Formative Years, 1822 to 1852.* Baltimore: Johns Hopkins University Press.

M'Mahon, B. 1806. *American Gardener's Calendar.* Philadelphia: B. Graves.

Meisel, M. 1926. *Bibliography of American Natural History.* vol. 2. New York: Premier.

Melville, H. 1980. *Moby Dick.* New York: Signet.

Merrill, E. D. 1935. Botany at Harvard University. *Scientific Monthly* 41:468–73.

———. 1937. Botanical Collections. In Report of the President of Harvard College and reports of the departments for 1935–36, *Official Register of Harvard University* 34(11):292–95.

[Merrill, E. D.]. 1938a. Current Arboretum activities of general interest. *Bulletin of Popular Information,* 4th ser., 6:51–57.

Merrill, E. D. 1938b. Preliminary report on the storm damage to the Arboretum on September twenty-first. *Bulletin of Popular Information,* 4th ser., 6:70.

———. 1945. The Arnold Arboretum during the fiscal year ended June 30, 1945. *Journal of the Arnold Arboretum* 26:484–99.

———. 1946. The Arnold Arboretum during the fiscal year ended June 30, 1946. *Journal of the Arnold Arboretum* 27:486–510.

———. 1948. *Metasequoia,* another "living fossil." *Arnoldia* 8:1–8.

———. 1953. Autobiographical: Early years, the Philippines, California. *Asa Gray Bulletin,* n.s., 2:335–70.

Michaux, F. A. 1819. *The North American Silva.* Paris: C. D. Hautel.

Miller, W. 1904. The world's greatest tree garden. *Country Life in America* 15:386–91, 444.

Morison, J. H. 1881. George Barrell Emerson, LL.D. *Unitarian Review* 16:59–69.

Morison, S. E., ed. 1930. *The Development of Harvard University since the Inauguration of President Eliot, 1869–1929.* Cambridge, Mass.: Harvard University Press.

Morison, S. E. 1965. *The Oxford History of the American People.* New York: Oxford University Press.

Morris Arboretum of the University of Pennsylvania, ed. 1933. *Proceedings at the Dedication of the Arboretum, June 2, 1933.* Philadelphia: Morris Arboretum.

Morse, A. P. 1929. John Robinson, botanist, of Salem, Massachusetts. *Rhodora* 31:245–54.

Muir, J. 1903. Sargent's Silva. *Atlantic Monthly* 92:9–22.

Mumford, L. 1971. *The Brown Decades. A Study of the Arts in America, 1865–1895.* New York: Dover.

Newton, N. T. 1971. *Design on the Land.* Cambridge, Mass.: Harvard University Press.

Old Dartmouth Historical Society. 1919. Proceedings of the special meeting, February 25, 1919 [excerpts from the diaries of J. Q. and C. F. Adams were read]. *Old Dartmouth Historical Sketches* 47:11–23.

Olmsted, F. L., and T. Kimball, eds. 1922. *Frederick Law Olmsted, Landscape Architect, 1822–1903. Early Years and Experiences.* New York: G. P. Putnam's Sons.

———. 1973. *Forty Years of Landscape Architecture: Central Park.* Cambridge, Mass.: MIT Press.

Ornduff, R. 1978. Using living collections—the problems. *Bulletin AABGA* 12:113–17.

Palmer, E. J. 1930. The spontaneous flora of the Arnold Arboretum. *Journal of the Arnold Arboretum* 11:63–119.

———. 1934. Indian relics of the Arnold Arboretum. *Bulletin of Popular Information,* 4th ser., 2:61–68.

———. 1935. Supplement to the spontaneous flora of the Arnold Arboretum. *Journal of the Arnold Arboretum* 16:81–97.

Paxton, J. 1835. Some account of the arboretum lately commenced by His Grace the Duke of Devonshire, at Chatsworth, in Derbyshire. *Gardener's Magazine,* n.s., 1:385–91.

Pease, Z. W. 1919. Arnold's garden. *Old Dartmouth Historical Sketches* 48:16–19.

———. 1924. The Arnold Mansion and its traditions. *Old Dartmouth Historical Sketches* 52:5–35.

Plimpton, P. A., ed. 1979. *Oakes Ames. Jottings of a Harvard Botanist, 1874–1950.* Cambridge, Mass.: Botanical Museum of Harvard University.

Potter, W. J. 1860. To the memory of Mrs. Sarah R. Rotch ["not published" pamphlet in the Archives of the Arnold Arboretum].

———. 1868. *To the Memory of James Arnold.* New Bedford, Mass.: Fessenden and Baker, Printers.

Pyle, R. 1925. Report committee of botanical gardens and arboretums. *American Nurseryman* 42:20.

Raup, H. M. 1935. Notes on the early uses of land now in the Arnold Arboretum. *Bulletin of Popular Information,* 4th ser., 3:41–74.

———. 1937. The Chinese collection. *Bulletin of Popular Information,* 4th ser., 5:25–28.

———. 1940. The genesis of the Arnold Arboretum. *Bulletin of Popular Information,* 4th ser., 8:1–11.

Raymo, C. 1986. Tree book as Yankee as cod. *Boston Globe* 28 July.

Rehder, A. 1898. Das Arnold Arboretum. *Mitteilungen der Deutschen Dendrologischen Gesellschaft* 7:89–93.

———. 1908. Notes from the Arnold Arboretum. *Horticulture* 7:717.

———. 1927a. Charles Sprague Sargent. *Journal of the Arnold Arboretum* 8:69–86.

———. 1927b. *Manual of Trees and Shrubs Hardy in North America.* New York: Macmillan.

———. 1930. Ernest Henry Wilson. *Journal of the Arnold Arboretum* 11:181–92.

[Rehder, A., J. H. Faull, and H. M. Raup]. 1936. Professor Ames and Professor Jack of the Arnold Arboretum. *Science* 38:200–201.

Rehder, G. 1972. The making of a botanist. *Arnoldia* 32:141–56.

Ricketson, D. 1858. *The History of New Bedford.* New Bedford, Mass.: published by the author.

Robbins, M. C. 1892. New England parks. The Arnold Arboretum. *Garden and Forest* 5:27–29.

———. 1893. A tree museum. *Century Magazine* 45:867–78.

Robinson, B. L. 1930. Botany. In S. E. Morison, ed., *The Development of Harvard University since the Inauguration of President Eliot, 1869–1929,* pp. 338–377. Cambridge, Mass: Harvard University Press.

Rockwell, S. F. 1932. *Davis Families of Early Roxbury and Boston.* North Andover, Mass.: Andover Press.

Roeding, G. C. 1924. On the eve of a trade semi-centennial. *American Nurseryman* 39:115–16.

Rollins, R. C. 1951. The end of a generation of Harvard botanists. *Taxon* 1:3–4.

———. 1963. Clarence Emmeren Kobuski 1900–1963. *Taxon* 12:213–15.

Roper, L. W. 1973. *FLO, A Biography of Frederick Law Olmsted.* Baltimore: Johns Hopkins University Press.

Sargent, C. S. 1876. *A Few Suggestions on Tree Planting* [from report of the Massachusetts State Board of Agriculture for 1875]. Boston: Wright and Potter, State Printers.

———. 1878. *Notes on Trees and Tree Planting* [from the twenty-fifth annual report of the Secretary of the Board of Agriculture]. Boston: Rand, Avery and Co., Printers to the Commonwealth.

———. 1879. Arboretum [report for the year ending 31 August 1878]. In *Annual Reports of the President and Treasurer of Harvard College, 1877–78,* pp. 100–104. Cambridge, Mass.: Harvard University.

———. 1880. Letter to Board of Park Commissioners. In *Fifth Annual Report of the Board of Commissioners of the Department of Parks for the City of Boston for the Year 1879.* Boston: City of Boston.

———. 1881. *Annual Report of the Director of the Arnold Arboretum to the President and Fellows of Harvard College for 1879–80.* Cambridge, Mass.: John Wilson and Son.

———. 1883. A national forest preserve. *Nation* 37:201.

———. 1884. *Report on the Forests of North America* [vol. 9 of the Tenth Census of the United States]. Washington: U.S. Government Printing Office.

———. 1885. To the President of the University [report for the year ending 31 August 1884]. In *Annual Reports of the President and Treasurer of Harvard College, 1883–84,* pp. 154–156. Cambridge, Mass.: Harvard University.

———. 1886. To the President of the University [report for the year ending 31 August 1885]. In *Annual Reports of the President and Treasurer of Harvard College, 1884–85,* pp. 147–52. Cambridge, Mass.: Harvard University.

———. 1887. The Arnold Arboretum [report for the year ending 31 August 1886]. In *Annual Reports of the President and*

Treasurer of Harvard College, 1885–86, pp. 122–24. Cambridge, Mass.: Harvard University.

———. 1888. Asa Gray. *Garden and Forest* 1:1.

Sargent, C. S., ed. 1889. *Scientific Papers of Asa Gray.* 2 vols. Boston: Houghton, Mifflin.

Sargent, C. S. 1893. The Arnold Arboretum [report for the year ending 31 August 1892]. In *Annual Reports of the President and Treasurer of Harvard College, 1891–92,* pp. 169–70. Cambridge, Mass.: Harvard University.

———. 1894. *Forest Flora of Japan.* Boston: Houghton, Mifflin.

———. 1895a. The Arnold Arboretum [report for the year ending 31 July 1894]. In *Annual Reports of the President and Treasurer of Harvard College, 1893–94,* pp. 188–89. Cambridge, Mass.: Harvard University.

———. 1895b. Untitled note about Peters Hill. *Garden and Forest* 8:292.

———. 1906. The Arnold Arboretum [report for the year ending 31 July 1905]. In *Annual Reports of the President and Treasurer of Harvard College, 1904–05,* pp. 246–48. Cambridge, Mass.: Harvard University.

———. 1911a. *A Guide to the Arnold Arboretum.* Cambridge, Mass.: Riverside Press.

———. 1911b. Preface. In A. Rehder, ed., *The Bradley Bibliography.* vol. 1, p. iv. Cambridge, Mass.: Riverside Press.

———. 1912a. The Arnold Arboretum [report for the year ending 30 June 1911]. In *Annual Reports of the President and Treasurer of Harvard College, 1910–11,* pp. 187–89. Cambridge, Mass.: Harvard University.

———. 1912b. Introduction. In E. H. Wilson, *Vegetation of Western China.* London: printed for the subscribers.

———. 1914. Arboretum. In L. H. Bailey, *Standard Cyclopedia of Horticulture.* vol. 1, pp. 347–52. New York: Macmillan.

———. 1917. The Arnold Arboretum. What it is and does. *Garden* 26:122–25.

———. 1918. Charles Edward Faxon. *Rhodora* 20:117–22.

———. 1922. The first fifty years of the Arnold Arboretum. *Journal of the Arnold Arboretum* 3:127–71.

———. 1926. The Arnold Arboretum [report for the year ending 30 June 1925]. In *Reports of the President and Treasurer of Harvard College, 1924–25,* pp. 217–21. Cambridge, Mass.: Harvard University.

———. 1927. The greatest garden in America. *Home Acres* 12:95, 112.

Sargent, E. W. 1923. *Epes Sargent of Gloucester and His Descendants* [with biographical notes by C. S. Sargent]. Boston: Houghton, Mifflin.

Sargent, H. W. 1855. Evergreen shrubs. *Horticulturist,* n.s., 5:205–7.

———. 1856. A third winter on the new evergreens. *Horticulturist,* n.s., 6:227–28.

———. 1865. Supplement. In A. J. Downing, *Treatise on the Theory and Practice of Landscape Gardening.* 7th ed. New York: Orange Judd.

Sax, K. 1947a. The Arnold Arboretum of Harvard University. *Arboretum Bulletin* 10(3):9–10, 24.

———. 1947b. The Bussey Institution. *Arnoldia* 7:13–16.

———. 1949. John George Jack, 1861–1949. *Journal of the Arnold Arboretum* 30:345–47.

———. 1950. Oakes Ames, 1874–1950. *Journal of the Arnold Arboretum* 31:335–49.

———. 1955. Plant breeding at the Arnold Arboretum. *Arnoldia* 15:5–12.

Schofield, E. A. 1987. A life redeemed: Susan Delano McKelvey. *Arnoldia* 47:9–23.

Schuyler, D. 1976. The evolution of the Anglo-American rural cemetery: Landscape architecture as social and cultural history. *Garden History* 4:291–304.

Seeley, J. G. 1969. The Cornell Plantations. *Longwood Program Seminars, 1968–1969* (University of Delaware, Newark) 1:48–52.

Seibert, R. J. 1956. Arboreta and botanical gardens in the field of plant sciences and human welfare. *American Journal of Botany* 43:736–38.

Shaw, E. A. 1986. Changing botany in North America: 1835–1860. The role of George Engelmann. *Annals of the Missouri Botanical Garden* 73:508–19.

Simo, M. L. 1981. John Claudius Loudon: On planning and design for the garden metropolis. *Garden History* 9:184–201.

Smallwood, W. M. 1941. *Natural History and the American Mind.* New York: Columbia University Press.

Smith, C. E. 1954. A century of botany in America. *Bartonia* 28:1–30.

Smith, J. J. 1847. Arboricultural gossip. *Horticulturist* 2:28–31.

[Smith, J. J.]. 1856. Visits to country places, No. 3. About New York. The north river. *Horticulturist,* n.s., 6:445–49.

Smith, J. J. 1857. Visits to country places, No. 7. Around Boston. *Horticulturist,* n.s., 7:65–69.

Sorensen, P. D. 1978. Dates of publication of Sargent's Silva of North America. *Journal of the Arnold Arboretum* 59:68–73.

Spongberg, S. A. 1979. The collections policy of the Arnold Arboretum. Taxa of infraspecific rank and cultivars. *Arnoldia* 39:370–76.

———. 1990. *A Reunion of Trees.* Cambridge, Mass.: Harvard University Press.

Stearn, W. T. 1972. From medieval park to modern arboretum: The Arnold Arboretum and its historic background. *Arnoldia* 32:173–97.

Steere, W. C. 1969. Research as a function of a botanical garden. *Longwood Program Seminars, 1968–1969* (University of Delaware, Newark) 1:43–47.

Stout, B. B., ed. 1981. *Forests in the Here and Now. A Collection of the Writings of Hugh Miller Raup.* Missoula: Montana Forest and Conservation Experiment Station.

Sutton, S. B. 1970. *Charles Sprague Sargent and the Arnold Arboretum.* Cambridge, Mass.: Harvard University Press.

———. 1971a. *The Arnold Arboretum: The First Century.* Jamaica Plain, Mass.: Arnold Arboretum of Harvard University.

Sutton, S. B., ed. 1971b. *Civilizing American Cities: A Selection of Frederick Law Olmsted's Writings on City Landscapes.* Cambridge, Mass.: MIT Press.

Sutton, S. B. 1972. The Arboretum administrators: An opinionated history. *Arnoldia* 32:3–20.

Teuscher, H. 1940. Program for an ideal botanical garden. *Memoirs of the Montreal Botanical Garden* (English edition) 1:1–32.

Trelease, W. 1929. Biographical memoir of Charles Sprague Sargent. *Biographical Memoirs* (National Academy of Sciences, Washington) 12:247–70.

van Melle, P. J. 1953. Andrew Jackson Downing, landscape gardener, 1815–1852. *Bulletin of the Garden Club of America* 41:15–19.

Van Rensselaer, Mrs. S. 1918. A living picture-book for artists. *Century Magazine* 96:223–28.

Verdoorn, F., ed. 1946. Merrilleana: A selection from the general writings of Elmer Drew Merrill, SC.D., LL.D. *Chronica Botanica* 10:127–394.

Wade, M. 1940. *Margaret Fuller. Whetstone of Genius.* New York: Viking.

Warner, S. B. 1977. *The Way We Really Live: Social Change in Metropolitan Boston since 1920.* Boston: Trustees of the Public Library of the City of Boston.

Waterston, R. C. 1884. *Memoir of George Barrell Emerson, LL.D.* Cambridge, Mass.: John Wilson and Son.

Waugh, F. A., ed. 1921. *Landscape Gardening [by] Andrew Jackson Downing.* 10th ed. New York: John Wiley and Sons.

Weaver, R. E., Jr. 1980. The restoration of Oak Path. *Arnoldia* 40:294–300.

West, A. G. 1935. Harvard calls a noted botanist to her gardens. *Boston Transcript* (magazine section, p. 3) 5 October.

Wheeler, W. M. 1930. The Bussey Institution. In S. E. Morison, ed., *The Development of Harvard University since the Inauguration of President Eliot, 1869–1929,* pp. 508–17. Cambridge, Mass.: Harvard University Press.

Whitcomb, H. M. 1897. *Annals and Reminiscences of Jamaica Plain.* Cambridge, Mass.: Riverside Press.

Whitehill, W. M. 1973. Francis Parkman as horticulturist. *Arnoldia* 33:169–83.

Wilder, M. P. 1881. *The Horticulture of Boston and Vicinity* ["substantially the same as the chapter prepared for the Boston Memorial Series volume 4"]. Boston: privately printed.

Williams, H. T. 1874. Shaw's Gardens, St. Louis. *Horticulturist* 29:225–28.

Wilson, E. H. 1916. Jackson T. Dawson, his work and his workshop. *Horticulture* 23:40–41.

———. 1925. *America's Greatest Garden.* Boston: Stratford.

———. 1928. *More Aristocrats of the Garden.* Boston: Stratford.

Winthrop, R. C. 1887. *Tributes of the Massachusetts Historical Society to Francis E. Parker.* Cambridge, Mass.: John Wilson and Son.

Wister, J. C. 1932. Report of the Arthur Hoyt Scott Horticultural Foundation. *Swarthmore College Bulletin.* Supplement to vol. 30, no. 3, pp. 1–42.

———. 1940. The Arthur Hoyt Scott Horticultural Foundation. A ten year history. *Swarthmore College Bulletin* 37(5):1–89.

———. 1956. A brief review of AABGA history. *AABGA Newsletter* 25:2–7.

Wyman, D. 1938a. The hedge demonstration plot at the Arnold Arboretum. *Bulletin of Popular Information,* 4th ser., 6:79–81.

———. 1938b. Hurricane damage at the Arboretum. *Bulletin of Popular Information,* 4th ser., 6:71–74.

———. 1938c. The Larz Anderson collection of Japanese dwarf trees. *Bulletin of Popular Information,* 4th ser., 6:31–41.

———. 1939. The past winter at the Arboretum. *Bulletin of Popular Information,* 4th ser., 7:1–4.

———. 1946. William H. Judd, propagator. *Arnoldia* 6:25–28.

———. 1947. The past year at the Arnold Arboretum. *Arnoldia* 7:1–8.

———. 1949. *Shrubs and Vines for American Gardens.* New York: Macmillan.

———. 1951a. The Arnold Arboretum. *Journal of the Royal Horticultural Society* 56:225–30.

———. 1951b. *Trees for American Gardens.* New York: Macmillan.

———. 1954. *The Arnold Arboretum Garden Book.* New York: D. Van Nostrand.

———. 1959. *The Arboretums and Botanical Gardens of North America.* Jamaica Plain, Mass.: Arnold Arboretum of Harvard University.

———. 1960. How to establish an arboretum or botanical garden. *Arnoldia* 20:69–83.

———. 1968. The new dwarf conifer collection. *Arnoldia* 28:9–27.

———. 1970. Horticulture at the Arnold Arboretum, 1936–1970. *Arnoldia* 30:81–99.

Zaitzevsky, C. 1982. *Frederick Law Olmsted and the Boston Park System.* Cambridge, Mass.: Harvard University Press, Belknap Press.

ILLUSTRATION CREDITS

Frontispiece: Photograph by the J. Horace McFarland Company [circa 1931], courtesy of the Archives of American Gardens, Smithsonian Institution.

1. Photograph by George R. King, 1917. Photographic archives of the Arnold Arboretum.
2. From *Gardener's Monthly*, 1859. Library of the Arnold Arboretum.
3. Photograph by George L. Goodale, courtesy of the archives of the Gray Herbarium.
4. From Phillip Miller, *Gardeners and Botanists Dictionary* (London, 1807). Library of the Arnold Arboretum.
5. Photograph by J. A. Whipple, 1865, courtesy of the archives of the Gray Herbarium.
6. Photograph by Gustavo Romero.
7. Photograph by Albert Bussewitz.
8. From *Magazine of Natural History*, 1830. Library of the Arnold Arboretum.
9. Photograph courtesy of the Society for the Preservation of New England Antiquities.
10. Photograph courtesy of the Museum of Comparative Zoology, Harvard University.
11. From *Horticulturist* (1848). Library of the Arnold Arboretum.
12. From *Woods and Forests* (1884). Library of the Arnold Arboretum.
13. Photograph by A. H. Folsom, 1884, courtesy of the Society for the Preservation of New England Antiquities.
14. From *Horticulturist* (1852). Library of the Arnold Arboretum.
15. From *American Journal of Horticulture and Florists Companion,* 1868.
16. "View from Mt. Auburn," engraving by James Smillie, 1847, courtesy of Mount Auburn Cemetery archives.
17. From F. L. Olmsted and Calvert Vaux, *Greensward* (1868 reprint). Library of the Arnold Arboretum.
18. Photograph by the Boston Park Commission, 1904. Photographic archives of the Arnold Arboretum.
19. Courtesy of the Harvard University Portrait Collection, Harvard University Art Museums. Bequest of Benjamin Bussey, 1894.
20. Archives of the Arnold Arboretum.
21. From *Century Magazine* (1892). Archives of the Arnold Arboretum.
22., 23. From Andrew J. Downing, *Treatise on the Theory and Practice of Landscape Gardening* (1844). Library of the Arnold Arboretum.
24. From *Magazine of Horticulture* (1848). Library of the Arnold Arboretum.
25. Photograph courtesy of the Harvard University Portrait Collection, Harvard University Art Museums. Gift of Arthur Rotch to Harvard for the Arnold Arboretum.
26. From Robert C. Waterston, *Memoir of George Barrell Emerson, LL.D.* (1884). Library of the Arnold Arboretum.
27. From Lorin L. Dame and Henry Brooks, *Typical Elms and Other Trees of Massachusetts* (1890). Library of the Arnold Arboretum.
28. From George B. Emerson, *Report on the Trees and Shrubs Growing Naturally in the Forests of Massachusetts* (1874). Library of the Arnold Arboretum.
29. From John C. Loudon, *Self Instruction for Young Gardeners* (1847). Library of the Arnold Arboretum.
30. From John C. Loudon, *The Derby Arboretum* (London, 1840). Library of the Arnold Arboretum.
31. From R. C. Winthrop, *Tributes of the Massachusetts Historical Society to Francis E. Parker* (1887). Library of the Harvard Divinity School.
32. Photograph courtesy of the Society for the Preservation of New England Antiquities.

33., 34. Photographic archives of the Arnold Arboretum.

35. Photograph by LeJune, Paris, 1868. Photographic archives of the Arnold Arboretum.

36. Archives of the Arnold Arboretum.

37. Photographic archives of the Arnold Arboretum.

38. Photograph by Alfred Rehder, 1899. Photographic archives of the Arnold Arboretum.

39. Photograph by T. E. Marr. Photographic archives of the Arnold Arboretum.

40. Photographic archives of the Arnold Arboretum.

41. Photograph, circa 1859, courtesy of the National Park Service, Olmsted National Historic Site.

42., 43., 44. Archives of the Arnold Arboretum.

45. Photograph by T. E. Marr. Photographic archives of the Arnold Arboretum.

46. Photograph by the Boston Park Commission, 1889. Photographic archives of the Arnold Arboretum.

47. Photograph by Alfred Rehder, 1900. Photographic archives of the Arnold Arboretum.

48. Photograph by the Boston Park Commission, 1890. Photographic archives of the Arnold Arboretum.

49. Archives of the Arnold Arboretum.

50. Photograph by T. E. Marr, 1904. Photographic archives of the Arnold Arboretum.

51. Photograph by T. E. Marr, 1903. Photographic archives of the Arnold Arboretum.

52. Photograph by Oakes Ames, 1931. Photographic archives of the Arnold Arboretum.

53. Photograph by Hardy, 1898. Photographic archives of the Arnold Arboretum.

54. Photograph by T. E. Marr, 1903. Photographic archives of the Arnold Arboretum.

55. From *Century Magazine* (1892). Archives of the Arnold Arboretum.

56. Photograph by G. R. King, 1916. Photographic archives of the Arnold Arboretum.

57., 58. Photographic archives of the Arnold Arboretum.

59. Photograph by J. G. Jack, 1899. Photographic archives of the Arnold Arboretum.

60. Pencil drawing by C. E. Faxon. Archives of the Arnold Arboretum.

61. Plate from C. S. Sargent, *Silva of North America* (1890). Archives of the Arnold Arboretum.

62. Photograph by H. H. Richardson, 1916. Photographic archives of the Arnold Arboretum.

63. Photograph by Alfred Rehder, 1921. Photographic archives of the Arnold Arboretum.

64., 65. Photographic archives of the Arnold Arboretum.

66. Photograph by Istvan Racz, 1988. Photographic archives of the Arnold Arboretum.

67. From George R. Shaw, *The Pines of Mexico* (1909). Library of the Arnold Arboretum.

68. Photograph by E. H. Wilson, 1916. Photographic archives of the Arnold Arboretum.

69. Photograph by Istvan Racz, 1991. Photographic archives of the Arnold Arboretum.

70. Photograph by Christopher Burnett. Archives of the Arnold Arboretum.

71. Photograph by Istvan Racz, 1989. Photographic archives of the Arnold Arboretum.

72. Photographic archives of the Arnold Arboretum.

73. Photograph by Istvan Racz, 1991. Photographic archives of the Arnold Arboretum.

74. Photograph by T. E. Marr, 1904. Photographic archives of the Arnold Arboretum.

75. Photograph by Istvan Racz, 1990. Photographic archives of the Arnold Arboretum.

76. Courtesy of Jon Katherine McKelvey. Photographic archives of the Arnold Arboretum.

77., 78. Photographs by T. E. Marr, 1903. Photographic archives of the Arnold Arboretum.

79. Photograph, 1927. Photographic archives of the Arnold Arboretum.

80. Photograph by E. H. Wilson, 1915. Photographic archives of the Arnold Arboretum.

81. Photograph by T. E. Marr, 1906. Photographic archives of the Arnold Arboretum.

82. Photograph by Istvan Racz, 1988. Photographic archives of the Arnold Arboretum.

83. Photograph by G. R. King, 1917. Photographic archives of the Arnold Arboretum.

84. Photograph by E. H. Wilson, 1923. Photographic archives of the Arnold Arboretum.

85. Photograph by T. E. Marr. 1913. Photographic archives of the Arnold Arboretum.

86. Photograph by Alfred Rehder, 1899. Photographic archives of the Arnold Arboretum.

87. Photographic archives of the Arnold Arboretum.

88. Photograph by E. H. Wilson, 1915. Photographic archives of the Arnold Arboretum.

89. Photographic archives of the Arnold Arboretum.

90. Photograph by Heman Howard, 1954. Photographic archives of the Arnold Arboretum.

91. Photograph by Christopher Burnett. Archives of the Arnold Arboretum.

92. Photograph by T. E. Marr, 1908. Photographic archives of the Arnold Arboretum.

93. Photographic archives of the Arnold Arboretum.

94. Archives of the Arnold Arboretum.

95. Photograph by N. R. Graves. Photographic archives of the Arnold Arboretum.

96. Photograph courtesy of the Scott Arboretum of Swarthmore College.

97. Photograph by Arthur G. Eldredge. Photographic archives of the Arnold Arboretum.

98. Photograph by E. H. Wilson, 1924. Photographic archives of the Arnold Arboretum.

99. Photograph by E. H. Wilson, 1925. Photographic archives of the Arnold Arboretum.

100., 101. Photographs by George E. Lawrence Co., 1927. Photographic archives of the Arnold Arboretum.

102. From *Through the Arnold Arboretum* (1934). Library of the Arnold Arboretum.

103., 104. Photographic archives of the Arnold Arboretum.

105. Photograph by Hugh M. Raup, 1933. Photographic archives of the Arnold Arboretum.

106. Photograph by E. H. Wilson, 1923. Photographic archives of the Arnold Arboretum.

107. Photograph by E. H. Wilson, 1929. Photographic archives of the Arnold Arboretum.

108. Photograph by W. R. Merryman, 1926. Photographic archives of the Arnold Arboretum.

109. Photograph by E. H. Wilson, 1925. Photographic archives of the Arnold Arboretum.

110. Photographic archives of the Arnold Arboretum.

111. Plant records of the Arnold Arboretum.

112. Photograph by Ernest J. Palmer, 1932. Photographic archives of the Arnold Arboretum.

113. Photograph by T. E. Marr. Photographic archives of the Arnold Arboretum.

114. Archives of the Arnold Arboretum.

115. Photograph by Donald Wyman, 1950. Photographic archives of the Arnold Arboretum.

116. Photograph courtesy of the botany libraries, Harvard University.

117. Photographic archives of the Arnold Arboretum.

118. Photograph by Donald Wyman, 1936. Photographic archives of the Arnold Arboretum.

119. Photographic archives of the Arnold Arboretum.

120. Photograph by Heman Howard, 1957. Photographic archives of the Arnold Arboretum.

121, 122. Photographs by Donald Wyman, 1938. Photographic archives of the Arnold Arboretum.

123., 124., 125. Photographic archives of the Arnold Arboretum.

126. Photograph by Heman Howard, 1947. Photographic archives of the Arnold Arboretum.

127. Photographic archives of the Arnold Arboretum.

128. Photograph by Heman Howard, 1947. Photographic archives of the Arnold Arboretum.

129. Photograph by Istvan Racz, 1989. Photographic archives of the Arnold Arboretum.

130. Photographic archives of the Arnold Arboretum.

131. Photograph by Istvan Racz, 1991. Photographic archives of the Arnold Arboretum.

132., 133. Photographic archives of the Arnold Arboretum.

134. Photograph by Istvan Racz. Photographic archives of the Arnold Arboretum.

135. Photographic archives of the Arnold Arboretum.

136. Photograph by Heman Howard, 1951. Photographic archives of the Arnold Arboretum.

137. Photograph by Heman Howard, 1959. Photographic archives of the Arnold Arboretum.

138. Photographic archives of the Arnold Arboretum.

139., 140. Photographic archives of the Arnold Arboretum.

141. Courtesy of Carroll E. Wood, Jr., and the Generic Flora of the Southeastern United States project, Arnold Arboretum.

142. Photographic archives of the Arnold Arboretum.

143. Photograph by Istvan Racz, 1990. Photographic archives of the Arnold Arboretum.

144., 145. Photographic archives of the Arnold Arboretum.

146. Photograph by Pam Bruns. Photographic archives of the Arnold Arboretum.

147. Photograph by Heman Howard, 1967. Photographic archives of the Arnold Arboretum.

148. Photograph by Kenneth R. Robertson, circa 1974. Photographic archives of the Arnold Arboretum.

149. Photographic archives of the Arnold Arboretum.

150. Photograph by Heman Howard, 1963. Photographic archives of the Arnold Arboretum.

151., 152. Photographs by Peter Chvany. Photographic archives of the Arnold Arboretum.

153., 154., 155. Photographic archives of the Arnold Arboretum.

156. Photograph by Istvan Racz, 1990. Photographic archives of the Arnold Arboretum.

157. Photograph by Istvan Racz, 1991. Photographic archives of the Arnold Arboretum.

158. Photograph by Istvan Racz, 1989. Photographic archives of the Arnold Arboretum.

159. Photographic archives of the Arnold Arboretum.

160. Photograph courtesy of Robert Howard, Photography Unlimited.

161. Photograph by Istvan Racz. Photographic archives of the Arnold Arboretum.

COLOR PLATES

1. Photograph by Jennifer Quigley, May 1979. Slide collection of the Arnold Arboretum.
2. Photograph, May 1960. Slide collection of the Arnold Arboretum.
3. Photograph by Gary L. Koller.
4. Photograph, May 1971, by Pam Bruns. Slide collection of the Arnold Arboretum.
5. Photograph, June 1960. Slide collection of the Arnold Arboretum.
6. Photograph, May 1989, by John H. Alexander.
7. Photograph, June 1981, by Richard E. Weaver. Photographic archives of the Arnold Arboretum.
8. Photograph by Albert Bussewitz.
9. Photograph by Albert Bussewitz.
10. Photograph, September 1987, by Ethan W. Johnson. Slide collection of the Arnold Arboretum.
11. Photograph by Albert Bussewitz.
12. Photograph by Albert Bussewitz.
13. Photograph, May 1993, by John H. Alexander.
14. Photograph, April 1985, by Corliss K. Engle. Slide collection of the Arnold Arboretum.
15. Photograph, May 1985, by Barbara O'Connor. Slide collection of the Arnold Arboretum.
16. Photograph by Albert Bussewitz.
17. Photograph, June 1989, by Corliss K. Engle. Slide collection of the Arnold Arboretum.
18. Photograph by Albert Bussewitz.
19. Photograph by Gary L. Koller. Slide collection of the Arnold Arboretum.
20. Photograph, August 1985, by Gary L. Koller. Slide collection of the Arnold Arboretum.

INDEX

Meadow Road *(cont.)*
  construction of, 96, 97
  cork trees along, 163, *296*
  golden-rain tree and, *Plate 20*
  natural woods near, *168*
  plans for Arboretum and, 88
  Rhus collections and, 163
Mechanization, 242–43, 252, 287
  equipment garage and, 293–94
Meconopsis integrifolia, 136
Medicinal Plants of East and Southeast Asia, The (Perry), 280
Melville, Herman, 47
Melvin, Mary Armstrong, 225, 227
Memorial rose, 258
Mendel, Gregor, 17
Mendum Street entrance, 100
Mercer, Martha Dana, 280, *285, 286*
Mercer Fellows, 280, 302
Merino sheep, 22, 40
Merrill, Dana, 232
Merrill, Elmer Drew, 228, *230,* 269
  appointment of Wyman, 235
  Asiatic botany and, 269
  background of, 232–34
  "The Botany of Cook's Voyages," 276
  consolidation of Harvard botanical units and, 234–35, 248–50
  death of, 276
  herbarium in Manila and, 232–33
  hurricane of 1938 and, 240–44
  living collections and, 239–46, 257–64
  research program and, 247–48, 276
  retirement as administrator, *249, 250*
'Merrill' magnolia, 264, *265*
Meservy, Julie, 316
Metasequoia glyptostroboides, Plate 17, 262, 263–64, 301, 308–9, *312*
Mexican Boundary Survey, 14
Mexico, *Pinus* species in, *132*
Meyer, Frank N., 185
Michaux, André, 56
Michaux, François André, 56
  North American Sylva, 22
Microclimate, 94, 223–24
Miller, Carl Ferris, 306
Missouri Botanical Garden, 185, 220, 228, 279, 309, 311
M'Mahon, Bernard, *American Gardener's Calendar,* 29
Mock orange, 159
  'Avalanche' variety, *159*
Möller's Deutsche Gärtner-Zeitung (German weekly), 126, 127

Monographic studies, 129–34. *See also specific monographs*
  Asian species and, 141, 142
Monograph of Azaleas, A (Rehder and Wilson), 148
Montgomery, Mrs. Robert H., 268
Montgomery, Robert H., 268
Monticello, 27, 29
Morgan, J. P., 200
Morris, John T., *226,* 227
Morris, Lydia Thompson, *226,* 227
Morris Arboretum, Chestnut Hill, Pa., *226,* 227
Morse, John, *208*
Morton, J. Sterling, 188
Morton, Joy, 188
Morton Arboretum, Lisle, Ill., 188–89
Moss Hill, *62, 67*
Motley, Thomas, 39, 63
Mountain andromeda, 294
Mountain ash, 262. *See also Sorbus*
  Korean, 143
Mountain camellia, *16*
Mountain laurel, 11. *See also Kalmia latifolia*
  Hemlock Hill Road and, *108*
  South Street bank and, *212, 213,* 218–19
Mount Auburn Cemetery, Cambridge, 23, *31, 32–33,* 41–42, 61
Mount Vernon, 27, 29
MSPA. *See* Massachusetts Society for Promoting Agriculture
Muir, John, 123, 124, 136–37
Mulching, 254, 287, *288*
  demonstration exhibit and, 295, 297
Mulligan, Brian O., *267*
Museum building, *103, 126. See also* Administration building
  exhibits at, 177, 294
  expansion of, 143, 145–47
  shrub plantings and, 101, *103*
Museum movement, 18–21
Myricaceae, 313

Nakai, Takenoshi N., 217
National Academy of Sciences, 123–24
National Botanic Garden, Washington, D.C., 189
National Geographic, 182
National Historic Landmark designation, *295,* 295–96
National Park Service, 319
National Shade Tree Conference (1946), 254

Native American artifacts, *224,* 295
Native plantings
  Bussey Hill and, 109
  roadside shrubbery and, 93–94, *95,* 105
Natural areas, 223. *See also* Natural woods
Natural history, in nineteenth century, 5–6, 29, 51, 54–55. *See also* Horticulture
Natural history specimens, 5–6, 8, 18–21. *See also* Cabinet; Herbarium; Type specimens
Naturalistic landscape gardening. *See* English landscape gardening style
Natural woods, *168,* 170, 176, 225. *See also* Natural areas; Old tree stands
Natürlichen Pflanzenfamilien, Die (Engler and Prantl), 129
Neillia sinensis, 166
Nevling, Lorin I., 279, 295, 305
New Bedford, Mass., 45–50, *49*
New Bedford Horticultural Society, 24, 47
New England Botanical Club, 232
New Guinean flora, 280, 309
New York Botanical Garden, 185, 186–87, 199, 228, 234, 247, 274
Nicholson, George, 113
Nomenclature
  binomial, 11
  horticultural taxonomy and, 283–84
  Rehder and, 129
  unified system of, 129, 264, 283
Northern Transcontinental Survey, 118
North Meadow
  drainage and, 154, 155
  lake plans and, 99
  shrub collection and, 158
North Woods, 75, 109, 223
Norway spruce, *149*
Nursery trade
  Arboretum acquisitions and, 239
  Arboretum resources and, 183–85, 214
  cooperating nursery program and, 246
  expedition support from, *183*
  hedge demonstration and, 238
  hybridization and, 168
  proposals for arboreta and, 59–60
Nuttall, Thomas, 7–8, 9
  Genera of North American Plants with a Catalogue of Species, 8
Nyssa sylvatica, Plate 15

THE ARNOLD ARBORETUM

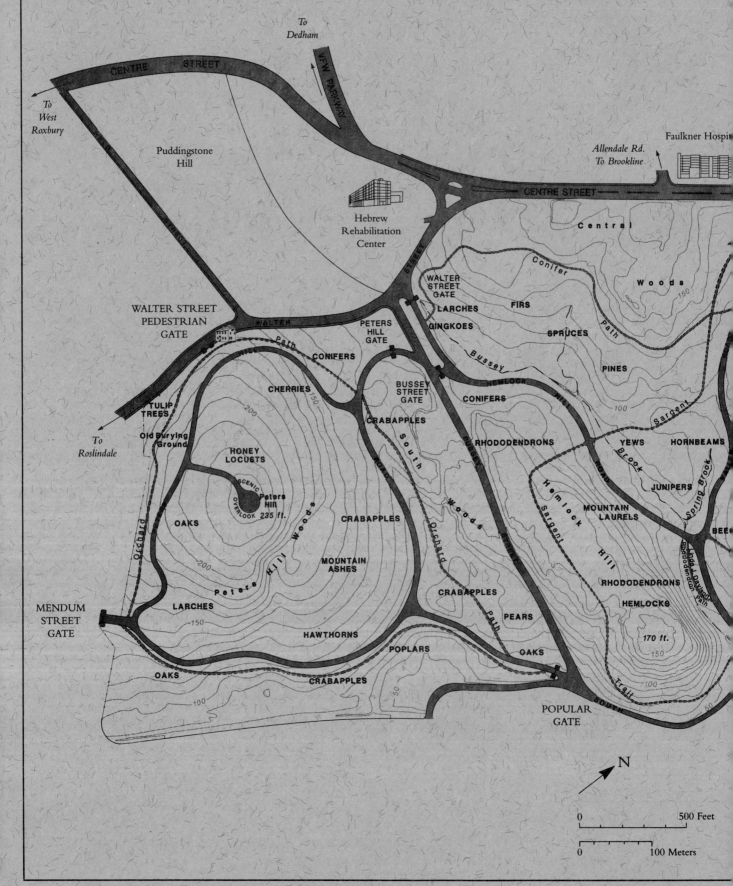